Evangelicals and Science

Recent Titles in
Greenwood Guides to Science and Religion

Science and Religion, 1450–1900: From Copernicus to Darwin
Richard G. Olson

Science and Religion, 400 B.C. to A.D. 1550: From Aristotle to Copernicus
Edward Grant

Science and Nonbelief
Taner Edis

Judaism and Science: A Historical Introduction
Noah J. Efron

Science and Islam
Muzaffar Iqbal

Science and Asian Spiritual Traditions
Geoffrey Redmond

Liberal Protestantism and Science
Leslie A. Muray

Evangelicals and Science

Recent Titles in Greenwood Guides to Science and Religion

Science and Religion, 1450–1900: From Copernicus to Darwin
Richard G. Olson

Science and Religion, 400 B.C. to A.D. 1550: From Aristotle to Copernicus
Edward Grant

Science and Nonbelief
Taner Edis

Judaism and Science: A Historical Introduction
Noah J. Efron

Science and Islam
Muzaffar Iqbal

Science and Asian Spiritual Traditions
Geoffrey Redmond

Liberal Protestantism and Science
Leslie A. Muray

Evangelicals and Science

MICHAEL ROBERTS

Greenwood Guides to Science and Religion
Richard Olson, Series Editor

Greenwood Press
Westport, Connecticut • London

Library of Congress Cataloging-in-Publication Data

Roberts, Michael (Michael B.)
Evangelicals and science / Michael Roberts.
p. cm. — (Greenwood guides to science and religion)
Includes bibliographical references and index.
ISBN: 978–0–313–33113–8 (alk. paper)
1. Religion and science—History. 2. Evangelicalism—History.
2. Creationism—History. I. Title.
BL240.3.R63 2008
261.5′5—dc22 2007041059

British Library Cataloguing in Publication Data is available.

Library of Congress Catalog Card Number: 2007041059
ISBN: 978–0–313–33113–8

First published in 2008

Greenwood Press, 88 Post Road West, Westport, CT 06881
An imprint of Greenwood Publishing Group, Inc.
www.greenwood.com

Printed in the United States of America

The paper used in this book complies with the Permanent Paper Standard issued by the National Information Standards Organization (Z39.48–1984).

10 9 8 7 6 5 4 3 2 1

Contents

Series Foreword

For nearly 2,500 years, some conservative members of societies have expressed concern about the activities of those who sought to find a naturalistic explanation for natural phenomena. In 429 BCE, for example, the comic playwright, Aristophanes parodied Socrates as someone who studied the phenomena of the atmosphere, turning the awe-inspiring thunder which had seemed to express the wrath of Zeus into nothing but the farting of the clouds. Such actions, Aristophanes argued, were blasphemous and would undermine all tradition, law, and custom. Among early Christian spokespersons there were some, such as Tertullian, who also criticized those who sought to understand the natural world on the grounds that they "persist in applying their studies to a vain purpose, since they indulge their curiosity on natural objects, which they ought rather [direct] to their Creator and Governor" (Tertullian, 1896–1903, p. 133).

In the twentieth century, though a general distrust of science persisted among some conservative groups, the most intense opposition was reserved for the theory of evolution by natural selection. Typical of extreme anti-evolution comments is the following opinion offered by Judge Braswell Dean of the Georgia Court of Appeals: "This monkey mythology of Darwin is the cause of permissiveness, promiscuity, pills, prophylactics, perversions, pregnancies, abortions, pornography, pollution, poisoning, and proliferation of crimes of all types" (Toumey, 1994, p. 94).

It can hardly be surprising that those committed to the study of natural phenomena responded to their denigrators in kind, accusing them of willful ignorance and of repressive behavior. Thus, when Galileo Galilei was warned against holding and teaching the Copernican system of astronomy as true, he wielded his brilliantly ironic pen and threw down a gauntlet to religious authorities in an introductory letter "To the Discerning Reader" at the Beginning of his great *Dialogue Concerning the Two Chief World Systems*:

> Several years ago there was published in Rome a salutary edict which, in order to obviate the dangerous tendencies of our age, imposed a seasonable silence upon the Pythagorean [and Copernican] opinion that the earth moves. There were those who impudently asserted that this decree had its origin, not in judicious inquiry, but in passion none too well informed. Complaints were to be heard that advisors who were totally unskilled at astronomical observations ought not to clip the wings of reflective intellects by means of rash prohibitions.
>
> Upon hearing such carping insolence, my zeal could not be contained.(Galileo, 1953, p. 5)

No contemporary discerning reader could have missed Galileo's anger and disdain for those he considered enemies of free scientific inquiry.

Even more bitter than Galileo was Thomas Henry Huxley, often known as "Darwin's bulldog." In 1860, after a famous confrontation with the Anglican Bishop Samuel Wilberforce, Huxley bemoaned the persecution suffered by many natural philosophers, but then he reflected that the scientists were exacting their revenge:

> Extinguished theologians lie about the cradle of every science as the strangled snakes beside that of Hercules; and history records that whenever science and orthodoxy have been fairly opposed, the latter has been forced to retire from the lists, bleeding and crushed, if not annihilated; scotched if not slain. (Moore, 1979, p. 60)

The impression left, considering these colorful complaints from both sides is that science and religion must continually be at war with one another. That view of the relation between science and religion was reinforced by Andrew Dickson White's *A History of the Warfare of Science with Theology in Christendom*, which has seldom been out of print since it was published as a two volume work in 1896. White's views have shaped the lay understanding of science and religion interactions for more than a century, but recent and more careful scholarship has shown that confrontational stances do not represent the views of the overwhelming majority of ether scientific investigators or religious figures throughout history.

One response among those who have wished to deny that conflict constitutes the most frequent relationship between science and religion is to claim that they cannot be in conflict because they address completely different human needs and therefore have nothing to do with one another. This was the position of Immanuel Kant who insisted that the world of natural phenomena, with its dependence on deterministic causality, is fundamentally disjoint from the noumenal world of human choice and morality, which constitutes the domain of religion. Much more recently, it was the position taken by Stephen Jay Gould in *Rocks of Ages: Science and Religion in the Fullness of Life* (1999). Gould writes:

> I... do not understand why the two enterprises should experience any conflict. Science tries to document the factual character of the natural world and to develop theories that coordinate and explain these facts. Religion, on the other hand, operates in the equally important, but utterly different realm of human purposes, meanings, and values. (Gould, 1999, p. 4)

In order to capture the disjunction between science and religion, Gould enunciates a principle of "Non-overlapping magisterial," which he identifies as "a principle of respectful noninterference" (Gould, 1999, p. 5).

In spite of the intense desire of those who wish to isolate science and religion from one another in order to protect the autonomy of one, the other, or both, there are many reasons to believe that theirs is ultimately an impossible task. One of the central questions addressed by many religions is what is the relationship between members of the human community and the natural world. This question is a central question addressed in "Genesis," for example. Any attempt to relate human and natural existence depends heavily on the understanding of nature that exists within a culture. So where nature is studied through scientific methods, scientific knowledge is unavoidably incorporated into religious thought. The need to understand "Genesis" in terms of the dominant understandings of nature thus gave rise to a tradition of scientifically informed commentaries on the six days of creation which constituted a major genre of Christian literature from the early days of Christianity through the Renaissance.

It is also widely understood that in relatively simple cultures—even those of early urban centers—there is a low level of cultural specialization, so economic, religious, and knowledge producing specialties are highly integrated. In Bronze Age Mesopotamia, for example, agricultural activities were governed both by knowledge of the physical conditions necessary for successful farming and by religious rituals associated with plowing, planting, irrigating, and harvesting. Thus religious practices and natural knowledge interacted in establishing the character and timing of farming activities.

Even in very complex industrial societies with high levels of specialization and division of labor, the various cultural specialties are never completely isolated from one another and they share many common values and assumptions. Given the linked nature of virtually all institutions in any culture it is the case that when either religious or scientific institutions change substantially, those changes are likely to produce pressures for change in the other. It was probably true, for example, that the attempts of Presocratic investigators of nature, with their emphasis on uniformities in the natural world and apparent examples of events systematically directed toward particular ends, made it difficult to sustain beliefs in the old Pantheon of human-like and fundamentally capricious Olympian gods. But it is equally true that the attempts to understand nature promoted a new notion of the divine—a notion that was both monotheistic and transcendent, rather than polytheistic and immanent—and a notion that focused on both justice and intellect rather than power and passion. Thus early Greek natural philosophy undoubtedly played a role not simply in challenging, but also in transforming Greek religious sensibilities.

Transforming pressures do not always run from scientific to religious domains, moreover. During the Renaissance, there was a dramatic change among Christian intellectuals from one that focused on the contemplation of God's works to one that focused on the responsibility of the Christian for caring for his fellow humans. The active life of service to humankind, rather than the contemplative life of reflection on Gods character and works, now became the Christian ideal for many. As a consequence of this new focus on the active life, Renaissance intellectuals turned away from the then dominant Aristotelian view of science, which saw the inability of theoretical sciences to change the world as a positive virtue. They replaced this understanding with a new view of natural knowledge, promoted in the writings of men such as Johann Andreae in Germany and Francis Bacon in England, which viewed natural knowledge as significant only because it gave humankind the ability to manipulate the world to improve the quality of life. Natural knowledge would henceforth be prized by many because it conferred power over the natural world. Modern science thus took on a distinctly utilitarian shape at least in part in response to religious changes.

Neither the conflict model nor the claim of disjunction, then, accurately reflect the often intense and frequently supportive interactions between religious institutions, practices, ideas, and attitudes on the one hand, and scientific institutions, practices, ideas, and attitudes on the other. Without denying the existence of tensions, the primary goal of the volumes of this series is to explore the vast domain of mutually supportive and/or transformative interactions between scientific institutions, practices, and knowledge and religious institutions, practices, and beliefs. A second goal is to offer the opportunity to make comparisons across space, time, and cultural configuration. The series will cover the entire globe, most major faith traditions, hunter-gatherer societies in Africa and Oceana as well as advanced industrial societies in the West, and the span of time from classical antiquity to the present. Each volume will focus on a particular cultural tradition, a particular faith community, a particular time period, or a particular scientific domain, so that each reader can enter the fascinating story of science and religion interactions from a familiar perspective. Furthermore, each volume will include not only a substantial narrative or interpretive core, but also a set of primary documents which will allow the reader to explore relevant evidence, an extensive bibliography to lead the curious to reliable scholarship on the topic, and a chronology of events to help the reader keep track of the sequence of events involved and to relate them to major social and political occurrences.

So far I have used the words "science" and "religion" as if everyone knows and agrees about their meaning and as if they were equally appropriately applied across place and time. Neither of these assumptions is true. Science and religion are modern terms that reflect the way that we in the industrialized West organize our conceptual lives. Even in the modern West, what we mean by science and religion is likely to depend on our political orientation, our scholarly background, and the faith community that we belong to. Thus, for example, Marxists and Socialists tend to focus on the application of natural knowledge as the key element in defining science. According to the British Marxist scholar, Benjamin Farrington, "Science is the system of behavior by which man has acquired mastery of his environment. It has its origins in techniques . . . in various activities by which man keeps body and soul together. Its source is experience, its aims, practical, its *only* test, that it works" (Farrington, 1953). Many of those who study natural knowledge in preindustrial societies are also primarily interested in knowledge as it is used and are relatively open regarding the kind of entities posited by the developers of culturally specific natural knowledge systems or "local sciences." Thus, in his *Zapotec Science: Farming and Food in the Northern Sierra of Oaxaca*, Roberto González insists that

> Zapotec farmers . . . certainly practice science, as does any society whose members engage in subsistence activities. They hypothesize, they model problems, they experiment, they measure results, and they distribute knowledge among peers and to younger generations. But they typically proceed from markedly different premises—that is, from different conceptual bases—than their counterparts in industrialized societies. (González, 2001, p. 3)

Among the "different premises" is the presumption of Zapotec scientists that unobservable spirit entities play a significant role in natural phenomena.

Those more committed to liberal pluralist society and to what anthropologists like González are inclined to identify as "cosmopolitan science," tend to focus on science as a source of objective or disinterested knowledge, disconnected from its uses. Moreover they generally reject the positing of unobservable entities, which they characterize as "supernatural." Thus, in an *Amicus Curiae* brief filed in connection with the 1986 Supreme Court case which tested Louisiana's law requiring the teaching of creation science along with evolution, for example, seventy-two Nobel Laureates, seventeen state academies of science, and seven other scientific organizations (1986, p. 24) argued that

> Science is devoted to formulating and testing naturalistic explanations for natural phenomena. It is a process for systematically collecting and recording data about the physical world, then categorizing and studying the collected data in an effort to infer the principles of nature that best explain the observed phenomena. Science is not equipped to evaluate supernatural explanations for our observations; without passing judgement on the truth or falsity of supernatural explanations, science leaves their consideration to the domain of religious faith.

No reference whatsoever to uses appears in this definition. And its specific unwillingness to admit speculation regarding supernatural entities into science reflects a society in which cultural specialization has proceeded much farther than in the village farming communities of southern Mexico.

In a similar way, secular anthropologists and sociologists are inclined to define the key features of religion in a very different way than members of modern Christian faith communities. Anthropologists and sociologists focus on communal rituals and practices which accompany major collective and individual events—plowing, planting, harvesting, threshing, hunting, preparation for war (or peace), birth, the achievement of manhood or womanhood, marriage (in many cultures), childbirth, and death. Moreover, they tend to see the major consequence of religious practices as the intensification of social cohesion. Many Christians, on the other hand, view the primary goal of their religion as personal salvation, viewing society as at best a supportive structure and at worst, a distraction from their own private spiritual quest.

Thus, science and religion are far from uniformly understood. Moreover, they are modern Western constructs or categories whose applicability to the temporal and spatial "other" must always be justified and must always be understood as the imposition of modern ways of structuring institutions, behaviors, and beliefs on a context in which they could not have been categories understood by the actors involved. Nonetheless it does seem to us not simply permissible, but probably necessary to use these categories at the start of any attempt to understand how actors from other times and places interacted with the natural world and with their fellow humans. It may ultimately be possible for historians and anthropologists to understand the practices of persons distant in time and/or space in terms that those persons might use. But that process must begin by likening the actions of others to those that we understand from our own experience, even if the likenesses are inexact and in need of qualification.

The editors of this series have not imposed any particular definition of science or of religion on the authors, expecting that each author will develop either explicit or implicit definitions that are appropriate to their own scholarly approaches and to the topics that they have been assigned to cover.

Richard Olson

Chronology of Events

1720s	Cotton Mather supports smallpox inoculation.
1730s	Beginning of Evangelical Revival in Massachusetts (Edwards) and England (Whitfield).
1738	Conversion of John Wesley.
1758	Death of Jonathan Edwards from smallpox vaccination.
1771	Francis Asbury goes to the American colonies and starts the Methodist church.
1795	Death of John Wesley.
1790s	Evangelicals blossom in Britain and America.
1790–1820s	Series of evangelical science professors at Cambridge.
1818	Rev. Adam Sedgwick elected Professor of Geology at Cambridge.
1812–1867	Michael Faraday at the Royal Institution, London, much experimental work and lectures.
1820s–1840s	Height of "evangelical" geologists—Sedgwick, Lewis, Miller in Britain and Hitchcock and Silliman in United States.
1859	Publication of Darwin's *The Origin of Species.*
1860s	Correspondence of Asa Gray and Darwin on design and evolution.
1880s	Height of "rapprochement" with B. B. Warfield and G. F. Wright.
1910	Publication of *The Fundamentals.*

1920s	Rise of anti-evolution, and splits over modernism.
1925	The Scopes Trial, Dayton, Tennessee.
1930s	Heyday of Harry Rimmer and George McCready Price.
1941	Formation of the *American Scientific Affiliation* in United States.
1944	Formation of what became *Research Scientists Christian Fellowship* (later *Christians in Science*) in London.
1949	First Billy Graham Crusade at Los Angeles.
1954	Publication of Ramm's *The Christian View of Science and Scripture.*
1961	Publication *The Genesis Flood.*
1962	Formation of *Creation Research Society.*
1972	Founding of *Institute of Creation Research* at San Diego.
1981	Trial at Arkansas.
1992	Formation of Evangelical Environmental Network (EEN).
1994	Formation of *Answers in Genesis* at Florence, Kentucky (with Australian roots).
2000	*Cornwall Declaration* opposing the EEN.
2005	Charles Townes, Nobel Laureate for MASER and LASER awarded Templeton Prize.
2006	American evangelicals divided over global warming.
2007	Opening of *Creation Museum* in Kentucky.

Introduction

"Evangelicals and Science!" The very thought will make many today smile at the sheer incongruity of it. After all evangelicals were the cause of the Scopes trial and the whole round of "monkey" legislation in almost every state in the United States. Many readers, like me, will remember that scene in *Inherit the Wind* where the local Tennessee fundamentalists were singing "The Old Time Religion." This has etched the perception that evangelicals are inherently hostile to science in the minds of many for the last half-century. Now there is some truth in this portrayal and I have been on the receiving end from latter day exemplars, who regard me as a heretic and compromiser, but the evangelicals have had a long and rich interaction with science, sometimes glorious and sometimes ignominious, and have produced some fine scientists. I prefer the tradition of Adam Sedgwick, in whose footsteps I have walked both literally and metaphorically (see Chapter 4).

The purpose of this volume is first, to give an overview of the way evangelicals have interacted with science since the start of the Evangelical Revival in the 1730s and this accounts for Chapters 3 through to 6. Second, it aims to analyze how the evangelicals of today interact with science and to put the controversy of creationism into its proper context and to avoid simplistic dismissals. Third, it aims to understand how and why evangelicals interact with science because of their various theologies. Fourth, it takes the two ethical issues of the environment and bioethics to see how evangelicals interact with aspects of science in the political arena.

The whole aspect of evangelicals and science is frequently misunderstood as it is assumed that evangelicals are entirely literalist and thus this book gives a wider perspective on that from both a historical and contemporary perspective. Of particular importance is that I demonstrate that with minor exceptions, and I mean minor, most evangelicals did not oppose science until the early twentieth century. It does, of course, explode the myth that there was virtual warfare between geology and Genesis in the early nineteenth century as I do over Sedgwick and others in Chapter 4. This continues my contribution to the *Geological Society of London Special Publication "Myth and Geology"* (Roberts, 2007), where I challenge the received account that the church was literalist over Genesis until 1800 and thus had serious problems over geological time. This myth renders many historical understandings of both evangelicals and science (and also Christianity in general and science) to be basically wrong, whether by evangelical creationists or secularists.

I also emphasize that the rise of creationism in the twentieth century is not a throwback to the seventeenth or eighteenth centuries, but rather an innovation caused by the synergism between the early twentieth-century fundamentalism, which had a strong anti-intellectual component, and new ideas of "Flood Geology" coming from McCready Price and the Seventh Day Adventists. Despite Numbers' excellent work *The Creationists,* many still hold to this.

Consequently the historical section is almost teleological in that it does not discuss every aspect of evangelicals and science from 1730 until now, but rather focuses on that which is important today, that is creationism. I can hear criticisms that I have left out this or that and that I should have given more space to miracles and evangelical scientists working in physics, chemistry, or astronomy. Of course, they are important, but when faced with a limitation on space, an author has either to deal with everything in extreme brevity or some things in depth. I have chosen the latter course, and have sought to select those, which seem to be more relevant today than those which may be of interest historically. I have not attempted to give a catalogue and chronology of scientific evangelicals in the last quarter of a millennium.

My overriding purpose is to put evangelicals and science *today* into historical and contemporary context and it is written both for the student, who may be studying the history of science or contemporary science studies, for clergy of all stripes, for educationalists, for the scientific community at large as well as the general reader.

I thus start with the earliest evangelicals in the 1730s and gradually move forward to the present day and focus mostly on British and American evangelicals. The book reflects my life-long interest in both the history of science in relation to evangelicalism and creationism, which began when I visited the evangelical study center at L'Abri in Switzerland in 1971. This, along with a historical study on Darwin's geology, has been my main academic interest during my ministry in Anglican parishes. I have also been to many creationist meetings and had the good fortune to attend the conference *Intelligent Design and its Critics* in Wisconsin in June 2000, when I met many leading Intelligent Designers and some of their critics.

Much of the book is not particularly original and has been said elsewhere, but my section in Chapter 4 on evangelicals and geology and "anti-geology" has come from my research into that period and presents evangelicals in a new light. An integral part of my research was deciphering the Welsh geological notebooks of Sedgwick, Buckland, and Sedgwick's pupil Darwin and spending many days walking the Welsh hills in all weathers retracing their steps. On one particular day in 1999, it took me over ten hours to retrace Sedgwick's route in the Carneddau, covering twenty miles of mountainous terrain. Above all, this book reflects nearly four decades of being personally involved in these issues and brings together my theological and scientific interests.

As this book deals with evangelicals both in history and today, the structure reflects this.

- Chapter 1 gives both a historical overview from the 1730s and contemporary snapshot of evangelicals in their sheer diversity from the most separatist fundamentalist to those who are nearly as liberal as Liberal Protestants.
- Chapter 2 considers the theological concerns of evangelicals and especially the nature and interpretation of the Bible and the iconic status of Inerrancy, along with the issue of the "Curse."
- Chapters 3 through 6 trace out evangelical attitudes to science from the beginning of the Evangelical Revival in the 1730s, to the publication of *The Genesis Flood* in 1961.
- Chapter 3 begins with a study of the conflict thesis of science and religion and how this distorts our understanding of evangelicals and science. I then deal briefly with the early years from 1730 to 1790, when evangelicals were a small group, who simply reflected the science of their day.
- Chapter 4 takes one from 1790 to 1850, which is when the evangelicals blossomed in numbers and when geology became the most popular science. Evangelicals were both in the forefront of geological research, and in the opposition to geology because of its implications on Genesis.
- Chapter 5 takes us though the heyday of evangelicalism in the late nineteenth century and the time of Darwinian controversies. It shows how evangelicals first "made peace" with Darwin and then with the rise of Fundamentalism broke the truce and listened to voices coming from the Seventh Day Adventists.
- Chapter 6 records the decline of evangelicalism and the rise of fundamentalism with its break from intellectual Christianity and the adoption of anti-evolution culminating in the Scopes trial of 1925, putting evangelicals and science in the doldrums. Then after the war, scientific concerns of evangelicals revived with the

American Scientific Association in the United States and the Research Scientists Christian Fellowship in Britain, but "creationism" was also quietly growing.

- The next three chapters bring us to today, with its unique mix of "Creationism" in all its forms and also evangelicals in the mainstream of science.
- Chapter 7 discusses the origin and influence of "creationism" both Young Earth Creationism (YEC) and Intelligent Design (ID) and their critics.
- Chapter 8 brings out the sheer diversity of today's evangelicals and science, and considers both evangelical scientists of all shades and how evangelicals at large, whether laity, clergy, or theologians understand science.
- Chapter 9 considers some of the ethical issues of science and how evangelicals deal with them. These are the environment (and global warming) and bioethics with attitudes to GM crops and stem cell research, over which evangelicals are sharply divided.
- Chapter 10 briefly sums up the book and attempts to crystal-gaze into the future, when evangelicals will be more prevalent in the Third World than they are today.

Evangelicals are split into three overlapping camps concerning science as described in Chapter 8. These are Young Earth Creationism, Intelligent Design and what is often called Theistic Evolution. Sometimes these co-exist, whether through desire or necessity, but at times this can lead to internecine warfare. The perception an individual will have of evangelicals and science will often depend on what type of attitude to science they have met among evangelicals, which varies from country to country, place to place and church to church.

I am very aware that my selectivity will not appeal to all informed readers, but I ask them to bear with me, particularly over my many omissions. I hope that through my recommended sources readers can read up for themselves what I have omitted. As McGrath wrote of himself in his writing of his massive three-volume work that he played to his strengths—biochemistry and historical theology—I have also done so and mine are geology and the history of science, and the former is often neglected. Geology and its history in relation to the Christian faith are essential to get a right perspective as so often it is glibly believed that in the early nineteenth century only liberal Christians accepted geology and when Darwin published *The Origin of Species* most Christians, whether educated or not, still accepted a six-day creation, whereas most educated Christians did not. To get this wrong, is to completely misunderstand the relationship of evangelicals and science and perpetuate the myth that good Bible-believers must accept Genesis literally.

I cannot claim to be a detached and disinterested outside observer on evangelicals, but then that creature does not exist. In one sense, the genesis of this book goes back thirty-six years when I first came across YEC at

Francis Schaeffer's centre L'Abri in Switzerland, where I first had to consider the relationship of evangelicals with science. I was originally helped by the Rev. Bob Ferris, now a minister of the Southern Baptist Church, and his wife Sue, who had studied under Russell Mixter. I was later given a second impetus when Canon Eric Jenkins invited me to his Science and Religion group in Liverpool Diocese in the early 1980s, which in turn led me to involvement with groups like the *Science and Religion Forum*, *Society of Ordained Scientists* and *Christians in Science*. Were it not for Eric's encouragement and friendship, this book would never have been written, but as he died in November 2006 he will not see it.

It is difficult to give all acknowledgments but I would single out the historians of science Jim Moore, Jim Secord, and John Brooke. Among clerical scientists I have already mentioned Eric Jenkins and now mention Arthur Peacocke, who passed away the same month as Eric. For many years I have benefited from membership of *Christians in Science* and have had particular help from Oliver Barclay, Sam (R.J.) Berry, Paul Marston and Bennet McInnes. In 2000 Angus Menuge made me very welcome at Wisconsin. I have benefited from the help of members of the *American Scientific Association*, especially Ted Davies, and Jack Haas, who made me so welcome at the ASA conference at Boston in 2001. My membership of HOGG (the History Of Geology Group of the Geological Society of London) has helped to literally ground my historical study, especially the one who welcomed me as a "bloody clergyman." John McKeown of the John Ray Initiative gave me a copy of a paper he gave on the environment, which has been invaluable. Most of my research has been done on a shoestring but over the years I have received small grants from the dioceses of Liverpool, St Asaph and Blackburn. I also express my appreciation to Kevin Downing, who has tried to keep me to deadlines and Richard Olson, the editor of this series.

And, finally, I give thanks to my wife, Andrea, who delights to refer to all I do as my "creationist rubbish" and now hopes that the study will regain some order and tidiness.

November 2007 Michael Roberts

Chapter 1

What Are Evangelicals?

One of the joys of television comes from watching old black and white classics late at night. Every few years one can see *Elmer Gantry* and *Inherit the Wind*. The first portrays the life of a prewar evangelist and is loosely based on the life of the anti-evolutionist, John Roach Straton. Of the same vintage is the screen version of Laurence and Lee's long-running Broadway play *Inherit the Wind*, with Spencer Tracy as the fictional version of Clarence Darrow and Fredric March in the lead role modeled on William Jennings Bryan. Both are fine films that continue to inform and misinform the public about the nature of Evangelicalism because of their portrayal of interwar fundamentalism (Larson, 1997). As we shall see, perceptions derived from these films can often mislead, particularly with respect to the relation of evangelicals to modern science. The popular image of evangelicals is of a subset of Protestants, who adopt biblical literalism, are anti-intellectual, and reject all science. That is, at best, a partial truth and misleading on its own.

I shall begin with a backward look at the history of evangelicals over the last quarter millennium, tracing the origins and beliefs of a very diverse movement. It is essential to understand what evangelicals stand for, before considering how they understand and interact with science. Unless one makes a careful, even if brief, study of the history of Evangelicalism, one is liable to set up a fundamentalist straw man and thus totally and utterly misunderstand evangelicals and their engagement, positive and negative, with science. As Americans have seen during the last three decades of the twentieth century, certain evangelicals have shown a special interest in science and have had considerable clout in both the teaching and practice of science at every level. On the one hand, evangelicals have been responsible for the Young Earth Creationism movement, which has generated widespread opposition in the scientific community. One the other hand, evangelicals have provided the highest leadership in the space sciences and the human genome project.

We are not dealing with a few religious people but a large minority in the world. Estimates vary from 150 million to over 400 million evangelicals in the world in 2000, that is between 2.5 and 6 percent of the world's population. If the present rate of growth continues there may be 600 million in 2025, that is 8 percent of the world's population.[1] Thus over the next few decades evangelicals are likely to become more significant on the world's stage than today.

WHAT ARE EVANGELICALS?

The simple answer to the question "What is Evangelicalism?" is that it is a movement of loosely affiliated enthusiastic groups of conservative Protestants, who have thrived since the 1730s and who put great emphasis on the truth of the Bible. Originally, evangelical beliefs were those of mainstream Trinitarian Protestantism coupled with enthusiasm; but diversity has crept in over the years. Scientifically, contemporary evangelicals range from Francis Collins, the director of the Genome Project, to Henry Morris, the founder of modern Young Earth Creationism. An immediate problem is that some evangelicals will not recognize the more "liberal" evangelicals as fellow evangelicals. This form of exclusion occurs, for example, when some Young Earth Creationists dismiss evangelicals who hold to "evolution" as either apostate or non-Christian. A further problem is that some groups, like Jehovah's Witnesses, call themselves evangelicals, but reject core beliefs such as the Trinity.

Evangelicals are a highly adaptive subspecies of Christians who can adapt to almost every type of society, yet still retain the core of evangelical belief. Within any one country there are an immense variety; Fundamentalists, Reformed, Baptists, Dispensationalists, various Methodists and Holiness groups, Afro-American groups, Pentecostalists, and many splinter groups. Some form separate denominations and some are evangelical ginger groups in mainline denominations like the Anglican, Methodist, and Presbyterian Churches, which are dominated by "liberals." In particular, evangelicals include a colorful cast of musicians, evangelists and tele-evangelists, faith healers, and some intellectuals. From the eighteenth and nineteenth centuries, there were the Wesleys, Edwards, Whitfield, Simeon, Finney, Asbury, Wilberforce. During the first half of the twentieth century, Amy Macpherson, Billy Sunday, Moody and Sankey, Warfield, Ryle, Spurgeon, Carl McIntire, Bob Jones, William Jennings Bryan, William

B. Riley, and the infamous Frank Norris were major figures. Since the Second World War, major evangelicals include Billy Graham, Jerry Falwell, Jim Wallis, Luis Palau, Tim Lahaye, John Stott, Archbishop Jensen of Sydney, Joni Erikson, Oral Roberts, Carl Henry, and two Ugandan bishops—Archbishop Janani Luwum, who was murdered by Idi Amin in 1977, and Festo Kivengere, who fled the country at that time.

One of the most useful definitions of Evangelicalism has been that of David Bebbington in his historical study *Evangelicalism in Modern Britain*, in which he defines the characteristics as (Bebbington, 1987, pp. 2–17)

1. Conversionism
2. Activism
3. Biblicism
4. Crucicentrism

Though he does not state this explicitly, Bebbington deals only with self-styled evangelicals who are Trinitarian. Bebbington discusses each characteristic at length and his morphology has formed the basis of much academic study on Evangelicalism throughout the world. Broadly speaking, an evangelical emphasizes the need for personal *conversion*, and the importance of being *active* in their faith, which is based four-square on the *Bible*. Moreover, they insist that their theology is centered on the *Cross* of Christ. Within these parameters, variations abound. Some insist on a dateable conversion, usually at a revivalist rally, others consider that conversion may take years. There is a great diversity of views on the Bible. The center of all evangelical belief is the atoning death of Christ, which is interpreted in different ways. A constant mark of evangelicals is their activism in evangelism and good works.

Bebbington has given scholars of Evangelicalism a good basis to work with, but he leaves out two important aspects of evangelical life—enthusiasm and fellowship. Evangelicals are more enthusiastic about their faith than most liberals, whether they are more restrained Anglicans or exuberant Charismatics. Though evangelicals are often individualistic and eschew fixed ecclesiastical structures, they usually form closely knit groups which may, on occasion, adopt rigid definitions of who is "in" and who is "out." This latter feature results in evangelicals having a unique form of *Fellowship*, which outsiders may see as oppressive and almost cultic. At times it is. Thus I would prefer to extend Bebbington's Quadrilateral to a Hexagon:

1. Activism
2. Bible

3. Conversion
4. Death of Christ
5. Enthusiasm
6. Fellowship

Bebbington's definition has been criticized for being too theoretical and focusing on the doctrinal rather than experiential aspects of Evangelicalism. This is correct, as some evangelicals are long on experience and short on thought and doctrine. It is an attempt to bring order to a highly diverse group of Christians. Though evangelicals will always claim to base their outlook on their doctrine on Scripture, there are many instances in which the belief is totemic rather than thought-through. It must be said in defense of evangelicals that they are often more articulate in the faith than many from mainstream churches.

THE DIVERSITY OF EVANGELICALISM

Although many will think of Elmer Gantry as the archetypal evangelical, evangelicals show an immense variety, which almost defies any summary. That is not surprising, as there are nearly half a billion scattered through every continent. On each of Bebbington's four points or my six, evangelicals are very diverse. They range from childish naiveté to a well-reasoned belief, from a simplistic crudity of a mechanical doctrine to a doctrine which is sensitive to the whole human predicament, from a talismanic approach to the Bible to one which uses every hermeneutical aid possible and from Young Earth Creationism to Harvard Professors of Astrophysics (Owen Gingerich) and Vice Presidents of the Royal Society of London (Sir Brian Heap).

Though the authority of the Bible is a common ground for all, the perception of its nature varies. The conservative wing puts great emphasis on its inerrancy and holds that the Bible contains no error of any kind whether moral or historical. Though many Inerrantists allow the Bible to *accommodate* itself to the thought of its day, some, especially those influenced by Young Earth Creationism, insist that it also teaches science. The more liberal evangelicals, especially those in Britain, eschew inerrancy, speak of the Bible's supreme authority in more general terms, and accept minor historical errors in the text. Some hardline Inerrantists have a tendency to demonize those who either question inerrancy or a Young Earth. Evangelicals are divided over what belief in the Bible entails.

Attitudes to conversion also vary. Before the Evangelical Revival began in the 1730s, "sudden conversions," even lasting a few months, were a rarity, but the early evangelicals speeded up the process and by the nineteenth and twentieth century some evangelicals were expecting a matter of hours or less. Some insist on a conversion, which is an identifiable event, others are flexible regarding whether conversion is sudden or gradual. On this issue, the latter almost slide over into the belief of mainstream Protestants regarding Christian nurture. The former can be very judgmental and doubt whether someone can be a Christian if they cannot put a time to their conversion. Billy Graham is often regarded as the epitome of an evangelical who emphasizes the need for a conscious conversion, due to his practice of calling for conversions during his evangelistic rallies. Graham, himself, was converted shortly after his sixteenth birthday in 1934 during an evangelistic rally. In contrast, his wife Ruth, who was born into a missionary doctor's family in China, had no conversion experience.

The death of Christ as the atonement for sin forms the heart of all evangelical religion. Simple presentations adopt a crude substitutionary atonement as the only understanding of the death of Christ. Here Jesus' death is portrayed as a sacrifice to placate an angry God. In more popular Evangelicalism, other New Testament portrayals of the meaning of the death of Christ are ignored—that is those portrayals which emphasize reconciliation, the suffering servant, and even Christ as an exemplar. Some evangelistic appeals can be very simplistic and without reference to the resurrection; but to claim that this is the case for all evangelicals is very misleading. Whether one reads an older study by John Calvin, a nineteenth-century study by Benjamin Warfield or James Denney, or recent understandings by John Stott or Joel Green, one finds much more careful doctrinal statements regarding why the death of Christ is the center of evangelical belief.

Many associate evangelicals with very enthusiastic worship, with hands raised high and the use of emotional music. Though evangelicals take their faith seriously and with enthusiasm, that enthusiasm takes many forms, from rowdy Pentecostal worship, interspersed with "Praise the Lord" and "Amen" at every possible opportunity to the more restrained liturgical worship of Lutherans and Anglicans, with every possible variety between, enthusiasm can be either restrained or ecstatic.

To outsiders, evangelical churches often seemed closed and only welcoming to fellow-believers. Thus their fellowship may seem to be sect-like. At times this is the case and that is especially so among groups who are either separatist or have distinct emphases. A dogmatic adherence to Young Earth Creationism does at times make evangelical groups sect-like. To risk generalization, evangelicals do put great emphasis on fellowship, and at times this can be excluding rather than including. At times Christians who do not accept the totality of what a particular church holds may find themselves ostracized and excluded from fellowship. This pattern is illustrated by what Bob Jones Sr. said of Billy Graham in 1953, that he "was doing more harm to the cause of Jesus Christ than any other living man" Jones then warned students in his college at Greenville, South Carolina, that they would be expelled if they attended a Billy Graham rally (Martin, 1992, pp. 317–318). Both Jones and Graham are evangelicals and thus our definitions must include both.

A BRIEF HISTORY OF EVANGELICALS

Church historians have long marginalized evangelicals and in the past have failed to either study them or realize their importance, resulting in a dearth of good histories of Evangelicalism. This may be due to the dominance of liberal Christians among protestant church historians, who unwittingly concentrate on the liberal wing of Protestantism. Since 1980 more historians have studied Evangelicalism, resulting in some very fine studies and now a five-volume *History of Evangelicalism* edited by David Bebbington and Mark Noll, of which three volumes have now been published. Put briefly, Evangelicalism began in the 1730s and grew slowly until 1790, when it began to expand rapidly on both sides of the Atlantic, with outposts in the British Empire. From about 1850 Evangelicalism was dominant in Britain and America until decline and liberalism set in about 1900. Thenceforth, Evangelicalism declined and retreated into the fortress of Fundamentalism, and it was regarded as a spent force. After World War II Evangelicalism revived again, and from about 1960 it expanded globally. The dominant figure in this expansion has been Billy Graham.

The Rise of Evangelicalism, 1730–1790[2]

Evangelicalism originated within the orthodox segments of British and American Protestantism in the 1730s. There was a sense that the church had become stagnant and needed revitalizing, as the churches had become less fervent after the rationalizing excesses of the mid-seventeenth century (Olson, 2004, pp. 83–110). It is difficult to pin the rebellion against "rational" Christianity down to one event, but in the 1730s, numbers of pastors were "enthused" on both sides of the Atlantic. Jonathan Edwards in New England, George Whitfield, who was commuting between England and the American Colonies, John and Charles Wesley in England, and Daniel Rowlands and others in Wales almost simultaneously pushed for a more emotionally involved religion. Edwards was a Congregationalist, and most of the English and Welsh evangelicals were Anglican. Thus the incipient revival began as within well-established churches, including the British established church—The Church of England. The outline of these events is well known, and in recent years scholars have begun to study them in far greater depth, producing a good and burgeoning literature.

The Evangelical Revival was not an unprecedented sudden irruption in the life of the English-speaking church, as there were precursors. These were initiated among Christians who belonged to churches of the Reformation and thus had protestant emphases on the Bible and Christian doctrine. Some of these proto-evangelicals belonged to the original English reformation church—the Church of England—and others to the "dissenting" bodies of Presbyterianism or Congregationalism, which came out of the Puritan movement. Thus the eighteenth-century evangelicals could look back for inspiration to the reformers and Puritans. Their more immediate precursors were Cotton Mather in the American colonies as well as the British Congregationalists Phillip Doddridge and Isaac Watts, and Anglican High Churchmen like William Law, who refused to swear an oath to William and Mary after the Glorious Revolution of 1688. In addition, the Moravians, who originated from Germany, had a great influence on the Wesleys and in much of North America. These disparate groups, coming from the Puritan tradition, High Anglicanism and Lutheran pietism, all contributed elements to form a new and distinctive tradition from an amalgam of older traditions.

Many history books oversimplify the origin of Evangelicalism and focus almost entirely on Wesley and the rise of the Methodist Church, as if that was the only product of the Evangelical Revival. Important as Wesley was, this simplistic account ignores the contribution of Edwards, Whitfield, and the many Anglicans. It also prevents Evangelicalism being seen as a multifaceted movement of great adaptability. Despite their geographical dispersion and their varied backgrounds, these revitalized Christians rapidly affiliated themselves to each other, either through personal contact or by letter.

It is essential to see the Evangelical Revival as Transatlantic in the literal sense rather than to focus on each country in turn. In fact, some significant events took place mid-Atlantic on board ship, most notably the developing spiritual angst of John Wesley and John Newton which resulted in their conversions. The most significant changes occurred in the middle of the decade. During 1734–1735, Jonathan Edwards was involved in a revival in Northampton, Massachusetts, where he was minister. That was no absolute novelty, as revivals had taken place in 1712 and 1718 while Edwards' grandfather, Solomon Stoddard was minister. Edwards was probably the greatest theologian of the eighteenth century and a prolific writer. His prodigious output included works on Original Sin, Free Will and "Some thoughts on the present revival of religion."

The revival associated with Edwards began to decline at the end of the 1730s, but in 1740 the revival in New England gained new life from an unexpected source when George Whitfield, an Anglican priest, came to Boston, the heartland of anti-Anglicanism. Whitfield had come to Pennsylvania in October 1739 and then worked southward to Savannah. After quarrelling with Anglican clergy he went back north and arrived in New England in September. He was not welcome in Anglican pulpits, but Dissenters made up for that, and thus the ecumenical side of Evangelicalism came into being. (Evangelicals often get on better with evangelicals from another denomination than with non-evangelicals in their own.) During the early 1740s Whitfield visited Northampton and preached for Edwards. A bond was forged and thus Evangelicalism became a transatlantic movement with preachers sharing pulpits or at least corresponding with each other, creating broad Methodist networks. (I use Methodist in the original sense as it was used indiscriminately of all evangelicals in the mid-eighteenth century.)

On the other side of the Atlantic, and within a few weeks of each other in South Wales, Howell Harris of Brecon and Daniel Rowland of Llangeitho had undergone experiences of forgiveness, which changed their lives. Also in the spring of 1735 the Oxford undergraduate George Whitfield had been converted. Two years later he stirred his hearers in London and Bristol, exhorting them to seek the new birth. In 1735, John Wesley traveled to Savannah for the Society for the Propagation of the Gospel and met with limited success. Both in Georgia and in London in 1738, Wesley had become acquainted with Moravian pietists, and on May 24, 1838, in Aldersgate Street in London his "heart was strangely warmed," which he considered to be his conversion.

When the birth of the Evangelical Revival is presented as the story of men such as Edwards, Wesley, and Whitefield, as it very often is, the impression is given that there was a sudden change in the life of the English-speaking church. Change there was, but it was slow, and for the next fifty years the evangelicals, or Methodists as they were often called, were a marginal group on the fringe of the mainstream churches. They were often vilified by the common man and the Anglican Bishops. Bishop Butler said to Wesley, of *enthusiasm*, the most hated feature of Evangelicalism, it was "a very horrid thing."

Suffice it to say that the ministry of Wesley and Whitfield was very peripatetic. Yet slowly, some Anglican clergy like William Grimshaw at Howarth (later of Bronte fame) and Samuel Walker, adopted the same faith but remained in their parishes. Thus Evangelicalism slowly grew in the Church of England. Some Anglican evangelicals had no contact with the Wesleys and Whitfields and their evangelical convictions were sui generis. Wesley, though an Anglican, worked as much outside the church structures as within and organized his circuits and classes with an iron fist without regard to the advice of any bishop. He sent workers to America, most notably Francis Asbury. In 1795, shortly after Wesley's death, his successors decided to leave the Church of England and form their own Methodist church, which was never Wesley's intention.

It is tempting to assume that evangelicals, with their religious fervor and biblical outlook, were both anti-intellectual and anti-Enlightenment. Evangelicalism is often cast as a reactionary movement; but as Bebbington has demonstrated, it was to a significant degree an expression of Enlightenment principles. Many evangelicals were close followers of Locke's *Essay on Human Understanding*, which argued that one's personal experience was basic for obtaining reliable knowledge. Thus evangelical religion was not merely the mental adoption of orthodox theology, but, most of all, an *experience* of what was believed, so evangelicals could say with Wesley, "My heart was strangely warmed." Evangelicals shared the Enlightenment's trust of the affections; however, they retained a historical allegiance to Scripture and traditional protestant orthodoxy. Mark Noll concludes his *Rise of Modern Evangelicalism* by pointing out that (Bebbington, 1989, pp. 42–74; Noll, 1992, pp. 140–141), "failing to pay full attention to the early evangelicals' exploitation of Enlightenment thought-forms would short-circuit explanations for the rise of Evangelicalism as much as failing to recognize how deftly Evangelicalism spoke to the shifting social landscape of the period." Part of the misreading of early evangelicals reflects liberal historians' disdain for the anti-intellectualism of much later fundamentalism.

The Expansion of Evangelicalism, 1790–1850[3]

After 1790, evangelicals increased almost exponentially for the next three decades. The growth of evangelicals, especially in Britain, coincided with the Wars with France, when the whole of British society swung rapidly in a conservative direction. During this decade radicals like Joseph Priestley were often vilified and subject to violence. By the end of the decade the evangelicals were poised to become the most significant church grouping throughout Britain, both through the leadership of clergy of all denominations and through politicians such as William Wilberforce. Growth in North America was slower. Evangelicals were a minor force in the fledgling republic and had had no influence in the setting up of the American Constitution, which had more to do with radical dissent and the Enlightenment than with Evangelicalism. The Founding Fathers looked more to Thomas Paine than to Wesley or Whitfield. (That irony is often lost today.)

Even though the 1790s was a crucial decade for evangelicals on both sides of the Atlantic, in 1800 evangelicals were still a small and often despised group within the churches. As the nineteenth century wore on, however, they came to prominence, almost dominance, throughout the English-speaking world, and Evangelicalism took root throughout much of the world. Evangelicals also spread their moral influence throughout British and American society.

From the 1790s, Anglo-Saxon Christianity flourished as never before, particularly its evangelical wing. In the United States this flowering of enthusiastic religion became known as the Second Great Awakening, with the Methodist ministers, Francis Asbury and Charles Finney as two of the major leaders. Whereas the First Great Awakening in America was dominated by Congregationalists and Presbyterians, the Second was dominated by Methodists, Baptists and Disciples, with the result that Methodists and Baptists began to outnumber the older denominations, initiating a trend that has continued until today. When the English born Asbury arrived in America in 1771, for example, there were four Methodist ministers, who had been ordained by Wesley against Anglican practice, and 300 laypeople. When he died as leader of the American Methodist Church in 1816, there were 2,000 ministers and 200,000 Methodists, making the Methodists the largest denomination in the United States.

In the generation after Asbury, Charles Grandison Finney (1792–1875) became the best-known revivalist. Finney's approach to revival was much more organized than that of the first evangelicals, who simply let the Holy Spirit take his course. Finney developed many practices, which became the norm in much evangelism, most notably the "anxious bench" where enquirers could sit at the front of the meeting place to be admonished on the state of their souls, and hopefully, to profess conversion. Finney originated from New York, but after his break with the Presbyterians he moved to Ohio, where he was heavily involved with Oberlin College. Though to him, evangelism was a priority, he also was an abolitionist, a pioneer in women's education, and a reformer in temperance, peace, sabbatarianism and care for the retarded.

During the First Great Awakening Baptists had been a small but significant minority, but the Second Awakening resulted in their rapid growth, particularly as a result of evangelizing the frontier population. Apart from an emphasis on believers' baptism, the Baptists' distinguishing mark was their emphasis on the independence of the local congregation. Despite their congregational autonomy, Baptists found ways to work together, setting up local associations and Missionary Societies. However, there were North–South tensions over slavery, especially over whether missionary candidates should own slaves. In 1845 this resulted in Baptists from the South withdrawing and forming the Southern Baptist Convention, which is now the largest protestant denomination in the United States.

Whatever one might think of the nation's religious life in antebellum America, it could not be described as dead. It was marked by growth, vitality and exuberance, and it spawned various separatist sects, including the Millerites and Mormons. Some observers of the American cultural scene, like Frances Trollope, mother of the novelist, Anthony, complained that "The whole people appear to be divided into an almost endless variety of religious factions" and she described Camp Meetings as "that most terrific saturnalia." In fact, the majority of the population professed church allegiance, and as Mark Noll points out, the nation's ethos, if not actual practice, was thoroughly evangelical (Noll, 1992).

In 1800, British evangelicals were about to become a major force religiously, politically and socially. The breakaway Methodist church was growing, and split within a decade. Anglican evangelicals had come to the end of a period of virtual stasis, during which they were loathed and opposed by the church hierarchy, and began to be accepted both by their church and society. In the Church of Scotland, evangelicals became a majority among the laity and the bias against the appointment of evangelical ministers resulted in the Disruption of 1843, when the evangelical leader Thomas Chalmers led out many to form the Free Church of Scotland. By 1810 Britain had an evangelical Prime Minister—Spencer Perceval, who was assassinated in the Houses of Parliament in 1812. By 1815 the first evangelical of many—Henry Ryder—was made an Anglican Bishop, and in 1848 an evangelical, John Bird Sumner, became an Archbishop. This period also saw the formation of many missionary societies both for missions overseas and at home, the British Foreign and Bible Society and other evangelical societies for good works including the (Royal) Society for the Prevention of Cruelty to Animals. The vast amount of evangelical activism, both evangelistic and in the encouragement of good works, resulted in the growth of all protestant churches and a reforming approach to society. There was a spate of church building during the early decades of the nineteenth century, and in 1851 a religious census in England and Wales showed that over 40 percent of the population were worshippers, half were Anglican and most the rest nonconformist with a small, but rapidly, growing number of Roman Catholics. A higher proportion of the population in Ireland and Scotland were worshippers. This was the highpoint of church attendance in Britain.

Evangelical Dominance and the Beginning of Decline, 1850–1910[4]

The successes of British and American Evangelicalism in 1850 seemed unstoppable. Evangelicalism was booming on the American frontier, in the North and in the South. Literature poured off the evangelical presses, whether tracts or weighty tomes. Colleges thrived. In Britain there was an Evangelical Archbishop of Canterbury and the newly formed Free Church of Scotland was growing rapidly. Methodists, Baptists, and

Congregationalists were at their peak. The Positivist novelist George Eliot even wrote about evangelicals sympathetically in *Adam Bede, Scenes from Clerical Life* and *Middlemarch.* Optimism on both sides of the Atlantic seemed well founded, as evangelical churches were growing, overseas and home missions were booming, and evangelicals had much influence in society. However the success and confidence gradually faltered and by the beginning of the twentieth-century Evangelicalism was beginning to decline.

In England the High Church or Anglo-catholics began winning over evangelicals and providing a more ritualistic worship after the middle of the nineteenth century, though their theology was conservative. An even greater challenge came from beyond the Rhine in the form of German biblical criticism. David Strauss and Ferdinand Baur where challenging the supernatural element in the Bible, and Friedrich Schliermacher was changing the whole basis of Christian theology by arguing that the Bible should be read, like other literary efforts, as a reflection of the society which produced it. Over the next half-century these new ideas became more and more accepted in mainline protestant churches, moving them away from their evangelical positions.

The 1860 publication of *Essays and Reviews,* which brought German critical perspectives to England, was the most significant religious event in Britain during the middle of the century, and the book was soon republished in America. In many churches there were heresy trials of clergy who adopted critical findings. Some—Joseph Colenso and two of the contributors to *Essays and Reviews* in England, W. R. Smith in Scotland, and Briggs in America were found guilty, but the liberalizing trend was unstoppable and the result was that most denominations moved in a liberal direction in the latter decades of the Victorian era.

Many evangelicals sought to counter this gathering together in a series of conferences in the 1890s at Niagara Falls to focus on what they called "The Fundamentals," which were in opposition to Modernism of any sort. These conferences brought together Revivalists and reformed scholars, but they did not stem the liberal tide, and by 1900 evangelicals were beginning to decline in their influence in most mainstream denominations.

During the whole of the nineteenth century there was a tremendous interest in prophecy and millenarianism throughout the Anglo-Saxon world. Proponents included serious Anglican theologians as well as wayside preachers. The end result was that many, if not most, evangelicals adopted some kind of millenarian scheme, often known as Dispensationalism, which almost became orthodoxy and is still widely held, if in a modified form, today. One minor millenarian group of the mid-nineteenth century has become very important in retrospect and that is the Seventh Day Adventist Church. Led by Ellen G. White, who claimed to receive divine messages from God, her followers regarded these messages as having parity with the Bible. She published a series of books, and in *Prophets and Patriarchs* she argued that Creation took place in Six Days and that Noah's Flood was a great catastrophe. These books, which were, and are, part of the foundation of Seventh Day Adventist belief, would be of little interest were it not for George MacCready Price, who wrote extensively on flood geology and anti–evolution in the first half of the twentieth century. Price looked to White for inspiration, and Price's ideas were later taken up and modified by Henry Morris, the leader of the Creation Science movement (Number, 1992).[5]

One result of the pragmatic emphasis of putting evangelism above theology was the development of the Holiness Movement, which was a strongly experiential, nonconfessional Evangelicalism, focusing on evangelism, holy living and, sometimes, ignoring academic theology. This movement had roots in the Pietist and Wesleyan traditions. Belief centered on Jesus, the need for conversion, and the living of a Christian life. Looking back to the early evangelicals, this movement can be seen as centering entirely on the "Religious Affections" of Jonathan Edwards, minimizing the intellectual affirmation of doctrine.

The de-emphasis on theology was accompanied by a commitment to *Dispensationalism*, which was a late nineteenth-century development of the premillenial and prophetic ideas of the early nineteenth century. These ideas involved an elaborate interpretation of Old Testament prophecy, especially Daniel and the Book of Revelation. Biblical prophecy was considered to be dominantly predictive and to be interpreted literally rather than symbolically. The scheme of future events began with the return of Christ to gather the believers *before* the Millennium, hence Premillennialism. In the early nineteenth century, prophetic writers included the highly erudite Evangelical Anglican G. S. Faber, whose scientific opinions we will discuss later, and Spencer Perceval, mentioned earlier, who thought Daniel 12 referred to Napoleonic France. In the 1830s, J. N. Darby developed the theme and it reached its apogee with the Dispensationalism of C. Schofield in the early twentieth century. Schofield's prophetic interpretation became almost fundamentalist orthodoxy, because of the wide use of the Schofield Reference Bible. Dispensationalism got its name because eight *Dispensations* were identified in the Bible, five in the Old Testament, the "New" Dispensation of Jesus, and in the future the Tribulation and the Kingdom. Though most Calvinists reject it, Dispensationalism has been the dominant prophetic belief of American evangelicals for over a century. It has been so pervasive that it permeates popular culture to the extent that the common perception of what Christianity holds about the end of the world and the return of Christ is a garbled version of Dispensationalism. One of the best-selling books of the 1970s was Hal Lindsell's *The Late Great*

Planet Earth, which sold over 15 million copies in the 1970s and was made into a film in 1979. Some of the best-selling novels over the last few years are the *Left Behind* series of Tim Lahaye and Jerry Jenkins, which develop dispensationalist themes in a neverending series of novels.

Apart from the fact that Dispensationalism in one form or another—pretribulationist, posttribulationist, etc.—has dominated American Evangelicalism, it is also significant for three other reasons. First, Dispensationalism takes the Bible literally, including the more poetic writings of the Prophets and Revelation. The literal hermeneutic is often, but not always, applied to the entire Bible, including Genesis. Second, there is a close correlation between those who hold some form of Dispensationalism and those who hold Young Earth Creationist views—that is who argue that the earth can be no more than about 10,000 years old, in spite of the modern scientific consensus that would make it several billions of years old. It is no accident that Lahaye and Henry Morris have been associated with each other for decades, and that Morris sees his Prophetic and Creationist beliefs as two sides of the same coin. Third, Dispensationalists believe the return of Christ, or rather the Rapture, is imminent and that the Lord "will not tarry" more than a few years. As this "tarrying" will be before oil reserves run out, there is little need the need for much environmental concern. Why worry if the lesser-spotted raccoon goes extinct if they would have done so in the Rapture a few years hence!

The combined influence of Holiness and Dispensationalism was innovative because it gave a new direction to Evangelicalism, but distanced it from the wider church and much intellectual endeavor, whether theological or scientific. It was coupled with a rather atomistic and literalist hermeneutic of the Bible, in which little consideration was given to whether any part of the Bible was poetic, narrative or doctrinal. With the detachment from any rigorous scholarly study of the Bible in favor of the claim that everyone is competent to study the Bible for themselves, the way was opened for flowery and fanciful interpretations of the Bible, of which Dispensationalism is just one. The argument for biblical interpretation by anyone can be seen as an extreme out-working of the doctrine of the Priesthood of all Believers, emphasized at the Reformation. In one sense it gives an inner cohesion to popular Evangelicalism but it has three negative consequences. First, it encourages schismatic and sectarian attitude, which has always been an evangelical failing. This attitude has resulted in the plethora of evangelical denominations and the non-denominational church. Second, it encourages a separatist attitude toward other Christians, who can be shunned for not being proper believers. Third, it encourages a fortress mentality, according to which evangelicals consider themselves to be under attack from secular forces and from mainstream Protestants, not to mention Roman Catholics (Marsden, 1987, 1991).

The Disruption of Evangelicalism, 1900–1940s[6]

The twentieth century saw a fall and subsequent renewal of evangelicals in both Britain and the United States. In 1900 evangelicals were a powerful force, but were being riven apart by various Modernist controversies. In reaction to the fragmentation of the late nineteenth and early twentieth centuries, two American businessmen funded the publication of the epynonymous *The Fundamentals,* which were published as twelve paper-covered little books in 1910. The authors were mostly not fundamentalist in the twenty-first century sense but were mostly competent conservative scholars writing on every aspect of Christianity, including science. They included two Anglican bishops, Handley Moule of Durham and J. C. Ryle of Liverpool, who had died some years before. B. B. Warfield of Princeton and James Orr of Glasgow each contributed several articles. Warfield was a proponent of Biblical Inerrancy, which Orr denied. None of these could be described as "Bible teachers," though several of the other contributors were. Frequently evangelical attitudes hardened into Fundamentalism a term suggesting those who generally accepted the views represented in *The Fundamentals* (in spite of their occasional inconsistencies)—after the First World War.

Fundamentalist Interlude, 1918–1946

In the United States, most evangelicals viewed Evolution along with Biblical Criticism and any form of Modernism or liberal theology as opposed to Christianity. The 1920s were thus marked by the Scopes trial and by battles for the control of denominations, in which the evangelicals usually lost to modernists. One result was that separatist denominations, seminaries and schools were founded. Another was that Modernists got control of the older denominations and seminaries in most denominations. By 1920 the older leaders in the classical evangelical tradition such as Warfield, Strong, and Wright were fading from the scene and were being replaced by a new breed with fewer denominational ties. Prominent were John Roach Straton, William Bell Riley, and Frank Norris, who achieved notoriety in 1926 for shooting and killing D. E. Chipps in his study after an altercation with the Roman Catholic mayor of Ft Worth. These new leaders were Dispensationalists, which they made the new evangelical orthodoxy after the First World War, particularly in North America.

The defining controversy of the 1920s was the Scopes trial of 1925 in Dayton Tennessee. This trial will be discussed in detail Chapter 6. However, it served was a symptom, illuminating the many issues behind the Fundamentalist–Modernist controversy, which rumbled on for decades.

The evangelicals who left their old, "impure," denominations to found a pure orthodox church, often split again after a few years. This happened with the Presbyterians. Following the rupture at Princeton Theological Seminary in 1929, Gresham Machen left to form the Westminster Theological Seminary. In 1936 Machen joined with other Presbyterian dissidents to set up an alternative church. The following year this church split into two groups, the Orthodox Presbyterian Church and the Bible Presbyterian Church, the former being confessionally Calvinist and the latter tinged with Dispensationalism and teetotalism. This schismatic nature of Fundamentalism resulted in a plethora of independent churches!

In Britain, unlike the case in the United States, there were no monkey trials, and the more conservative evangelicals waned while liberal evangelicals gained influence. This is no better seen than in the controversies of the Church of England in the 1920s. For several years there had been rumblings about missionaries from the evangelical Church Missionary Society teaching critical views of the Bible in Japan. After much controversy the more conservative Bible Churchman's Missionary Society was formed in 1922. Parallel with this was the dominance of liberal evangelicals and the formation of groups like the Anglican Evangelical group Movement (AEGM) which rapidly became part of liberal Anglicanism, leaving conservative evangelicals as a small remnant.

In spite of important differences, British and American evangelicals were brought together by two movements from the nineteenth century. The first was the Holiness Movement which found a quintessential English expression in the annual Keswick Convention held every July. The Holiness Movement has its roots in the teachings of Phoebe Palmer and Hannah Whittall Smith in the late nineteenth century. Theirs was a subjective and pietistic movement stressing the Inner Life. By focusing on Inner Life and a believer's personal relationship to God, the whole theological emphasis became claiming the power of Christ's blood, that is the forgiveness due to Christ's death, and the work of the Holy Spirit in sanctification. The second movement was Dispensationalism discussed above.

The effect of Holiness and Dispensationalism together was to present a form of Christianity which was not just long on personal faith and short on theology, but also one which was largely disengaged from the wider world. The social concern of early evangelicals such as Wilberforce, Finney, and Shaftesbury was quietly sidelined until it was revived in the 1970s. As a result, by 1939, churches in the United States were polarized into the liberal mainstream and the separatist evangelical or fundamentalist churches. Until recently, historians have regarded interwar fundamentalism as a dying relic and have ignored the fact that it retained the loyalty and trust of a good proportion of the American population. As a result, a numerically large base for evangelical Christianity was secured, while more liberal churches began to decline. Whatever can be said about the excesses of prewar fundamentalism, they retained a form of Evangelicalism from which the evangelical movement rose almost phoenix-like after the War.

The Renaissance and Global Expansion of Evangelicalism, 1940s–2000s[7]

The situation changed rapidly after the end of the Second World War, with a remarkable turnabout for evangelical fortunes. This occurred independently both in Britain and America. In the United States, an unknown evangelist, Billy Graham, began to organize evangelistic crusades, first in America and then throughout the world. Though Graham was no intellectual, he and his colleagues encouraged a more scholarly approach to their faith, and became associated with scholars like Carl Henry. This resulted in an evangelical renaissance with an emphasis on theological scholarship. Undoubtedly it is wrong to give Graham[8] the sole credit for this rebirth of Evangelicalism, but he was a major focal point.

Over the next few decades the intellectual renaissance grew with numbers of young theologians including Carl Henry, a systematic theologian, George Ladd, a New Testament scholar, and Bernard Ramm, a theologian of great importance to evangelical science, transforming the reputation of evangelical theologians. Henry and Ladd were associated with Fuller Theological Seminary, which was founded in 1947 by the radio evangelist Charles Fuller. Fuller Seminary offers a microcosm of the evangelical renaissance, as Marsden (1987) shows in his book *Reforming Fundamentalism*. It started as a conservative college, with a conservative faculty—Wilbur Smith, Everett Harrison, Carl Henry, Harold Ockenga, and Harold Lindsell, who were Inerrantists; but matters began to change when E. J. Carnell joined the staff. Carnell questioned strict inerrancy in his 1959 *The Case for Orthodox Theology*. Then, Daniel Fuller, son of the founder and Dean of the seminary, argued for a limited doctrine of inerrancy during the 1960s.

Soon the more conservative evangelicals began to be concerned lest their doctrines, especially inerrancy, should be diluted by theological liberalism. In 1976, Harold Lindsell made this manifest in his book *The Battle for the Bible*, which bewailed the drift from inerrancy and devoted a whole chapter to "The Strange Case of Fuller Theological Seminary" In 1979 a number of evangelical theologians met in Chicago and produced the Chicago Declaration on Inerrancy, which will be discussed in a later chapter. During the 1970s and 1980s, controversy also struck the Lutherans and the Southern Baptists. With the Lutherans two main churches resulted; the broad ELCA (Evangelical Lutheran Church of America) and the conservative LCMS (Lutheran Church Missouri Synod). The LCMS was both very conservative theologically and officially adopted Six-Day Creationism as its foundational principle.

Up to the 1970s the Southern Baptists had allowed moderate critical positions regarding interpretations of the Bible and did not insist on inerrancy But then they swung to the right and numbers of seminary professors were obliged to leave to be replaced by more conservative theologians who were Inerrantists. This has resulted in the largest Protestant denomination in the United States being highly conservative and sympathetic to Young Earth Creationism. The emphasis on Inerrancy has encouraged many evangelicals to apply their version of literal inerrancy to the cosmological and "scientific" aspects of the Bible and thus to adopt a Young Earth position despite the Chicago Declaration.

The burgeoning of Evangelicalism in the United States from 1950 onwards has had several apparently contradictory results. Many might have expected that as the clergy became more intellectual there would be a liberal drift. Such a drift has happened, but it has been paralleled by a new conservative swing, which has considerable intellectual content in its conservative theology. Gone are the days of the caricature of anti-intellectual supporters of old-time religion as portrayed in *Inherit the Wind.*

As the prime medium for delivering religion to the populous at large changed from radio to TV, many evangelists developed that means to spread their faith, and, at times, to line their pockets. Despite a series of scandals concerning tele-evangelists, their influence is considerable and vitality undiminished. The more conservative and Young Earth evangelicals have dominated this medium. Since the early 1970s many evangelicals have also been espousing right wing political causes. Leaders of this political movement among evangelicals have included Pat Buchanan, Jerry Falwell, and Pat Robertson. For a time their followers were known as the Moral Majority and then as the Religious Right. In addition to their general right wings views, political evangelicals have typically emphasized antiabortion.

In Britain, Evangelicalism also revived after World War II with John Stott, Rector of the Anglican church of All Souls, Langham Place and Martyn Lloyd Jones, the minister of Westminster Chapel as the two dominant leaders (Dudley-Smith, 1999, 2001). Steady growth was boosted by the Billy Graham rallies of the 1950s and 1960s. Many Anglican clergy look to these rallies as what made them Christian in the first place, though they are fast approaching retirement age. Postwar evangelical growth in Britain was not without the schismatic problems which faced Americans. At an evangelical conference in 1966, Lloyd Jones exhorted Anglicans and others in mainstream churches to leave their denominations and form a pure evangelical super church. Due to Stott's leadership, this did not happen; but since then there has been a slight rift between Anglican evangelicals and Independent evangelicals. From 1970, evangelicals have been the dominant group in the Church of England, in marked contrast to the Episcopal Church of the United States of America, the American branch of Anglicanism, in which only a tiny minority are evangelicals. Many non-Anglican evangelicals in Britain have united under the umbrella of the FIEC (Federation of Independent Evangelical Churches). While Anglican evangelicals tend to reject inerrancy and play down hell, many from the FIEC emphasize both. It is no accident that migrating English Anglicans tend to settle at the Fuller theological seminary.

Evangelicalism Goes Global, 1960 Onwards

Evangelicalism began as an Anglo-Saxon movement in Britain and North America (with some German influence) and remained limited to English-speaking regions until the nineteenth century. During that century, evangelical missionaries went from their mother countries to many parts of the world and set up evangelical churches there, though often these were part of mainstream denominations. This is particularly so in the old British Empire. But missionaries also went to all parts of the world, especially China. By 1900 there were small indigenous evangelical churches in most countries but they had limited indigenous leadership. By 1960 there were possibly over 100 million evangelicals worldwide, but Western evangelicals outnumbered non-Western by two to one. Since 1960 the growth of non-Western evangelicals has been exponential and now they outnumber evangelical westerners by three to one—there are 300 million non-Western evangelicals and just over 100 million Western evangelicals. Assuming the same rate of growth, by 2010 non-Western evangelicals will outnumber westerners by ten to one and there will be more evangelicals in China than in the United States. Since 1960 there has been intense evangelistic activity, much emanating from then United States; but increasingly, Third World nations are sending out missionaries to other countries including those of Europe.

Increasingly, non-Western evangelicals are making their presence felt, as is happening in the Anglican Communion at the present moment. In both Britain and North America, for example, evangelicals from Africa, Asia, and South America are opposing moves among Anglicans and Episcopalians to tolerate homosexuality. There are 32 million African Anglicans, most of whom are evangelical, as opposed to 1.5 million in the United States and 1.5 million in Britain (though 25 million claim to be Anglican in the United Kingdom). Elsewhere, Third World evangelicals are making their mark as well. These include Luis Palau from Argentina; the New Testament scholar, Seyoon Kim, from Korea, and many others. Though at present the United States may dominate the evangelical scene through sheer weight of resources and missionary involvement, the balance of power and involvement is shifting to the non-Westerners.

This shift from North America may be seen in the following table, drawn from *Operation World*.[9]

Continent	Population ($\times 10^6$)	Evangelicals ($\times 10^6$)	Percentage	% Growth
Africa	784	116	14.8	4.6
Asia	3692	133	3.6	7.6
Europe	729	17.4	2.4	1.8
Latin America	519	55	10.6	4.0
North America	310	94	30.3	2.0
Pacific	31	4.7	15.2	0.7

Even if these figures are not totally accurate they illustrate the trend. The majority is no longer WASP, and non-WASP evangelicals are growing at least twice the rate of North American evangelicals. The growth rate in Europe is small at 1.8 percent, but that is largely growth in a population which adheres to Catholicism, Orthodoxy, or a mainstream Protestant church, all of which are in decline.

Increasingly those from outside Anglo-Saxon countries are gaining in influence on the worldwide stage and these persons are less influenced by Enlightenment ideas and thus more skeptical of science, especially Darwinism. African evangelicals, for example, are far more convinced of the existence of spirits and demons than their Western counterparts. From a Western perspective they may thus be seen as less scientific, or naturalistic, in outlook. Evangelicals from China and the old Soviet bloc are also skeptical of Enlightenment views because the ideology of the Communist Blocs has been anti-Christian and part of the Marxist ideology was to use Darwinism as a weapon against Christianity. Reaction against Marxist uses of Darwininian evolution has already provided fertile ground among Chinese and former Soviet evangelicals for accepting a Young Earth Creationism.

THE WORLD DISTRIBUTION OF EVANGELICALS TODAY

Much of this chapter has traced out the development of evangelicals as a largely Anglo-Saxon phenomenon, but the last section has described how evangelicalism is now a global movement, which will be very significant in the foreseeable future. Furthermore, as the majority of evangelical missionaries originate from the United States, there is a very large American evangelical influence throughout the world. Missionaries are to be found in France, the Philippines, Africa, Latin America, Asia, and more and more in the former Soviet bloc countries. To this must be added the missionaries from almost every country in the world being sent to other countries. This survey of Evangelicalism in the non-Western world might initially seem superfluous in a book on evangelicals and science; but just as we will consider the impact of evangelicals on science and science teaching in the United States in recent decades, we also need to look to the future and consider how evangelicals may impact science and the teaching of it in countries as diverse as Argentina, Nigeria, and Russia. By understanding the dynamics of the evangelical movement as it spreads throughout the world, we may be able to anticipate what will happen in various countries, as evangelicals become numerous enough to have some political clout. Already there are indications that this will follow the American pattern.

Evangelicals in the United States of America at the Beginning of the Twenty-First Century[10]

Evangelicals make up at least a third of the American population. Though many of these may be cultural evangelicals rather than regular church attenders. Most evangelicals belong to monochrome evangelical churches, whether large denominations like the Southern Baptist Convention or the Lutheran Church Missouri Synod, but there is also an array of smaller separatist churches and independent churches. Within the mainline Presbyterian, Congregationalist and Baptist churches dating from the seventeenth century there are a minority of evangelicals as there are in the Methodist churches. Evangelicals have always been a small minority in the Episcopal Church. In the twentieth century several originally continental confessional churches like the Lutheran Church Missouri Synod and the Christian Reformed Church (Dutch Reformed) have affiliated with the evangelicals. However, prohibition minded evangelicals found it problematical that "hymn singing and theological discussion were often lubricated by lager and pursued through clouds of cigar smoke" (Noll, 1992, p. 482). There is a considerable range in theological opinion but the balance of evangelical opinion in the United States is strongly in favor of Biblical Inerrancy and with a bias to some kind of Dispensationalism. On the periphery of Evangelicalism, and highly significant to science, are the Seventh Day Adventists, who add to evangelical belief the works of Ellen White and also worship on a Saturday.

Evangelicals in the United Kingdom and the Old British Dominions

Compared to the United States Britain is not a church-going nation, with less than 10 percent regular worshippers. About half of these are evangelical. The other major difference is the high proportion of Anglican (Episcopal) evangelicals, which is now about 35 percent. This situation stems from the 1730s when the Evangelical Revival began in Anglican churches of England and Wales. Since the older protestant churches went "liberal" in late nineteenth century there has always been a proportion of evangelicals in the Methodist, Congregationalist, Baptist, and Scottish Presbyterian churches. These probably account for a half of British evangelicals and the rest belong either to Pentecostal or independent evangelical churches, who have grown dramatically since 1970. Because of the numbers from mainstream churches British evangelicals place less emphasis on Inerrancy and Dispensationalism, creating a very different ethos. There is also less of a taboo on alcohol than is common among American evangelicals.

In the eighteenth and nineteenth centuries Britain exported its churches, along with evangelicalism to its "White colonies" Canada, Australia, and New Zealand, so that the church situation there is similar to that of Britain, though the proportion of evangelicals is somewhat higher.

Western Europe[11]

Half a century ago there were very few evangelicals in Europe. In Catholic countries there was at times a small, marginalized, Protestant church, with little or no evangelical ethos. However that is changing, largely due to evangelistic groups from the United States. For example, in 1940 there were 40,000 evangelicals in France. In 2000, there were 477,000 evangelicals, as opposed to 300,000 in the traditional Protestsant Eglise Reformee. Thus in every medium-sized town in France there is now at least one evangelical church. The situation in Italy, Spain and Portugal shows a similar growth of evangelicals, particularly in Portugal where evangelicals are 3 percent of the population compared to 8 percent in Britain. The change in mainland Europe is in part due to evangelism from mission groups such as Operation Mobilisation and France Mission. The situation in northern Europe is different because Protestant churches took root at the Reformation. There the bulk of the population is affiliated to one of the older mainstream, but declining, Protestant churches and evangelicals can claim up to 5 percent of the population. In most Western European countries evangelicals are increasing at the rate of 2–5 percent a year and will thus become even more significant in years to come. From a European perspective continental evangelicals will soon be more numerous than British who are growing at only 0.3 percent per annum.

Eastern Europe, Including Russia

Until the fall of Communism in 1989, Christians were persecuted or harassed by the Soviet regimes, whether they were Orthodox, Catholic, or

Protestant. Vast numbers of churches were closed and often clergy were killed or imprisoned. In Russia, the Russian Orthodox Church had to compromise with the authorities to survive and in Poland the Roman Catholics adopted a silent opposition until the election of Pope John Paul II gave them a new confidence and possibly hastened the collapse of the Soviet system in 1989. After years during which being an acknowledged Christian meant having no prospects and during which the state promoted scientific atheism, the change was immense. Since 1989 many evangelistic groups have descended on Eastern Europe to take advantage of the evangelistic opportunities. "Taking advantage" is the right word, as some Western evangelicals have shown little respect for the indigenous churches. They have met with opposition from the older churches and in some countries their activities are being curtailed. Many are from the conservative side of Evangelicalism and thus bring with them a commitment to Young Earth Creationism.

Africa, Asia, and Latin America

It is these continents which have seen both a dramatic rise in the number of Christians and evangelicals in the last half-century. In 1980 over half of all Christians came from these continents, and the figure now stands at two-thirds. However they contain three-quarters of all evangelicals. Part of the growth has been due to Western missionaries, especially in the early years, but the greater part has been due to indigenous evangelical movements, which started within a few decades of the arrival of Western missionaries. Protestant missionaries, mostly of an evangelical persuasion first came to each of these continents by the mid-nineteenth century. The churches are now indigenous, with almost entirely local clergy and leadership, but there is still considerable influence from the United States, which is often financial.

EVANGELICALS AND SCIENCE

The common perception is that evangelicals are opposed to any kind of science, but especially that associated with evolution or geology. This perception is given credence by the monkey trials of the 1920s and *Inherit the Wind* as well as by the influence of the Young Earth Creationist Movement during the last forty years. As a result of the common acceptance of the *discredited* conflict thesis of science and religion, this perception is difficult to contradict. Too often it is assumed that the conflict seen in recent decades over Young Earth Creationism is simply a continuation of past confrontations. This assumption is used by different groups of people both to denigrate evangelicals and to support Young Earth Creationism. But the interactions between evangelical Christianity and the scientific community have been complex, and they have as often been mutually supportive as they have been.

This chapter has tried to give a brief outline of the development of Evangelicalism, breaking its history into six important periods. In the remaining chapters we will consider the historical relationships between Evangelicalism and science, during those same periods. Then we will turn to explore several contemporary ethical concerns raised by scientific practices, articulating evangelical responses to them. Finally, we will conclude by trying to look into the future of evangelical/science interactions.

FUNDAMENTALISM

Earlier I used the word "Fundamentalism" and discussed its rise. The term is very generally used to label, and even denigrate, hard-line adherents of most religions, especially Muslims and Christians. The way fundamentalism is used to describe extreme forms of Islam is not our concern here, but only when it is used for Christians. Initially in the early twentieth-century fundamentalism had no bad connotations and merely described those who adhered to the fundamentals of protestant Christianity. These were often summed up in five headings:

1. The Authority of the Bible
2. The atonement
3. The Virgin Birth
4. The Resurrection
5. The work of the Holy Spirit in the believer

After the First World War it was used to define those American evangelicals, who objected to the Modernist drift of the mainstream protestant churches. It rapidly acquired negative connotations and thus George Marsden defines Fundamentalism in this context as "militantly anti-modernist Protestant evangelicalism" Some American evangelicals happily called themselves Fundamentalists and aggressively lived out Marsden's definition. Following the rebirth of evangelicalism after the Second World War the word changed in meaning as those who followed Billy Graham and other leaders insisted that they were Evangelicals and that Fundamentalism was "a bad thing." As the word has now evolved it used to express the user's negative vibes toward Evangelicalism (if they are not evangelicals) or to more hard-line forms if they are. Thus someone like Bishop John Spong will dismiss all evangelicals as "Fundamentalist," but moderate evangelicals will use it only to label those whom they consider to be immoderate. Wags may define a Fundamentalist as a Christian who is more conservative than they are! Frequently use of the term simply adds heat rather than light; thus I shall only use the term very sparingly indeed, either in its original connotation, or in the negative sense. I hope my meaning will be clear from the context.

NOTES

1. *World Christian Encyclopedia* by Barrett and *Operation World* by Johnstone and Mandryk are the two authorities on statistics. I have used *Operation World* as it is more available.
2. *The Rise of Evangelicalism* by Noll deals with a similar period.
3. *The Expansion of Evangelicalism* by Wolfe deals with the same period.
4. *The Dominance of Evangelicalism* by Bebbington deals with this period.
5. Also in Chapters 6 and 7 in this book.
6. Marsden's *Fundamentalism and American Culture* and *Understanding Fundamentalism and Evangelicalism* has formed the basis of much of my treatment.
7. *The Global Diffusion of Evangelicalism* by Stanley will deal with this period when published.
8. Graham has attracted too many biographies to list, both hagiographical and critical!
9. Based on Johnstone and Mandryk's *Operation World.* Their figures are consistently higher than Barrett.
10. Partly based on *A History of Christianity in the United States and Canada* by Noll.
11. No studies are presently available and this is based on personal knowledge and Johnstone and Mandryk.

Chapter 2

Evangelicals, the Bible, and Science

With evangelicals being people of the Book, it is inevitable that they wish to reconcile science with the Bible. At a popular level, evangelicals adopt a *default* literalism and "follow" the plain meaning of scripture rather than "interpret" it. Thus many interpret early Genesis literally and accept YEC (Young Earth Creationism) by default. A similar process occurs with prophetic books and Revelation, whereby many adopt Dispensationalism, evidenced by the popularity of the *Left Behind* series of novels on the Rapture.

The second concern is "suffering and death before the Fall," and one of the main attractions of YEC is that it denies even animal death before the Fall (Morris and Whitcomb, 1961, pp. 454–489; Sarfati, 2004, pp. 195–225). As God surely made Creation "very good" with an absence of suffering and death then these entered the world as a result of the sin of Adam and Eve in Eden (Genesis 3). Thus suffering is part of the "curse" due to the Fall. Further many believe that salvation comes through Jesus's death on the cross, when He conquered death, which is the result of sin. Hence billions of years of death before Adam nullifies the Christian Faith, especially the atonement. Many YECs make this the basis of the biblical and evangelistic appeal of YEC.

As Evangelicalism is centered on the Bible, much of the controversy of Evangelicalism and science soon returns to these two points. Thus to understand this, we need to consider how evangelicals approach the Bible, and have done since 1730, in respect of its authority, its nature as inerrant or not, its interpretation and how evangelicals interpret the Bible today, with a focus on Genesis 1–11.

THE AUTHORITY OF THE BIBLE

As Bebbington (1987) demonstrated, the Bible is one of the four distinguishing marks of Evangelicalism, as the Bible is the authoritative Word of God. To say that the Bible *is* the Word of God emphasizes that evangelicals consider the Bible to be revelation and thus communication from God. Here they differ from liberal Christians who see the Bible as *containing* the Word of God, or a record of man's experience of God. In doctrine and ethics, evangelicals appeal to the Bible. The problem comes in assessing what the Bible says, especially with regard to science. The popular perception is that evangelicals are literalist and insist on a six-day creation. The prominence of YECs gives credence to that, but in fact, until twenty years ago most evangelicals have not been literalist. Since 1730 there has been a great variety in the ways that different evangelicals appeal to the authority of the Bible. All emphasize its historical reliability particularly on miracles, but they differ on whether the Bible is *inerrant* or should be interpreted literally. Early Genesis has been a focus of attention.

THE PROTESTANT DOCTRINE OF SCRIPTURE INHERITED IN THE 1730s

As evangelicals emerged from orthodox Protestantism in the eighteenth century, they adopted the same understandings of the Bible. These included its final authority in any matter of faith and doctrine, with church tradition subordinate to Scripture, hence *sola scriptura. Sola scriptura* was a reformation slogan to emphasize the Bible's final authority, rather than the *only* authority. Some evangelicals take this to mean the *only* authority and thus reject any interpretation based on "non-biblical knowledge." The Reformers with their Humanist/Renaissance heritage had a wider perspective and as heirs of the catholic tradition also valued church tradition and the early church Fathers like St. Augustine. This is particularly seen in the work of John Calvin.

As most early evangelicals were of English origin, their main doctrines, except for ministry and church order (and baptism for Baptists), derived ultimately from the Anglican teaching of Cranmer and the Edwardian Reformation of 1547–1553. The official view of the Anglican Church from which all Anglo-Saxon churches were hewn, is found in the *Thirty-nine Articles* of 1559, which were reaffirmed in 1662 in the *Book of Common Prayer.* These articles summarize the Christian Faith and Article VI deals with the Bible.

> Holy Scripture containeth all things necessary to salvation: so that whatsoever is not read therein, not may be proved thereby, is not to be required of any man, that it should be believed as an article of Faith, or be thought requisite of salvation.

The fact of its inspiration and authority were assumed, as it was common ground with Roman Catholics.

The following century under Cromwell, the Church of England was suppressed and the Westminster Confession written in 1644. After 1660 this became the doctrinal standard of English speaking Presbyterian churches. The confession was produced in question and answer form and Question 3 reads:

Q. 3. *What is the Word of God?* A. The Holy Scriptures of the Old and New Testament are the Word of God, the only rule of faith and obedience.

Neither statement makes any reference to inspiration or inerrancy. In view of later developments it is often thought that the Reformers laid the foundation of inerrancy and a literalist interpretation, with the Old Testament as of great importance as the New. Thus an obscure law in the Old Testament (e.g. Lev. 19.27) is as important as the Sermon on the Mount. In fact, the Reformers recognized both *Discontinuity* and *continuity* between the Testaments and that the New is superior to the Old (McGrath, 2001, p. 164).

The thirty-nine articles mentioned Creation in passing but the Westminster Confession was far more explicit, referring to Creation in six days:

Q. 15. *What is the work of creation?* A. The work of creation is that wherein God did in the beginning, by the Word of His power, make of nothing the world, and all things therein, for Himself, within the space of six days, and all very good.

The first confession of faith to mention explicitly the six days of creation were the Irish Articles of the (Anglican) Church of Ireland of 1615. These 104 articles were drawn up by James Ussher (1581–1656), later the Archbishop of Armagh, while he was Professor of Divinity at Trinity College, Dublin. They were largely based on the thirty-nine articles and later formed the basis of the Westminster Confession. The eighteenth article deals with creation;

Article 18. In the beginning of time when no creature had any being, God by his word alone, in the space of six days, created all things, and afterwardes by his providence doth continue, propagate, and order them according to his own will.

The most well known of Ussher's works was *Annales Veteris Testamenti* (1650), which was a solid piece of chronological scholarship in which he argued from historical grounds that Jesus was born in 4 BC. But he is remembered for his date of creation—4004 BC. Despite popular representations, he did not arrive at this figure from arithmetic applied to dates of patriarchs and other Old Testament figures. To Ussher there were six Chiliastic days of 1,000 years apiece followed by the seventh day of the millennium. There were four Chiliaistic days before Christ and thus Creation took place in 4004 BC, on the night before October 23. Adam was created on October 28. However, this is good seventeenth century historical scholarship. He dated the sack of Jerusalem by the Babylonians described in II Kings 25 in 588 BC, close to today's accepted date of 587/6 BC. His dates earlier than King David are rejected by all except the most conservative evangelicals, for example Exodus in 1491 BC, the departure of Abraham from Ur in 1921 BC and the Noachian Deluge in 2468 BC.

In 1704 Ussher's dates were printed in many editions of the Bible. This gave the impression that these were the official biblical dates. It has also resulted in Ussher becoming a figure of fun. Yet this was not part of the Reformers' heritage, nor was it the belief of most educated Christians from 1660 to the end of the eighteenth century when geologists made the vast age of the earth manifest (Roberts, 2007).

The Reformers' interpretation of Scripture developed from the new learning of the Humanists, such as Erasmus, and the way they studied all ancient texts, whether divine (hence *divinity*) like the Bible, or human (hence *humanist*) like Classical literature. Calvin's own study of the Bible was based on his humanist background as is seen in his commentary on Seneca. The Reformers appealed to the *literal* sense of Scripture and rejected the mediaeval methods of allegorization. The word *literal* means the plain sense of scripture rather than slavish literalism, which plagues popular Evangelicalism. It was a *literal* exegesis rather than allegorical. An irony is that the popular hermeneutic of today stemming from Dispensationalism, which claims to be literal, ends up as *allegorical* as any medieval exegesis.

ACCOMMODATION

The refusal to adopt a slavish literalism is seen best in Calvin's understanding of the *accommodation* of Scripture, which he based on earlier theologians including Augustine. In 1554, eleven years after Copernicus published *De revolutionibus*, Calvin published his commentary on Genesis. Calvin made no reference to the Copernican theory and stressed that *Genesis* was not written to teach astronomy. As he dealt with the Mosaic description of the firmament of Genesis 1 vs 6 he wrote, "He, who would learn astronomy and other recondite arts, let him go elsewhere" (Calvin, 1847, p. 79). He considered the firmament not to be the solid dome, which is implied by Egyptian astronomy, but a representation of rain clouds, because "nothing is here treated of but the visible form of the world." Calvin was wrong at this point as most ancients considered the *firmament* to be a solid dome (Seely, 1991, 1992). In other words Moses *accommodated* himself to the limitations of human thought and as Calvin commented on chapter 1 vs 15, "For as it became a theologian, he had respect to *us* rather than the stars." Calvin approached his task with Ptolomean assumptions of a spherical earth and did not question a 6,000-year-old earth or a universal flood. His accommodating interpretation eased the path for many to accept Copernicanism, which some Roman Catholics called the "Calvino–Copernican" theory. In the following centuries Calvin's doctrine of *accommodation* allowed Protestants to accept the findings of science without rejecting the authority of scripture (Hooykaas, 1972, pp. 117–124). McMullen argues that *accommodation* goes back to Augustine and that Galileo took over his ideas at a time when both Protestant and Catholic had become more literalist (McMullen, 1998).

In the eighteenth century many writers, both Catholic (Fr J. Needham and Buffon) and Protestant (de Luc), utilized a form of *accommodation* to incorporate long geological time in Genesis One (Roberts, 2007). A recent study of de Luc by Martin Rudwick (Rudwick, 2001) demonstrates that de Luc developed Ussher's historiography to include geological time. Lutherans were more literalist than Calvinists and less inclined to adopt *accommodation* and thus more opposed to Copernicanism.

THE CHAOS OF GENESIS 1 VS 2

Alongside this literal exegesis, tempered by *accommodation*, many writers considered the earth being created "without form and void" to be paralleled by many classical writers, such as Ovid and Heisiod, who wrote of the formation of an original *chaos*. Calvin, who surprisingly for a classical scholar, made no reference to these writers, referred to the earth being a "rather shapeless chaos (on Genesis I vs 2). Many Reformation and Renaissance writers were more explicit and Williams states that "[f]ew commentators could refrain from quoting the opening lines of the [Ovid's] *Metamorphoses* . . ." (Williams, 1948, p. 49). Thus Genesis One was interpreted as God first creating *chaos* (i.e. without form and void) and then subsequently re-ordering this chaos in six days. This, in fact, opened up the way for a longer time span of creation, as the duration of Chaos was undefined. Up to about 1650 Genesis could not be interpreted in the light of geological evidence concerning the age of the earth.

The Church Fathers of the first five Christian centuries had varied approaches to Genesis. Some like Theophilus of Antioch in about 180 AD interpreted it literally at calculated the Creation at 5515 BC (Theophilus, 1970). Augustine of Hippo was ambivalent seemingly both hold a figurative sense of day and a literal one. Other Fathers were clearly figurative (Van Till, 1996). The period of the Reformation resulted in a more rigorous biblical interpretation with an emphasis on the literal rather than allegorical meaning. This inclined most theologians to understand the day of Genesis as of twenty-four hours and thus many writers, Protestant and Roman Catholic, reckoned the earth to have been created in about 4000 BC.

THROUGH THE SEVENTEENTH CENTURY INTO THE 1730s

During the Renaissance all knowledge was considered a unified whole and thus "Biblical History" was related to other spheres of knowledge both classical and modern. Genesis was considered with reference to classical writers who spoke about chaos. This is seen in many writers of the early seventeenth century: Grotius, Mersenne, Bacon, and Descartes. Grotius in *The Truth of the Christian Faith in Six Books* argued that "the most antient tradition among all Nations [Phonecian and Greek] is exactly agreeable to the Revelation of Moses" (Grotius, 1719, section XVI) and his work was translated and used throughout Europe. Later writers, like Chalmers, cited Grotius in support of a chaos of undefined duration.

A few decades later in 1656 Ussher published *Annales Veteris Testamenti*. Although his chronology from the first humans in 4004 BC to the time of Christ was widely accepted, his strict understanding of six days was not, and was almost a minority opinion over the next 150 years. His influence on the churches is grossly exaggerated and during the next century many writers passed over him in silence. But the date of 4004 BC in the margins of many English Bibles gave the impression that this was orthodox belief on the date of creation.

Toward the end of the seventeenth century many *Theories of the Earth* were published in Britain by such writers as Burnett, Whiston, Woodward, Ray, and Hobbes. These are often considered to be an attempt to rationalise the early history of the earth into six days, but, in fact, all writers allowed an indefinite time for *chaos* and melded Genesis, classical writers, scientific observation, and speculation into a seamless whole. Burnett wrote of indefinite chaos, "so it is understood by the general consent of commentators" and the commentator Bishop Simon Patrick wrote of the duration of chaos that "(I)t might be a great while"(Roberts, 2002, pp. 145–148).

The chaos-restitution interpretation was adopted by most commentators in the eighteenth century—including the evangelicals Wesley, Gill, and Pantycelyn (writer of *Guide me o thou Great Redeemer*) (see Chapter 3). Many did not refer to the duration of chaos. This interpretation formed the basis for a libretto for a planned oratorio by Handel, which Haydn acquired in 1792 and used in *The Creation*. The whole schemata of original chaos followed by development was an essential part of the eighteenth-century worldview, whether Christian or not, and was bequeathed to the evangelicals and forms the historical background to the ways evangelicals and other Christians have understood Genesis and geology since 1800, particularly in the development of the "Gap Theory."

By 1730 heliocentricity was long resolved and almost all educated Protestants were Copernicans except for a few eccentrics the matter was closed, as all had accommodated their understanding of scripture to heliocentricity.

This summarizes the consensus of Protestant biblical understanding in the 1730s when the Evangelical Revival began. Obviously, evangelicals strenuously rejected the interpretations of the Deists and the scatological biblical criticism of the coffee shop atheists of London. Many early evangelicals were highly educated, whether from Yale or Oxford, so that their background was orthodox Protestantism, whether Congregationalist or Anglican and full of the learning of Newton and Locke. Thus, the early evangelicals accepted both the Bible as the ultimate authority *and* Newtonian science, which provided the second book of revelation, God's Works, augmenting the book of God's Word. This commanded almost universal acceptance, except for the few, who adopted Hutchinson's Mosaic science. To my knowledge no evangelical until the late twentieth century questioned Copernicanism, except for late nineteenth century American German Lutherans, and some devotees of Frances Turretin, the seventeenth-century Calvinist.

EVANGELICAL INTERPRETATION FROM 1750 TO 1850

As the growth of Evangelicalism resulted in theological diversification it was inevitable that as new sciences impinged on biblical interpretation, it would open up diverse interpretations of the Bible. Some were more accommodating than others leading to some controversy over geology in the nineteenth century. During the eighteenth century evangelicals, like most Protestants, were either literalist or semiliteralist in interpretation, but allowed some "accommodation." After 1770 geologists like de Luc, de Saussure, Hutton, Smith, Werner, Cuvier, and others started to demonstrate the antiquity of the earth, Christians rethought their position (Rudwick, 2004; Roberts, 2007). Because of the popularity of the Chaos-Restitution and Day-Age interpretations a radical rethink was unnecessary for those familiar with the natural philosophy of the day. Those without the knowledge of natural philosophy were more liable to take the Bible at face value, and adopt a literal hermeneutic. However, attacks on the possible heresy of those who accepted a vast age were rare and attacks were usually confined to the possible deism of writers like Hutton, rather than against a more elastic view of Genesis. Thus de Luc and Kirwan attacked Hutton not for accepting deep time but almost allowing the earth to be of an infinite age.

I can think of no example of evangelicals before 1800 criticizing geology on theological grounds, though some, for example Thomas Scott, Charles

Simeon, Andrew Fuller, and John Newton simply ignored the question. Some did defend "Genesis" against the geologists, notably the Roman Catholic Chateaubriand (1768–1848) in France[1] and the High Church Hutchinsonians in England, who dissociated themselves from evangelicals, but these were minority views among educated Christians. Taking the sequence of the *Theories of the Earth* with God first creating Chaos, re-ordering Creation with man being created in about 4000 BC and then the Deluge evolved into a vastly extended Chaos, which encompassed a multiplicity of Deluges, with Noah's Flood being the last of many, theologians quietly slipped geology into the Chaos. The first theologian who is *known* to have done this was Thomas Chalmers at St. Andrews in the winter of 1802 (Hanna, 1852, vol. 1, pp. 79–80).[2] This was before Chalmers became an evangelical in 1811 but his understanding of geology and Genesis remained unchanged. Many writers, notably Henri Blocher and Weston Fields, wrongly credit Chalmers with a novel interpretation. Blocher wrote that Chalmers "was seeking to reconcile Genesis with the new discoveries about the age of the earth" (Blocher, 1984, p. 41), but Fields reckoned that "Chalmers deemed it *necessary* to harmonise the Scriptures and science in order to save Christianity from the onslaught of atheism!" (Fields, 1976, p. 40).

A few years later in 1816 a future Archbishop of Canterbury, the evangelical John Bird Sumner (1780–1862) published *A Treatise on the Records of Creation*. Much was on political economy but the appendix was on the relationship of Christianity and Geology, which followed the Chaos-Restitution interpretation (Sumner, 1833, vol. II, pp. 339–359). Chalmers and Sumner were largely responsible for forging a new geologico-theological consensus by modifying older interpretations. Most Protestants on both sides of the Atlantic accepted this reconciliation of Genesis and geology in the 1810s and 1820s, especially by clerical geologists. Conybeare and Phillips' (1822) *Outline of the Geology of England and Wales* was the most widely read book on British geology of that era.[3] The introductory chapter, presumably by Conybeare (1787–1857), later Dean of Llandaff, who was on the fringes of Evangelicalism, contains a long section on the theological implications of geology. Conybeare wrote that

> Two only points can be in any manner implicated in the discussions of Geology.
>
> I. The Noachian Deluge
>
> II. The Antiquity of the Earth (Conybeare and Phillips, 1822, p. lvi).

As a Diluvialist the former was no problem to Conybeare. On the latter, Conybeare followed Sumner. Human antiquity was the 6,000 years indicated by a strict reading of the Bible—something which was not questioned for a decade. He gave three hypotheses "With regard to the time requisite for the formation of the secondary strata." The first is a literal six days, which he does not expressly exclude, the second the Long Day, which was forcibly expounded by the evangelical G. S. Faber (Faber, 1823, pp. 111–165) (1773–1854), and the third Chaos-Restitution. Conybeare avoided recommending any of the three, but his preference is implicit in a long footnote citing Sumner on the *Records of Creation* (Conybeare and Phillips, 1822, p. lxi).

William Buckland (1784–1856) devoted part of his Inaugural Lecture *Vindiciae Geologicae* at Oxford in 1819 to the relationship of geology and "the Mosaic Records," adopting the Chaos-Restitution hypothesis citing Sumner, Horsely, and Buffon for support (Buckland, 1820, pp. 25–28). Buckland returned to this in his Bridgewater Treatise (Buckland, 1836) where the second chapter considered the *Consistency of Geological discoveries with sacred History*. That chapter offended a few for espousing an ancient earth, and thus his Bridgewater was followed by several anti-geologies condemning "infidel" geology in the late 1830s. Buckland rejected any notion of "a detailed account of geological phenomena in the bible," and rejected that all strata were laid down in the Flood and had reservations over a "Long Day." To support his case Buckland referred to Chalmers, Pusey, Burton, Horsely, Sumner, and others, a cross-section of conservative theologians. He also cited Adam Sedgwick's (1785–1873) *Discourse on the Studies at the University of Cambridge* (Sedgwick, 1834/1969) and the long discussion on geology in the *Christian Observer* in 1834.

A good example of a slow shifting away from a nondogmatic literalism can be seen in the writings of G. S. Faber. Faber was a prolific evangelical writer, writing many volumes on prophecy and other theological themes. Among these are many references to geology and Genesis. His Bampton Lectures *Horae Mosaicae* refers once to geology "while the bowels of the earth are ransacked to convince the literary world of the erroneousness of the Mosaical Chronology." (Faber, 1802, p. viii.), implying hostility. By 1816 Faber demonstrated his acceptance of geology in *The Origin of Pagan Idolatry*. Tucked away in volume two are a few pages referring to de Luc's geology. He continued his interest in geology in *A Treatise of the Three Dispensations* of 1823 and *The Difficulties of Infidelity* of 1824, and cited Cuvier, Dolomieu, and de Luc in support of a devastating Deluge, thus indicating the influence of Buckland. According to Rupke, the Oxford geologist William Buckland had three theological advisors and supporters, Faber, J. B. Sumner and Shute Barrington, the first two being evangelicals (Rupke, 1983, p. 14).

A survey of contemporary theological writings show that Chaos-Restitution was the most widespread "reconciliation" of geology and Genesis in the period 1810–1850 and that the biblically literalist anti-geologies, such as Cockburn, Fairholme, Fitzroy, and others were in the minority, even among evangelicals (Roberts, 1997, pp. 247–250). It is easy to regard the Chaos-Restitution interpretation of Genesis as special pleading and a forced exegesis, but it was widely held until mid-century. Hugh Miller (1802–1856) questioned it in a footnote in *Footprints of the Creator* (Miller, 1881, p. 332), his antievolutionary critique of the *Vestiges* in 1847. This he expanded in his posthumous *The Testimony of the Rocks* (Miller, 1857), both in the Preface and in two chapters on Genesis and geology. He explained why he felt it necessary to reject Chalmers' Gap Theory, which had been widely held for fifty years in favor of his concept of *The Mosaic Vision of Creation.* In the preface, Miller spelled out the geological reasoning behind this change. He wrote, "I certainly did once believe with Chalmers and with Buckland that the six days were simply natural days of twenty-four hours each . . . and that the latest of the geologic ages were separated by a great chaotic gap from our own" (Miller, 1857, pp. x–xi). This was reasonable to Catastrophists, who reckoned that each geological era was closed off by a catastrophe. Miller explained that there was no problem with "the Palaeozoic and Secondary rocks," but there was with recent strata. He continued, "During the last nine years (written in ca. 1856), however I have spent a few weeks every autumn in exploring the later formations." From his study of the Pleistocene, he concluded that many of our "humbler contemporaries" especially molluscs existed long before man. Thus "No blank chaotic gap of death and darkness separated the creation from which man belongs from that of the old extinct elephant . . . and hyaena, or for familiar animals . . . lived throughout the period which connected their times with our own." As a result Miller rejected the whole idea of Chaos then Restitution and adopted the view of six *prophetic* days of creation. Chalmers' ideas were more congenial to Catastrophism than to Uniformitarian geology, with its seamless geological development throughout time.

Within a few years Gilbert Rorison was arguing for a totally poetical exegesis of Genesis in Wilberforce's *Answers to Essays and Reviews* (Wilberforce, 1861, pp. 281–286) and the Chaos-Restitution interpretation rapidly went out of fashion. Archdeacon Josiah Pratt of Calcutta (Pratt, 1871) was one of the last serious writers to expound it. The Day-Age interpretation gained ground among the more "intellectual" conservatives, most notably by J. W. Dawson (Dawson, 1877). By the end of the nineteenth century the Chaos-Restitution interpretation was given a Dispensationalist twist as the "Gap Theory" first by George Pember and then enshrined in the *Scofield Reference Bible.* This allowed "fundamentalists" from 1890 to 1970 to accept geological science, if not evolution. However following the rise of YEC after 1961, the Gap Theory has now been almost completely discarded (Fields, 1976). YECs like Morris, Ham, and Sarfati reject it with virulence. From a later vantage point, whether the late nineteenth or even early twenty-first century, it is difficult to conceive that this interpretation *made sense at the time* whether for theological or scientific reasons, or both. It was considered to be a careful well-thought theological understanding and biblical interpretation, which both took earlier understandings into account (tradition) and understandings of science especially geology (reason).

LATE NINETEENTH CENTURY DISPENSATIONALIST AND POPULAR EVANGELICAL INTERPRETATION

Form the late nineteenth century until about 1950, moderate evangelicals tended to adopt a Day-Age interpretation, which allowed for a generalized *concordism* parallel between Genesis One and geology. The more conservative evangelicals, who tended to be Dispensationalist, were more literalist and preferred the Gap Theory which enabled them to accept geological time. Dispensationalism had many attractions. It gave a seemingly coherent and accessible scheme of biblical interpretation, and appealed to the popular evangelical belief anyone can interpret the Bible. It gave great authority to the Bible and interpreted it in the most obvious, and literal, way. "Interpret" is not the best word, as Dispensationalists claimed to take the Bible at its word rather than interpret it. This has resulted in the ambivalence of the heirs to fundamentalism since 1945. Many evangelicals almost have a "default literalism" and this may explain why many evangelicals rejected the Gap Theory as the YEC movement got under way in the 1970s. Since about 1970, a literal interpretation of Genesis has become more common among evangelicals throughout the world, thus reversing the trend of two centuries since 1770.

EVANGELICAL BIBLICAL INTERPRETATION FROM 1950

The postwar revival of Evangelicalism resulted in a renaissance of theological scholarship. Many students went to mainstream universities in the United States and Europe for a doctorate. A proportion broadened out theologically and often were regarded to have gone liberal. Marsden has made a good case study on this in his study of Fuller Theological Seminary *Reforming Fundamentalism* (Marsden, 1987). In 1949 the *Evangelical Theological Society* was founded in the United States, which insisted on inerrancy for membership and at about the same time the *Theological Students Fellowship* and the *Tyndale Fellowship* were founded in Britain, which significantly did not.

Whereas in 1950 there were few notable evangelical biblical scholars, there are now considerable numbers as well as innumerable Ph.Ds in theology. As a result an immense volume of evangelical theology of varying quality is published. Some theologians drifted away from Evangelicalism and at times made a name in the scholarly world, as have James Barr and Maurice Wiles in Britain and Bart D. Ehrman in the United States. Of the many who remained in the evangelical fold, their theological perspectives vary greatly. The more liberal, who adopt a conservative critical approach, are often indistinguishable from conservative non-evangelical scholars. An example is Bishop Tom Wright the New Testament scholar. At the other extreme some have scarcely moved from a fundamentalist perspective with an insistence on fanciful typology, an Ussher chronology and a literal Genesis. Hence today there is a great diversity of Old Testament interpretation, with an immense diversity on how Genesis should be understood.

CONTEMPORARY EVANGELICAL BIBLICAL SCHOLARSHIP

Over the last fifty years there has been a growing number of competent evangelical biblical scholars. New Testament scholars have been far more numerous than Old Testament scholars and include F. F. Bruce, I. H. Marshall, R. Baukham, R. P. Martin. J. D. G. Dunn and N. T. Wright (now Bishop of Durham) from Britain and G. E. Ladd, Ward Gasque, Joel Green, G. Fee, and T Schreiner from across the Atlantic. Fewer Old Testament scholars have gained distinction in Old Testament studies. This is because the text and the history of the Old Testament are not as straightforward as the New Testament. The Old Testament text itself is often unclear and any translator or exegete has to cope with that, along with questions of historicity and authorship. This means that it is harder to regard the text as authoritative and inerrant. Consequently Old Testament scholars often find that they cannot subscribe to an evangelical basis of faith about the Bible. Even so there are numbers of evangelical Old Testament scholars publishing competent work. These contribute to the commentary series such as the Tyndale and Word Old Testament commentary series and the multivolume IVP Dictionary of the Old Testament.

There is no one evangelical perspective of the Old Testament. The most conservative insist on the Mosaic authorship of the Pentateuch, that the OT history is precise and retain Ussher's 1656 chronology of the Old Testament with the Flood occurring in 2473 BC, the unity of Isaiah, etc. At the other extreme the more liberal evangelical accept that the Pentateuch was compiled centuries after Moses, the OT is only generally historical, the Flood was local if it occurred. Needless to say that there is an inerrancy divide here. Apart from implications on how archaeology impinges on the Old Testament, there are very different understandings on how science relates to early Genesis and thus I consider a selection of writers on this.

I have put these writers into three cohorts based entirely on their acceptance or not of geological time. Because of the evangelical understanding of scripture, evangelicals do not take the position of many liberal scholars, for whom Genesis is not historical. Barr, for example, argues that though the original authors of Genesis thought that the days of Genesis One were solar days, the Bible is clearly wrong on this point, but it still has theological value. This is anathema to the conservative evangelical. This is John Whitcomb's argument in *The Genesis Flood* (Morris and Whitcomb, 1961, chap. 6) where he posits "a scriptural framework for historical geology."

At the popular level many expositions of Genesis argue for six solar days as the "true" interpretation. More serious studies are rare. One of the most influential is Douglas Kelly's *Creation and Change* (Kelly, 1997). In 2003 John Currid, Carl McMurray Professor of Old Testament at the Jackson campus of the Reformed Theological Seminary, published a two volume commentary on Genesis for the Evangelical Press Study Commentary Series. This is an academic commentary making much use of the Hebrew text. Currid argues for six solar days and a global Flood as the best interpretation, but unusually for a conservative maintains that the firmament of Genesis 1 vs 6–8 was a solid dome, and that is what the author of Genesis (Moses) thought along with a belief in a flat earth, which was typical of ancient Egyptian cosmology. In this he followed the work of Seely discussed below. No Old Testament theologian has done more to encourage evangelicals to accept a literal Genesis than the coauthor of *The Genesis Flood*, John Whitcomb.

At the other end of the evangelical spectrum, some reject both literalism and concordist interpretations. Instead they adopt a "framework" interpretation, which understands the six days as thematic rather than chronological. Arie Noordzij of the University of Utrecht first used it as an interpretive tool for Genesis 1 in 1924, and Meredith Kline of Westminster Theological Seminary developed it in "Because it had not rained" (Kline, 1958). Kline wrote as an exegete rather than an apologist, that a chronological six-day creation does not fit with Genesis 2 vs 5 (because it had not rained). As he was loath to admit to contradiction between "two creation accounts" or that early Genesis was legendary or mythical, he recognized "the figurative NATURE of the several chronological terms of Genesis 1" and argued that the author "used the imagery of a chronological week to provide a figurative framework" for the creation acts. Kline later developed this in *Space and Time in the Genesis* Cosmogony (Kline, 1996). Kline's thesis has been widely accepted by many evangelicals, especially those convinced of geological time, but has been criticized by more conservative theologians like Wayne Grudem (Grudem, 1994, pp. 302–305), where he summarizes the framework hypothesis and its alleged problems. Grudem is also critical of evolutionary theory and inclines to YEC. Ken Ham is also critical in his AIG booklet *Six Days or Millions of Years?* However Kline did not intend to open the floodgates for evolution and many orthodox Presbyterians (especially members of the Orthodox Presbyterian Church and the Presbyterian Church of America) hold to the framework hypothesis but deny evolution.

Despite many criticisms, a form of framework hypothesis is followed by several leading evangelical scholars in their commentaries on Genesis: Gordon Wenham (Word Commentary), Bruce Waltke (Genesis: A Commentary, Zondervan), John Walton (NIV Application Commentary, Zondervan), and Conrad Hyers in various writings. One of the most accessible expositions is by Henri Blocher in his work *In the Beginning* (Blocher, 1984, chap. 2). For an evangelical who is inclined to accept evolution, the Framework theory is attractive as there is no need to devise any chronological concordance with six days.

Declining numbers still hold to either the "Gap Theory" or a "Day Age Theory," but they are between a rock and a hard place as they claim to be literalist. Both appear to have been largely eclipsed after the rise of YEC. The main scholar who holds to the Day age is Gleason Archer. Two non-theologians who argue for this are Glenn Morton, whose apostasy from YEC is discussed in Chapter 7, and Hugh Ross of *Reasons to Believe*, who has expounded the day-age theory at length in his recent book *A Matter of Days: Resolving a Creation Controversy* (Ross, 2004) and has received virulent criticism from *Answers in Genesis*. Davis Young held to a Concordist Day Age view in the 1970s (*Creation and the Flood* [Young, 1977, pp. 81–134]), but now has adopted the framework theory. The physicist turned theologian Robert Newman and Herman Eckelmann Jr. also argue for this (Newman and Eckelmann, 1977) in as does John Wiester (Wiester, 1983). Older writers include Peter Stoner and Edwin Gedney in Chapters 2 and 3 of *Modern Science and Christian Faith* by members of the American Scientific Affiliation (Everest, 1950, pp. 9–57). Few today argue for the Gap Theory and the last significant evangelical to do so was Arthur Cunstance. The Gap Theory is strongly criticized by YEs, most notably by Weston Fields. It is significant that the Day Age theory today is held by evangelicals, who are OEC rather than YEC or TE.

These three views still cause considerable debate among American evangelicals and a useful discussion is to be found in the book *The Genesis Debate: Three Views on the Days of Creation* (Duncan et al., 2000). Three pairs of authors put forward their case and respond to the others. J. Ligon Duncan III and David Hall for solar days, Gleason Archer and Hugh Ross for the Day Age, and Lee Irons and Meredith Kline for the Framework Theory. In the blurb, Geisler stated "The Genesis Debate is a worthwhile volume that will help you better understand the biblical doctrine of creation."

Even so, it is wrong to assume that adherence to a literal Genesis blinds the scholar to critical study of the Bible. David Fouts of the YEC Bryan College argues that the large numbers in the Old Testament are polemical hyperbole and thus are not to be taken literally (Fouts, 1997). However he still maintains that Creation occurred in 144 hours.

As a rider these three views they also tend to reflect "three views" on science. Those who accept evolution tend to accept the Framework Hypothesis as do many in the ASA or Christians in Science, Day Age appeals to Old Earth creationists who reject evolution, and the Solar Day, not surprisingly, appeals to YECs. The Gap Theory is in eclipse.

THREE REPRESENTATIVE EVANGELICAL THEOLOGIANS: KELLY, LUCAS, AND BLOCHER

I have chosen these three as their work is at a serious lay level. The three are well-regarded theologians in America, France, and Britain. They also indicate the range of scientific understanding by evangelicals; Kelly is a convinced YEC and also argues that YEC is correct and reflects a changing paradigm of science. Lucas, a scientist-theologian, accepts evolution, though he flirted with YEC over thirty years ago. Blocher firmly rejects YEC but is hesitant about evolution.

Ernest Lucas graduated in chemistry from Oxford and obtained first a Ph.D. in chemistry from the University of North Carolina and then a Ph.D. in biblical studies. Henri Blocher, a French protestant, was appointed to the Gunther H. Knoedler Chair of Theology at Wheaton College in 2003. Since 1965 he had been Professor of Systematic Theology at the Faculté Libre de Théologie Evangélique in Vaux-sur-Seine. He was educated at the Sorbonne, London Bible College, Gordon Divinity School, and Faculté Libre de Théologie Protestante of Paris and has written many theological books, including *In the Beginning, Evil and the Cross* and *Original Sin*. Douglas Kelly originally studied in the States and earned a Ph.D. in systematic theology in Edinburgh under T. F. Torrance.

The subtitle of Kelly's book *Genesis 1.1–2.4 in the light of changing scientific paradigms* (Kelly, 1997) makes his thesis clear. Kelly adopts Kuhn's paradigm shifts in his *The Structure of Scientific Revolutions* and argues that as a paradigm shift, that is the success of YEC, has occurred in science, there needs to be a related paradigm shift in theology away from previously-held old earth interpretations of scripture. After putting forward his arguments for accepting YEC and a literal hermeneutic of Genesis, he concluded, "There is only one way for massive intellectual, moral and cultural healing to occur, and it entails a revolutionary 'paradigm shift' from mythological evolution to a Scripturally revealed and scientifically realistic paradigm of special, divine creation" (Kelly, 1997, p. 245). His arguments on science reflect conventional YEC understandings of science, but his theological arguments need considering.

Douglas Kelly

Kelly begins with a chapter entitled *Creation: Why it Matters*, which is strongly based on the Scottish theologian Tom Torrance and standard writers on the history of science and Christianity—Hooykaas, Jaki, etc. though none doubts the "evolutionary paradigm." However he does move on to the challenge to evolution posed by Johnson and Behe and concludes the chapter by saying "God provides us with such information in the first three chapters of Genesis . . . from the One was the eye-witness . . . " (Kelly, 1997, p. 30). It is difficult not to conclude that Kelly, like many YECs, draws his science from the Bible.

The center part of the book are discussions of the various days of creation. Kelly argues against those who reject a 24-hour day for *yom* (Heb day), and questions all alternative interpretative schemes whether "Gap Theory," Day-age or Framework. His weakest argument is to claim that there are fifty-seven references to Genesis 1–11 in the New Testament and that "[I]n none of these references . . . is there the slightest indication of anything other than the literal, chronological understanding of the six days of creation . . . " (Kelly, 1997, p. 134).However, most of these references have no bearing on a literal Genesis. At the end of the book Kelly argues for no death before the fall (Kelly, 1997, p. 228f) from Genesis 1 vs 31, "And God saw that it was very good," stating that "very good" means no suffering or death and that this is in accord with Genesis 3, Romans 5 vs 12, Romans 6, and I Corinthians 15 vs 21.

On scientific questions Kelly accepts the consensus of YEC arguments. These include the moon dust argument, the circular reasoning of the Geological Column, catastrophic deposition at Mt. St. Helens and others. All have been shown to be fundamentally wrong. Whether or not one finds Kelly's arguments convincing, it is probably the best theological argument for a YEC "paradigm."

Ernest Lucas

Lucas argues that standard science, whether cosmology, geology, or evolutionary biology, are conformable to evangelical belief. He writes for the well-informed layman and presents his case eirenically. The title *Can We Believe Genesis Today?* (Lucas, 2001) is rhetorical. The first third of the book deals with Biblical interpretation and stresses the variety of literary forms before moving onto Genesis itself, as well as considering scientific matters. Various young earth arguments like the decay of the magnetic field are found wanting. Despite Lucas' scientific credentials, he deals far more with theological questions, and moves easily between science and theology. Unlike Kelly, Lucas uses science to inform his interpretation of scripture, and draws a parallel with the use of archaeology. He also stresses how archaeological evidence has assisted in the understanding of the Greek of the New Testament.[4] In the New Testament there are several Greek words, which were not found in classical texts and thus their meaning was obscure, until some Greek documents were found in the nineteenth century in Egypt, using these words, revealing the meaning. Lucas (2001, pp. 61–62) argued that the principle behind this is the same as using science to illuminate the meaning of scripture as it is using the best available knowledge. His next two chapters apply this approach to Genesis.

Lucas is the antithesis of Kelly and comes to a diametrically opposed conclusion. These two books are highlight the theological division within Evangelicalism and are instructive as both are accessible to both the nonscientist and nontheologian.

Henri Blocher

Blocher's books represent evangelical theology at its best. *In the Beginning* is an extended study on the first three chapters of Genesis. His aim is theological but makes reference to science and refers to scholars from both sides of the Atlantic, Protestant and Roman Catholic. His approach is thematic and provides a useful appendix on *Scientific hypotheses and the beginning of Genesis*. On the creation week he outlines the various interpretations and favors the Framework hypothesis. In his discussion of evil in Genesis 3 he is reluctant to posit that the Fall had any physical effects. Blocher was familiar with YEC, but rejects it (Blocher, 1984, p. 214), preferring to accept standard science with reservations about evolution. Philosophical extensions of science have no appeal for him. Blocher's book has been widely used by evangelicals but has come in for much criticism by YECs, like Douglas Kelly (Kelly, 1997, pp. 115–120), who likens Blocher to a Mediaeval Nominalist. The fact that Blocher was appointed to a chair at Wheaton College in 2003 demonstrates his acceptability to American Evangelicalism.

INERRANCY

Most evangelicals today hold that the Bible is *Inerrant*. This means that the Bible is absolute truth and does not err in its statements. It is easy to conclude that evangelicals, who believe in biblical inerrancy, equate it with literalism and thus YEC. Though this is often the case, there are many exceptions. Evangelicals who espouse YEC adopt both literalism and inerrancy and this is often written into credal statements of evangelical churches and colleges, as well as YEC groups like AIG and ICR. However to leave it at that would be misleading.

It is a matter of debate whether inerrancy has been the main protestant doctrine of the Bible since the Reformation or not. In 1979, at the height of the inerrancy debate centered on the writings of Harold Lindsell, Rogers and McKim (Rogers and McKim, 1979) argued that inerrancy was introduced by the Haldane brothers in 1828 and developed by the Princeton theologians Hodge and Warfield after 1860. Calvin along with most Reformers and Doddridge, Thomas Scott and others in the eighteenth century allowed some error in the Bible, without questioning its absolute authority.

The classic nineteenth-century expression of inerrancy is in Hodge's *Systematic Theology* of 1870 (Hodge, 1870) and Warfield's (1851–1921) many writings (Warfield, 1951) on the authority of scripture. Hodge likens the errors in the Bible to tiny specks of sandstone in the marble of the Parthenon (Hodge, 1870, vol. 1, p. 170). Both theologians accepted geological ages and Warfield reckoned himself a Darwinian. Thus in its classic formulation, Inerrancy embraced a nonliteral interpretation of Genesis. Biblical inerrancy became a central belief among the early twentieth-century American fundamentalists, often with an acceptance of geological time. With the growth of the "New Evangelicals" after 1950, some, like E. J. Carnell and others from Fuller seminary, began to question inerrancy. D. P. Fuller put forward the case for a *limited inerrancy*, in which the Bible is not inerrant on matters of history and science (Marsden, 1987). This came to a head in the 1970s with Lindsell's books, notably *The Battle for the Bible* (Lindsell, 1976), followed in 1978 by the *International Council on Biblical Inerrancy* which met in Chicago in October 1978.

The International Council on Biblical Inerrancy

The *Chicago Statement on Biblical Inerrancy* was signed by nearly 300 noted evangelical scholars, including James Boice, Norman L. Geisler, Carl F. H. Henry, Harold Lindsell, John W Montgomery, J. I. Packer, and Francis Schaeffer. Most of these accepted geological ages and Packer accepted evolution.

Article 12 of the Chicago Statement refers to earth history:

> We affirm that Scripture in its entirety is inerrant, being free from all falsehood, fraud, or deceit.
>
> We deny that Biblical infallibility and inerrancy is limited to spiritual, religious, or redemptive themes, exclusive of assertions in the fields of history and science. We further deny that scientific hypotheses about earth history may be properly used to overturn the teaching of Scripture on creation and flood.

In 1982 the council met again to discuss the hermeneutics of the Bible and produce a second report—the *Chicago Statement on Biblical Hermeneutics.* This contained twenty-five articles and the twenty-second dealt with the early chapters of Genesis.

WE AFFIRM that Genesis 1–11 is factual, as is the rest of the book.

WE DENY that the teachings of Genesis 1–11 are mythical and that scientific hypotheses about earth history or the origin of humanity may be invoked to overthrow what Scripture teaches about creation.

Since the historicity and the scientific accuracy of the early chapters of the Bible have come under severe attack it is important to apply the "literal" hermeneutic espoused (Article XV) to this question. The result was a recognition of the factual nature of the account of the creation of the universe, all living things, the special creation of man, the Fall, and the Flood. These accounts are all factual, that is, they are about space-time events which actually happened as reported in the book of Genesis (see Article XIV).

The article left open the question of the age of the earth on which there is no unanimity among evangelicals and which was beyond the purview of this conference. There was, however, complete agreement on denying that Genesis is mythological or unhistorical. Likewise, the use of the term "creation" was meant to exclude the belief in macro-evolution, whether of the atheistic or theistic varieties.

This affirmed the factuality of Genesis and denied that it could be either mythical or that "scientific hypotheses" could "overthrow what Scripture teaches about creation." The article seems to point to a literal Genesis, but Norman Geisler made it clear in his commentary that "The article left open the question of the age of the earth on which there is no unanimity among evangelicals" but "the use of the term 'creation' was to exclude macro-evolution." In the volume *Hermeneutics, Inerrancy and the Bible* produced for the Council, Walter Bradley and Roger Olsen claimed that Progressive Creation was the best combination of "the biblical and scientific particulars," thus giving semiofficial support to the refusal to espouse YEC.[5] However responding to Bradley and Olsen, Henry Morris called progressive creation an "old time-worn, compromising hermeneutical system" and refused to sign the declaration.

Thus on early Genesis the 1982 Council failed to resolve anything, as evolution was stated to be contrary to inerrancy but old-earth ideas were not excluded. This, in itself, marked a considerable hardening of the definition of inerrancy from that of Warfield a century earlier and also James Packer, who wrote a classic defense of inerrancy in the 1950s. Though the statement was equivocal, it undermined those who accepted evolution and gave YECs confidence. Since then, if not before, YECs have insisted that the only right view of the Bible is inerrancy and inerrancy implies YEC. This is a powerful debating tactic and gives immediate advantage to the YEC, who can then charge any "Old Earther" as "Liberal"

Inerrancy Today

In the United States, the majority of evangelicals hold to inerrancy today, which makes the total acceptance of geology and evolution extremely difficult.[6] Where the Chicago Statements are regarded as authoritative, evolution is out. There are some evangelicals who hold to both evolution and inerrancy but that goes against the general opinion. For many evangelicals, to accept evolution is to reject inerrancy and thus to have a weakened belief in the Bible. This outlook is increasingly being accepted throughout the world, including Britain.

Definitions of inerrancy vary considerably. At the popular level inerrancy is assumed to imply literalism and a young earth. Thus scientific evangelicals may reject inerrancy for scientific reasons, being oblivious of more nuanced treatments. Among those who have gone through Evangelical seminaries, there is a considerable range of opinion but most will recognize the literary nature of the Bible. Even so, seminary professors may disturb students' notions of inerrancy by pointing out that there are many grammatical errors in the Greek of Paul's letters. After all, if the Bible is inerrant, the grammar must be also![7]

Today Inerrancy is held in a variety of forms. Some evangelicals continue in the tradition of Hodge and Warfield, which recognizes the variety of literary forms in the Bible and accept evolution. These include both theologians like Jim Packer and John Stott and scientists like Oliver Barclay and Denis Alexander. This is not by shared by many YECs who argue that acceptance of an old earth is "theological compromise" as it destroys inerrancy. As the *correct* hermeneutic of the Bible is to read in it a literal way this means that Flood must be universal and that Creation took place in six 24-hour days. However as no one can deny that the earth is spherical, then all references in the Bible to the shape of the earth must be inerrant. Thus every biblical passage in the Old Testament, which can possibly be taken to imply a flat earth, *must* be taken to support the earth's sphericity, or else inerrancy would be denied. Thus the natural meaning of passages like Genesis 1 vs 6–8, Exodus 20 vs 4, and Isaiah 40 vs 22 is ignored (see below) and taken to support sphericity contrary to the usage of Hebrew words. This is the logical conclusion of attempting to extend inerrancy to "scientific" matters and not recognizing that the Biblical writers were limited to the "scientific" understanding of their day and in the words of Calvin "Moses wrote in a popular style" for "the unlearned and rude as of the learned."

Because of these types of questions, some evangelicals avoid the use of inerrancy and prefer to speak of the supreme authority of Scripture. Others simply reject inerrancy altogether and happily affirm that the Bible though authoritative contains minor errors. That in turn elicits opposition from those who adopt the extremer forms of inerrancy and so the internecine conflict between evangelicals continues. Because of the voices for inerrancy, especially in America, the large number of evangelicals who either reject it are often not heard. Howard Marshall, professor emeritus of theology at the University of Aberdeen, discussed inerrancy at length and rejected it as unhelpful as it tends to make people expect the Bible to be "literally" true. (Marshall 1982, p. 49ff) Gerald Bray, a British scholar at Beeson divinity school in Birmingham, Alabama, has similar reservations (Bray, 1996, pp. 539–563). It is also true to say that most evangelicals in Britain reject or avoid inerrancy. Risking oversimplification evangelicals can be divided into three groups:

1. Those who do not accept inerrancy and prefer to speak of the trustworthiness of scripture. This includes a large minority of evangelical scholars, who would not be found in the most conservative schools.
2. Those who accept a nuanced form of inerrancy and allow for *accommodation*. This would include most evangelical scholars in more conservative schools.
3. Those whose inerrancy is decidedly not nuanced and dependent on the scientific accuracy of the Bible. This is the stance supported by colleges affiliated to TRACS and includes many "popular" evangelicals.

The most strident defenders of Inerrancy come from the third group, who as Noll says often have "lush but eccentric interpretations" (Noll, 1994). Some will be discussed in the chapter on Young Earth Creationism. They are probably the largest group in the United States. It is important to realize the differences among evangelicals to understand the "biblical" reasons evangelicals have for adopting particular attitudes to science.

The whole subject of inerrancy may seem to a side-show on evangelicals and science, but it is crucial in the understanding of controversies over evolution, issues of medical ethics (like stem cell research) and the nature of what it is to be human and whether a body–soul dichotomy is tenable. It is surely no accident that the earliest attempts at ID from Olsen and Bradley came shortly after their attempts to harmonize the Chicago Statement, which tentatively allowed an old earth but not evolution.

DOES THE BIBLE TEACH A FLAT EARTH? A CASE STUDY

I have never met a flat-earther but some Afrikaner farmers in the remoter parts of South Africa were in the 1970s. My source was a fellow geologist,

Dr. Piet Joubert, also an Afrikaner, who regaled friends about it. When local farmers asked about his work, Piet happened to mention that the earth was spherical, to which they retorted, "Ek is plat!" About the same time other geologists working in Zimbabwe told their African laborers that men had walked on the moon and were told, "Yes, baas, Neil Armstrong and Edwin Aldrin." With these exceptions there can be few who accept that the world is flat and hence it is a good example to discuss the Bible in relation to science.

Before doing so, the myth of a flat earth must be dispelled. Most people in the West believe that until the time of Columbus most Europeans believed in a flat earth and it was the voyages of Columbus and Magellan, which disproved a flat earth around about 1,500 and is still repeated by some (Moore, 2002, p. 148). It is one of the instances where the Church opposed science and Andrew White waxes eloquent on the subject. The myth of the flat earth was wonderfully exploded by J. B. Russell (Russell, 1991), who demonstrated that few theologians believed in a flat earth in contrast to the majority like Augustine and Aquinas who took the earth's sphericity for granted.

However the cosmogony of the Bible tells a very different story. The New Testament makes no clear reference to cosmogony but Rudolf Bultmann claimed that it teaches a three-decker universe, but this cannot be substantiated. In his classic argument presented in 1941 Bultmann in *New Testament and Mythology* (Bultmann, 1984, pp. 1–2), the world-picture of the New Testament as something highly mythological:

> The world is like a three-storied building. In the middle is the earth; above it is heaven, below it is the subterranean world. Heaven is the dwelling-place of God . . . the lower world is hell, the place of torment.

He argued that "modern man" cannot accept Christianity without "demythologizing" the biblical world view. Though few still adhere to Bultmann's "demythologizsation," many still believe that the New Testament writers held to a flat earth. That would be highly unlikely, especially for the Greek-educated Luke and Paul, as the Greeks had demonstrated the earth's sphericity in 500 BC. It is entirely reasonable to regard the apparently mythological descriptions of "the heavens" in the Lukan and Pauline writings as metaphorical. However, it is possible that Gallilean fishermen and carpenters could have adhered to a three-decker universe. In the centuries before Christ, astronomers considered the earth to be spherical but the stars were tiny and fixed on the celestial dome. The small size of the stars is probably reflected in Matthew 24 vs 29, "and the stars will fall from heaven."

However, the Old Testament was a very different world, going back to 2000 BC.[8] The dates of the actual composition of Old Testament books are in dispute. Many liberal scholars hold that all were written after the Exile and thus are post 500 BC, with Genesis being of a Babylonian origin. The most Conservative Evangelicals reckon that Job was written before 1500 BC, and that Moses wrote the Pentateuch in about 1450 BC. Moderate evangelicals argue that little of the Old Testament was committed to writing before 1000 BC.

The differences are more than those of theology as if the biblical books were written before 500 BC. then the authors could not have known that Greek astronomers had demonstrated the earth's sphericity, and thus would have held to the conventional beliefs of their societies, viz., that the earth was a flat disc, with the hemispherical firmament above, and the underworld below. To those who consider that the Bible will reflect the world view of the writers' day that presents no problem. But to those who hold fast to a strong form of Inerrancy, then the "science" in the Bible *must* be accurate. Thus some evangelicals argue that biblical writers believed that the earth was spherical. As Moses and Isaiah lived in the fourteenth and eighth century BC this was before the time of Plato (427–348/7 BC) when most educated Greeks began to accept that the earth was spherical.

To maintain that the Israelites believed that the earth was spherical (often with the implication that this had been revealed to them by God) it is necessary to interpret several Biblical passages contrary to their "plain and literal" meaning. Take Isaiah chapter 40 vs 22, "It is he [God] who sits above the circle of the earth, and its inhabitants are like grasshoppers." Most commentators take the word for circle *khug* to mean a flat disc or the dome of the firmament. However Mark Eastman in his article, "Science and the Bible," states:

> Despite contrary assertions, the fact of a spherical earth was clearly proclaimed in the Bible by the prophet Isaiah nearly twenty-eight centuries ago . . . "It is He who sits above the circle of the earth, and its inhabitants are like grasshoppers [etc.]" Isaiah 40:22 (NKJ). When Isaiah wrote this verse he used the Hebrew word "khug" to describe the shape of the earth. Although this word is commonly translated into the English word "circle," the literal meaning of this word is "a sphere."[9]

Jonathan Sarfati in *Refuting Evolution*, argues in a similar way, as does Henry M. Morris, who in *The Biblical Basis of Modern Science* asserts that *khug* in Isa. 40:22 often translated "circle" means a "sphere" (Morris, 1984a, pp. 245–246). All of these writers claim that the Hebrew *khug*—or *hûg* of Isaiah 40:22a means "sphericity." No biblical scholars support this; in the nineteenth century Delitzsch translated it as the "vault of heaven" which is supported by Allen in DOTTE. Arguments that the Bible teaches the earth's sphericity are to be found in many YEC writings and Web sites.[10] Sarfati also argues that Luke 17:34–36 implies that Jesus believed the earth to be spherical, because Jesus "stated that different people on earth would experience night, morning and midday at the same time!"

This raises a fundamental question: Just how should one interpret the Bible in light of modern scientific knowledge? The YECs Nelson and Reynolds state that one should not read meanings into biblical texts that are not there in order to make them conform to modern scientific knowledge (Moreland and Reynolds, 1999). Some YECs do not follow their advice. Besides the earth's sphericity, Eastman finds references to such modern scientific knowledge as ocean currents (Isaiah 43:16; Psalm 8:8), elementary particles (Hebrew 11:3), and nuclear explosions (2 Peter 3:10). Such fanciful eisegesis as this is matched by Morris' readings into the text of Job, whom he credits with knowledge of the hydrological cycle (28:24–27), and the rotation of the earth (38:12–14). He also claimed that Job describes dinosaurs in Job 40 vs 15ff and, according to Henry Morris (Morris, 1984, pp. 356–359) from Job chap. 41 vs 20–21 some dinosaurs were like dragons and breathed fire. No one can fault their devotion to the Bible, but by reading modern science *into* the Bible, they make mockery of it by ignoring the historical context of the Bible.

Ernest Lucas emphasizes that the thought world of the Ancient Near East of Babylon and Egypt demonstrates that the Cosmology of the Hebrews was similar to that of its neighbors, with some kind of flat earth and heavens above and the underworld beneath (Alexander and Baker, 2003, p. 137). This is manifest in Exodus 20 vs 4, which of the "heaven above, or that is in the earth beneath or that is in the water under the earth." This is similar to Egyptian cosmology (Alexander and Baker, 2003, p. 134). In parts of the Old Testament there are still references to Babylonian and Egyptian mythology associated with their cosmology.

There have been many interpretations of the *firmament* of Genesis chap. I vs 8. Calvin writing in his commentary on Genesis in 1553 claimed that this was a description of clouds carrying rain, no doubt because although no Copernican he was a well-informed Renaissance man and knew that the heavens or the *firmament* was not a solid dome. During the next three centuries most commentators evaded the question of what the firmament was, partly because Copernicanism was unquestioned. With the rise of more detailed biblical studies in the nineteenth century and research into other ANE cultures, scholars began to see that this fitted into typical Egyptian cosmogony. Conservative exegetes objected, as did Delitzsch in his commentary of 1852 (Keil and Delitzsch E. T., nd, p. 53) presumably to allow Genesis not to contradict modern astronomy. He also argued for six Solar Days and a global flood and questioned the reliability of geology. Several decades later he took a far more open line in a later commentary on Genesis.

Many recent commentators ignore the question, but Paul Seely, a graduate of Westminster Theological Seminary, has demonstrated conclusively that the firmament was a solid dome (Seely, 1991) and this has been incorporated into the commentaries by Walton and Currid.

To conclude then, the writers of the Old Testament clearly accepted a flat earth as part of the common ANE cosmogony. This is of no concern to many Christians, who accept that the writers were children of their time. However if Inerrancy extends to history and science, then it is inevitable that some Inerrantists would feel obliged to demonstrate that the Bible taught a spherical earth. The question of a flat earth in the Old Testament highlights the problems some evangelicals face in relation to science and the Bible.

CONCLUSION

This chapter has looked briefly at how evangelicals have related science and scripture in the last three centuries and the main issues today, which are centered on the interpretation of Genesis. The "scientific reliability" of the Bible is of much greater concern to those who hold to Inerrancy, and it is difficult not to conclude that the emphasis on a strict inerrancy during the last thirty years has encouraged a more literal hermeneutic and an inclination to YEC. As we now consider the relationship of evangelicals and science since 1730, it is important to keep the various evangelical understandings of scripture in mind.

NOTES

1. *Génie du Christianisme,* a Catholic literary tour-de-force re-acting against the French Revolution. He rejected Buffon's long timescale commenting, "Dieu a dû creér, et sans doute créér le monde avec toutes les marques de vétusté." This can be translated "created the world with all the marks of antiquity and decay;" thus the world may *appear* ancient but *is* actually a recent creation. Gosse took this up in 1857.

2. There is no detailed study of either Chalmers' Gap Theory or subsequent developments. My suspicion is that others anticipated Chalmers and the documentary evidence is somewhere in Scotland.

3. Samuel Wilberforce owned a copy which is now privately owned in Australia.

4. The Greek of the New Testament is not classical Greek but a popular or *koine* Greek.

5. We shall come across Bradley and Olsen again in Chapter 7 as pioneers of Intelligent Design.

6. This is because YECs describe conventional geology as evolutionary uniformitarianism. A knowledge of the history of geology will show how that is wrong.

7. I do not joke, as this has happened recently in a theological seminary in Britain.

8. The more liberal like van Seters and Thompson doubt whether the Old Testament is historical in any sense. There is variation among Evangelicals as the more conservative insist on a historicity which demands the reliability of the vast ages of the Patriarchs of Genesis 4–11. The less conservative like Wenham and Lucas still insist on the basic historicity of the stories of Abraham, Isaac, Jacob, and Joseph in Genesis.

9. www.marshill.org/Apologetics%20Pages/science_and_the_Bible.htm

10. "The Bible and the Earth's Sphericity" posted on the Creation Research Society web site: www.creationresearch. org/creation_matters/ *Astronomy and the Bible*, www.answersingenesis.org/docs/400.asp.

Chapter 3

Eighteenth-Century Evangelicals and Science: From Jonathan Edwards to John Wesley

So far we have considered what evangelicals are and how evangelicals have understood the relationship of the Bible and science. Over the next four chapters we shall see how evangelicals have inter-acted with science from 1730 until 1960. The first of these dates marks the start of the evangelical movement and the second is the year before the publication of *The Genesis Flood*, which changed the face of Evangelicalism's relationship to science. The time divisions adopted are the same as those in the historical overview of Evangelicalism for two reasons. First for convenience in having the same time periods, but secondly, and more importantly, each of the periods has one dominating scientific theme. During the eighteenth century, evangelicals simply shared the general Protestant acceptance of Newtonianism, the early nineteenth century (from 1780) was dominated by the rise of geology, the late nineteenth century by evolution, and the early twentieth century by anti-evolutionism.

THE PROBLEM OF THE CONFLICT THESIS OF SCIENCE AND RELGION

If care is not taken, each of these can be presented as case studies in the conflict of science and religion, so beloved by militant skeptics and in an inverse way by YECs. It is all too easy to shoehorn these interactions between science and faith into an ideological interpretative scheme. The dominant scheme since the late nineteenth century has been the warfare model so powerfully put forward by J. W. Draper (Draper, 1874) and Andrew Dickson White (White, 1896), who have molded opinions on science and religion, and according to Larson (1997) even molded the events of the Scopes trial, and set the scene for antievolution in the twenty-first century. Over the thirty years most historians of science have repudiated the warfare model (Lindberg and Numbers, 1986; Brooke, 1991), but it is still widely held as many assume that religion, especially Evangelicalism, and science are in conflict.

The Church Historian Sir Owen Chadwick (Chadwick, 1975, p. 161) pithily summed up this attitude in relation to the Huxley–Wilberforce confrontation over *The Origin of Species* at Oxford in 1860:

> Science versus religion—the antithesis conjures two hypostatized entities of the later nineteenth century; Huxley St George slaying Samuel smoothest of dragons; a mysterious undefined ghost called Science against a mysterious indefinable ghost called Religion; until by 1900 schoolboys decided not to have faith because Science, whatever that was, disproved Religion, whatever that was.

Most accounts tell us that Huxley trounced the good bishop and made him look stupid. It is quoted frequently to show how the church has always opposed science with bigoted obscurantism. Even the BBC produced a re–enactment for television and the book *Evolution, the Triumph of an Idea* (Zimmer, 2001, pp. 52–54), which accompanied the PBS series on Evolution, repeats a similar story. But it is not the whole truth. The story of Huxley trouncing Wilberforce was not told until thirty years after the event and it seems that Huxley's memories played tricks on him as he wrote his memoirs in the 1890s. In fact, Huxley could hardly be heard and his friend Hooker had to take the bishop to task. Even so Wilberforce made some telling criticisms of evolution and was supported by scientists including Sir Benjamin Brodie—the President of the Royal Society (Brooke, 2001; James, 2005).

Huxley was not alone in putting forward the conflict of science and religion. In 1896 White wrote *The Warfare of Science with Theology*, which describes the way the church has opposed science from the time of Christ to 1895. Colin Russell described the book as a "polemic tract masquerading as history." Yet for over a century it has encouraged people to believe that there has always been a conflict. This resurfaces frequently in a wide variety of writings—"pop" history of science, popular science, college, and even evangelical, surveys on the history of ideas and many works by theologians and church historians of all persuasions. Young Earth Creationists often adopt a similar historical scenario with the roles of "goodies" and "baddies" reversed!

Usually these conflict scenarios focus on one or more of three issues:

(a) The Churches' opposition to Copernicus and Galileo.

(b) The Churches' opposition to Geology around 1800.

(c) The Churches' opposition to Darwin in 1860.

The first predates Evangelicalism so does not concern us. In a later chapter I will discuss the rise of geology and that many early geologists were Christians. Yet the common view is that the church opposed geology at every turn. A minority Christians did oppose geology from 1800 to 1850 but did not represent the mainstream of the churches or Evangelicalism. Some years ago Simon Winchester, a journalist with an Oxford degree in geology, wrote a life of William Smith entitled *The Map that Changed the World* (Winchester, 2001). It became a best seller, but on many pages lambasted the church for opposing geology. He wrote on page twenty-nine, "The hunch that God might not have done precisely as Bishop Ussher had suggested [creation in 4004 BC], . . . , was beginning to be tested by real thinkers, by rationalists, by radically inclined scientists who were bold enough to challenge both the dogma and the law, the clerics and the courts" (Winchester, 2001, p. 29). Winchester seemed oblivious to the fact that Smith's main advisors were clergymen and he does not mention what British law forbade people to reconsider the age of the earth (assuming there was one!). Winchester is not alone as similar inaccuracies are repeated by many other writers.

My favorite story about the response to Darwin in 1860 is what the Bishop of Worcester's wife is supposed to have said, "Oh, my dear, let's hope that what Mr. Darwin says is not true. But if it is true, let us hope that it will not become generally known." The source of the story is unknown and is often regarded as an Urban Myth. Yet it appears on BBC documentaries about Darwin. Unfortunately it is the minority who do not repeat these myths. We may ask what the effect is on readers. It probably reinforces the popular perception that Christianity opposes science. There is also a reverse side to the conflict thesis, which makes some Christians believe that much of science is wrong and atheistic. The result is that some Christians, especially those sympathetic to YEC, believe the truth of any attack on science, which appears to contradict the Bible. Thus the geologist Charles Lyell is often presented as a godless anti-Christian, who developed his Uniformitarian geology to destroy Christianity. Mortenson develops this thesis in *The Great Turning Point* (Mortenson, 2004). That is incorrect as Lyell was a theist and a devout Unitarian, with many Christian friends. The conflict thesis of science and religion is deeply ingrained in many people's mind that many seem to assume that they were always in conflict. Hence it comes as a shock to discover that Darwin's most vocal opponents were physicists and geologists rather than clergy, and that some of his supporters were clergy, for example Charles Kingsley, Frederick Temple, and the Evangelical H. B. Tristram of whom more anon. Much historical writing during the last few decades has either modified or overturned the simple thesis of warfare between science and religion especially the two volumes edited by Lindberg and Numbers (1986, 2003). That is not to say that there was no conflict at all but rather there was no sustained war, as

Hedley Brooke emphasizes with great clarity. Throughout history there has never been one attitude to science by Christians, and thus there are examples when Christians were in the vanguard of science, for example the geologist Adam Sedgwick in the 1830s (see Chapter 4), when they were indifferent or when they were plain hostile, as with the anti-geologists of the 1820s and YEC of today. Even then, we need to ask WHY they were hostile, as there may be good reason.

Any over-arching system whether claiming conflict or harmony between science and religion is doomed to failure, whether it is the anti-religious approach of some skeptics or the massaging of history by YECs who wish to demonstrate that most branches of science were founded by a "creationist." In his book *A History of Modern Creationism*, Henry Morris attempted to demonstrate that "so many of the great founders of modern science were themselves devout creationists and Bible-believing Christians" (Morris, 1984b, p. 25ff) and is expounded at a popular level in *Men of Science, Men of God* (Morris, 1982), where Morris gives brief biographies of over fifty alleged "creationists" ranging from Leonardo da Vinci to James Clerk Maxwell. Of his sixty-nine "creationists" only one, Philip Gosse, could be termed a YEC. A good number of Morris's post-Darwinian "creationists" accepted evolution! Nearly all of those who lived after 1800 accepted geological ages. For good measure several rejected what evangelicals would regard as essential Christian dogma as Dalton and Agassiz were Unitarian.

Thus when studying the history of the evangelical engagement with science, we shall find a two-fold distortion. The one is the problem of the conflict thesis of science and religion and the second is the massaging of history by YEC historians. Both prevent an informed understanding of the past. When it comes to considering evangelicals and science the problem is amplified, as it is perceived that evangelicals are normally biblical literalists and thus there *had* to be a major conflict. As we shall see that was not the normal situation, at least until the 1980s.

In the first two chapters I stressed the variety of evangelicals and the diversity of ways they interpret the Bible. There is also a variety in the ways evangelicals have considered science, particularly geology and biology, which impinge on Genesis. Some have been indifferent to science as has been the case for most British evangelicals until recently, others are hostile when science impinges on "origins" that is the *historical* sciences and yet others not only accept all science with alacrity, but also are concerned for the development of science. With such diversity there is a great danger in a deliberate or accidental selection of certain evangelicals. To illustrate this, many years ago I gave a paper on evangelicals and geology in the early nineteenth century. I began by presenting those who opposed geology as it contradicted a literal Genesis. I gave an impressive list of people and apparently supported the conflict thesis. But then I dealt with the pro-geologists, who were more numerous. I concluded by explaining the diversity of evangelicals and that a wide selection could contain an in-built bias. However the weakness of this type of approach as it considers only those who *wrote* about science. The vast majority of evangelicals of any period—and any period—were either ignorant of or indifferent to science and probably considered there were far more important things in life, such as evangelism, supporting some good work or an overseas mission. If one is looking only for references to science all that will be overlooked, because the concern for salvation often displaced other priorities. Hedley Brooke also raised other questions. If an evangelical is interested in science that may have nothing to do with his faith. Frequently it is difficult to discern the relationship between a person's faith and their interest or lack of interest in science or anything else. There is also the problem of trying to categorize a person as a moderate or extreme representative of his tradition, which in turn raises questions over what is extreme or moderate Evangelicalism and as we consider the nineteenth-century attitudes to geology may not be a good test. For example Dean Francis Close was reckoned to an extreme evangelical of the mid-century and Henry Moule most moderate, yet Moule accepted geology with difficulty and Close was as progressive as Hugh Miller.

THE LEGACY OF ARCHBISHOP JAMES USSHER

One of the few things I can remember about my undergraduate lectures in geochronology was Ussher's date of the creation in 4004 BC. To pit 4004 BC against radiometric ages of the age of the earth as 4.6 by amuses any geologist. 4004 BC casts a long shadow over perceptions of how scientific understandings on the age of the earth have developed over the last 350 years. Ussher wrote his *Annales Veteris Testamenti* in 1656 before Steno and others began to look at strata, so that his work is classical humanist history and, in the charitable use of the word, pre-scientific. Ussher's dates for creation occupy only the first two pages of his book, and many of his dates for the Old Testament are fairly close to present estimates indicating his historical competence. Because the date of 4004 was included in many Bibles from 1702 it is often assumed that it was almost an official status. That was not the case and many did not accept it as binding and before the late eighteenth century many were more flexible on the date of creation, although most would have seen it in thousands rather than millions of years (Roberts, 2001). Those whom Bernard Ramm would have called the hyperorthodox have for centuries reckoned that Ussher was more or less correct in all his chronology and in 2003 *Answers in Genesis* published *Annales Veteris Testamenti* in an English translation and are attempting to rehabilitate Ussher as the scholarly voice of orthodoxy.

EVANGELICALS AND SCIENCE IN THE EIGHTEENTH CENTURY

The State of Science and Religion at the Dawn of Evangelicalism

Science in the eighteenth century was dominated by Newton and best epitomized by Alexander Pope's epitaph to Newton written in 1730:

Nature, and Nature's Laws lay hid in night.
God said, *Let Newton be!* and all was light.

This reflects the scientific progress, personified by Newton and the optimism so engendered. Physics, or rather natural philosophy, was beginning to usurp theology as the queen of the sciences, as the mysteries of the universe were being explained by mathematics. The decades after the foundation of the Royal Society in 1660 were incredibly fruitful, especially in physics and astronomy. Just mention of the names Newton, Halley, and Flamsteed is sufficient. Sir Christopher Wren designed St Paul's Cathedral to double up as an observatory. It almost seemed that Newton had completed the work that Copernicus, Kepler, and Galileo had started, and that the universe could be explained by math as predictable as a clock.

At the beginning of the eighteenth century, several Anglicans promulgated their fusion of Christianity and Newtonianism as *physicotheology* to provide a natural theology for a mechanistic era. The best known of these writers is William Derham, whose *Physico-Theology* (Derham, 1713) was widely read throughout the eighteenth century and laid the foundation for William Paley's *Natural Theology* at the start of the nineteenth century.

At the more earthly level was the work of Hooke and Boyle and other luminaries of the Royal Society. The classification of the organic world begun by John Ray in the late seventeenth century came to fruition with the binomial classification of Carl Linnaeus in *Systema Naturae* in 1758. The interrelatedness of the animal kingdom and the status of mankind were expounded in the many *Great Chains of Being*. Those studying the earth and its "fossils" (both minerals and organic remains) were beginning to grasp its vast age. The eighteenth century did not see a rapid advance in geology until about 1780, as observers continued the work of their seventeenth-century forbears. Geologically the most important question was how to work out the historical succession of strata and that occurred at the end of the century (Rudwick, 2004).

The late seventeenth century also saw the beginnings of a scientific study of the earth published in volumes known as "*Theories of the Earth.*" On a first reading these seem to be a literal reading of Genesis stories with a few semi-scientific glosses. A closer reading shows them to be more profound as they meld together the Bible, the classics, almost mediaeval "book" learning with the citing of endless authorities and scientific insight in a Chaos-Restitution interpretation of Genesis One. Here they shared the outlook of many theologians (except Ussher!) and literary writers such as Thomas Traherne and Alexander Pope. Instead of taking the Creation story to teach creation in six short days, writers, following an interpretation going back to the early Church Fathers, claimed from Genesis (chap. 1 verse 1) that God first created Chaos (*without form and void*) and after an interval recreated it in six days. The duration of Chaos was undefined. For Ussher it was twelve hours, but for most it was a long and unspecified duration. Thomas Burnet (1635?–1715), Edmond Halley (1656–1742), and William Whiston (1667–1752), reckoned the days to be more than twenty-four hours. Halley attempted a calculation of the age of the earth from the sea's salinity, but came to no firm conclusions other than it was tens of thousands of years old. Likewise theological writers of the day; Bishop Simon Patrick (1626–1707) reckoned that God first created Chaos and then later re-ordered it in six days. He said of the duration of Chaos, "It might be . . . a great while; . . . " (Patrick, Louth, Whitby, 1764, vol. 1, pp. 1–3). Few accepted Ussher's date of 4004 BC for the initial Creation, though most accepted that humanity first appeared in about the year 4000 BC, hence the general acceptance of the rest of Ussher's chronology. The extension of time by the "Theorists" and contemporary theologians was minute compared to the billions of years of geological time, but was, as Stephen Gould wrote of Whiston's argument that the day of Genesis one was a year long was, "a big step in the right direction." In Britain the way was open for accepting a longer time scale (Roberts, 2002, 2007).

Not until the late seventeenth century were "formed stones" or fossils recognized as imprints of dead creatures rather than formed as "sports of nature." Only then could "fossils" be used to demonstrate former life and it took a century before the succession of fossils was used to put strata into historical order. Possibly the first person who used the succession of fossils to demonstrate evolution was Charles Darwin in a notebook in 1838, shortly before he "discovered" Natural Selection. In the 1690s there were insufficient grounds to suggest "Deep Time" or the continual reworking of the earth's crust as understandings of erosion were rudimentary. Ray, Whiston, and others cannot be expected to have done otherwise.

Most of the writers had some "scientific" understanding and often spent as much time refuting each other as suggesting new ideas. Some were mostly speculative, as was Thomas Burnet's *The Theory of the Earth*. Despite his devotion to the Deluge, he sought to explain phenomena naturalistically and somewhat extended the duration of Genesis 1. John Ray's *Miscellaneous Discourses Concerning the Dissolution of the World* shows the beginning of careful observation on earth processes and questions over geological time. He nailed his colors firmly to the fence, and did not explicitly reject an Ussher chronology. However from his discussion of Chaos and other comments, it is fair to conclude that he accepted that the earth was more than 5,500 years old.

Two who broke loose from the *Theories of the Earth*, with their fairly limited time scale were de Maillet and Buffon. Benoit de Maillet (1656–1738) was a French diplomat with a sound grasp of the geography and geology of the Mediterranean and amplified Cartesian cosmogony. His work *Telliamed: Or Conversations Between an Indian Philosopher and a French Missionary* did not appear until 1748, though manuscripts had circulated from 1720. It was an odd work both accepting mermaids and reporting careful observation on marine deposition. Our main interest is that the author reckoned the earth to be over 2 billion years old and according to Albritton (1980) the work acted as a leaven among eighteenth-century geologists.

Buffon, born as Georges-Louis Leclerc (1707–1788) was the Keeper of the Jardin de Roi in Paris and in 1749 published the first volumes of *Histoire Naturelle*, but by his death in had published only thirty-five of the projected fifty volumes. He had little time for the *Theories of the Earth* and said, "I reject these vain speculations." However according to Roger, his biographer, Buffon borrowed more from Whiston than he was willing to admit. It also shows that the Theorists' longer time scale was well known on the continent. Buffon also carried out experiments on the cooling of red-hot globes of iron and then extrapolated his findings to a globe the size of the earth and estimated that the age of the earth to be about 75,000 years. Though vastly greater than 4000 BC, it was not drastically different from British writers in the previous century and gave some experimental data to support them. In unpublished manuscripts Buffon reckoned the earth to be 3 million years old. In 1751 he was censured by the theologians at the Sorbonne and responded by claiming that the first verse of Genesis should read; "In the beginning God created *the materials of* the heavens and the earth." Despite that a longer time scale was gradually gaining ground both in Britain and the Continent so that by 1780 most educated people accepted the earth was ancient, whether tens of thousands like Jean Andre de Luc or millions like Hutton, Buffon, and Soulavie.

The Evangelical Revival was born into this scientific milieu and along the vast majority of educated Christians, Protestant, or Catholic accepted the Newtonian world-view and the natural history of the day. Many Christians baulked at a great age of the earth but most accepted the slightly longer time scale of *The Theories of the Earth* in preference to Ussher's date of 4004 BC. for creation. The evangelicals were no exception and to my knowledge, no evangelicals except for a few Hutchinsonians questioned the truth of Newtonianism though all rejected Deism. Thus the early evangelicals if they wrote about science at all, and most did not, did so from as position of accepting eighteenth century Newtonian science. This was so for three of my examples of early evangelicals and science; Jonathan Edwards, John

Wesley, and William Williams, from New England, England, and Wales respectively. Writing about the position in the colonies in the early eighteenth century, Marsden wrote, "New Englanders had long been friendly to scientific advances and were confident that discoveries of God's ways of governing the natural world would only confirm what they knew from scripture." In New England one who did much to further science was the preacher the Rev. Increase Mather, who in 1683, helped found the Boston Philosophical Society, modeled on the Royal Society. His son, Cotton, carried on his father's tradition and was later elected to the Royal Society of London. However, despite their interest in science, many were enthralled by the preternatural, which in the 1690s led both to the Salem witch-hunts and Cotton Mather preaching on a prodigious cabbage root with one root branch like a cutlass, one like a rapier, and another like an Indian club. Back in England the pages of the *Transactions of the Royal Society* were also a mix of the scientific and the preternatural. The ambivalence of Cotton Mather, who was an immediate precursor of the American evangelicals, is to be seen both in his cabbage root and his daring stance during the smallpox epidemics of 1721 and 1722, after he had seen two wives and eleven children die of the disease, when he urged inoculation against the dreaded disease. For his efforts someone tried to bomb Mather's house and he was opposed by James Franklin in *New England Courant* as part of his anti-clericalism. Some clergy also opposed inoculation, but the example of James Franklin warns against putting too much on Christian reaction to science and its benefits.

HUTCHINSONIAN LITERALISM

Standing out against the prevalent Newtonianism was the clerical scientist John Hutchinson (1674–1737), his disciple Alexander Catcott (1725–1779), and other churchmen. In 1724 Hutchinson wrote *Moses' Principia* to oppose Newton. Both lay great store on Genesis and sought to correct the "errors" of Newtonianism. Far less is made of the Chaos than in the *Theories* and Hutchinson seems not to hold that the period of chaos or *tohu va bohu* was of any significant duration. In 1868 Catcott wrote his *Treatise on the Deluge*, which implied that Chaos was of short duration. As well as some detailed discussion of the biblical text, Catcott gave some fine descriptions of geomorphology. Some held Hutchinsonian ideas until the early nineteenth century and the last Hutchinsonian scientist seems to have been the entomologist and Highchurchman William Kirby (1759–1850), who argued for a six-day creation in his Bridgewater Treatise (Kirby, 1835). Hutchinsonianism was a *biblicist* rather than an *evangelical* reaction to Newtonianism.

Despite its biblical attraction Hutchinsonianism appealed to very few evangelicals. T. T. Biddulph (1763–1838), a vicar in Bristol, was one of the few to espouse it and John Wesley and William Romaine were Hutchinsonians for a short time in their early years. Romaine was vicar of St Georges. Hanover Square, in London from 1750 to 1755, when he was also Gresham Professor of Astronomy, and gave lecture series on Astronomy. The majority of Hutchinsonians were High Church Anglicans in the latter part of the eighteenth century, and thus do not concern us.

THE CONFORMITY OF EVANGELICALISM AND PROTESTANT ORTHODOXY ON SCIENCE

Because the Evangelical Revival began in mid-century and evangelicals were a beleaguered minority until the end of the century there is little material on evangelicals and science, but what there is shows a similarity with other Protestants. On the whole, evangelicals in the eighteenth century took little interest in science as their mission was to evangelize the world and lead godly lives. Most evangelical writings concentrate on the new birth, justification, sanctification and the whole gamut of the new evangelical teachings rather than intellectual matters. Most evangelicals were either too busy for science or just indifferent. In 1780 the Anglican Evangelical leader Henry Venn wrote to Francis Wollaston, a future Professor of Chemistry at Cambridge, not to let chemistry take over from his Christian ministry, "What comparison can there be between saving a soul and analysing a salt?" They also had very little to say on science, and what they did agreed with other protestants as they accepted orthodox Protestantism whether of a Calvinist or Arminian bent. The eighteenth-century outlook of acceptance of Newtonianism and a *fairly* recent inception of the universe was common ground and usually taken for granted. As there were no perceived or real flash points between science and religion, there was no point being diverted from the "real" work of the Evangelical Revival. As most of the leaders had been educated at leading universities like Yale, Oxford, Cambridge, or Edinburgh, they accepted the natural philosophy of the day and took it as read.

By the end of the century potential conflicts between science and religion were beginning to surface for several reasons. First was a change by some on the status of the bible. The earlier university-educated evangelicals emphasized *sola scriptura* in the traditional Protestant sense that the bible was the *only* ultimate source of religious authority but not the only source of knowledge. Wesley described himself as "a man of one book" but as we shall see of both his readings and publications, he most clearly went far beyond the good book. Wesley did not consider the bible as a source of science. Some later evangelicals almost scorned other learning and began to regard the bible as the source of all knowledge and that any knowledge perceived to be contrary to the bible must by definition be false. George

Bugg, whom we shall consider in the next chapter, expressed this most forcibly.

Second, Evangelicalism was attracting more people without university education. These tended to be self-taught and lacked a general knowledge of natural philosophy. Most of these came from those who attended neither the main universities nor the British dissenting academies.

Third, there was an increasing tendency for evangelicals to adopt biblical literalism, due to their emphasis on biblical authority. The biblical commentary of Thomas Scott, which seems to assume a literal hermeneutic and does not mention science in its exposition of early Genesis, is an example. Scott does not explicitly say that the days of Genesis are of a twenty-four-hour duration, but it is difficult to read it otherwise. Thomas Chalmers despite his accommodation to geological time with his Gap Theory presented his as the "one way of saving the credit of the literal history"

During the eighteenth century problems over Genesis did not surface as Copernicanism was long accepted and the vast age of the earth was only apparent in the 1790s. Thus the eighteenth century was a period of almost entirely harmonious relationships between Evangelicalism and science for the minority of evangelicals who took an interest in science. Most simply made no reference to science and were either disinterested or too busy in evangelistic work. Few have left any record of the interest in science, and are probably insufficiently numerous to allow anything but tentative conclusions to be drawn about the spectrum of eighteenth-century evangelical attitudes to science. It is also an area where very little historical research has been carried out. Serious research on evangelicals is only two decades old and hardly anyone, except Jack Haas on Wesley, has carried detailed research on evangelicals and science. This lack of research gives a provisionality to my discussion, as it is an open field for a future researcher.

However several of the founding fathers of Evangelicalism had both an interest and competence in science. These are Jonathan Edwards, Thomas Prince, John Wesley, William Romaine and William Williams. These come from both sides of the Atlantic, two from Yale, two from Oxford, and one from a Nonconformist academy and thus had the advantage of a broad eighteenth century university education. None may be considered even a minor contributor to science and by today's standards would be considered as amateurs. Except for Wesley on electrotherapy none had the stature of the Reverend Gilbert White, the Hampshire parson-naturalist, who wrote the classic *Natural History of Selborne.* When considered in relationship to eighteenth century science, they reflect what a well-educated man would do and understand. What is most conspicuously absent is any sense of conflict between science and religion, as these writers stood in the tradition of Newtonian physico-theology.

JONATHAN EDWARDS

Too often Edwards is known almost only for his sermon, *Sinners in the Hands of an Angry God*, which strong though it is, is not a rant by a hell-fire preacher. Today Edwards is being increasingly recognized not only as a great theologian but a philosopher as well but as Noll so aptly expressed it, "It remains an oddity that the greatest intellectual in the whole history of Evangelicalism was also its first great evangelical"(Noll, 2004, p. 243). He was born in Windsor, Connecticut, in 1703, raised as a Congregationalist, studied at Yale from 1716 and obtained an M.A. in 1722. For the next four years he was pastor to churches in New York and Boston before becoming a tutor at Yale. In 1726 we went to assist his grandfather, Solomon Stoddard, at Northampton in Massachusetts and became full pastor after Stoddard's death in 1729. He remained there until he was dismissed from the pastorate in 1750 and was leader of the awakening at Northampton from 1735. From 1751 he was pastor and missionary to Native Americans at Stockbridge and left in 1758 to become president of the College of New Jersey, only to succumb to a smallpox inoculation a few months later. He is best known as the theologian of the revival and his books on it are classics. But that is not our interest here. He was no simplistic revivalist preacher but a deep theological thinker of a philosophical bent with more than a passing interest in natural philosophy, which we now call science. Natural philosophy in all its forms was a lifetime's interest and ultimately it killed him as he died as a result of a smallpox inoculation, which his relative Cotton Mather had been promoting since 1721.

Edwards' scientific interests took third place to his theological and philosophical writings. His output was prodigious and included published sermons, studies on the New England Revival, serious theological works and theologico-philosophical studies on the Will. So far, Yale University Press has published twenty-two volumes of *The Works of Jonathan Edwards* since 1957, under the editorship of Perry Miller, John Smith, and Harry Stout. Volume 6 contains his scientific and philosophical writings, which contain his early "spider" papers and those of natural philosophy and "The Mind."

Edwards carried out most of his studies in natural philosophy while at Yale. While a teenager Edwards read Locke and Newton, largely thanks to the Dummer collection which was given to Yale in 1718. The Dummer collection included works by Locke, Newton, Boyle, Hooke, Whiston, and many other luminaries of the late seventeenth century. According to Perry Miller on reading Locke, Edwards "grasped in a flash" the implications of modern thought, but Marsden is skeptical of that. As a result of the Dummer collection the Yale curriculum included both Locke and Newton after 1718. Some have claimed that before 1718 the Ptolemaic rather than the Copernican system was taught at Yale, but this is unlikely according to Marsden. Harvard graduates of the late seventeenth century were familiar with Descartes and in the early 1700s sophisticated New Englanders subscribed to British journals and were familiar with Newton, Addison, and Steele. Cotton Mather had imbibed Newtonianism at Harvard or soon afterwards. Thus Edwards' education at Yale was not only extensive theologically but included much recent science and thought from Europe.

Edwards' writings during the three years from 1720 demonstrate the extent of his scientific studies and interests. His study of insects and his "*Spider*" letter of 1723 show more concern with the physics of the spider's flight. His other writings on *Natural Philosophy* and *The Mind* demonstrate his interest in matters scientific and his grasp of Newtonianism without a harking back to Aristotle. His Puritanism was evident by his strong providentialism. After the 1720s Edwards wrote little or nothing on science, because of his work as a pastor and involvement in the writing of more strictly theological works rather than any opposition to science, as is demonstrated by his death. In his large output of theology, Edwards did not discuss science and there is nothing in them, which indicates any hostility to science or questioning of the popular physico-theology of his day. He was no unquestioning preacher and in his writings on "*The Mind*" wrote:

> [12] BEING. It seems strange sometimes to me, that there should be being from all eternity, and I am ready to say, "What need was there that anything should be?" I should then ask myself if there should be either something or nothing. If so, 'tis not strange that there should be; for that necessity of there being something or nothing implies it. (Edwards, 1980, vol. 6, p. 343)

Edwards is remembered for many things. He was the leader of the First Great Awakening in the American Colonies and an evangelist to both the American colonists and the Native Americans. Above all he was a great theologian and philosopher who sought to present a post-Newtonian statement of classic Augustinian and Reformed theology. Apart from Edwards there have been no profound evangelical thinkers, though there were, and are, many competent ones. There are many reasons for that, especially the evangelical emphasis on *Activism* for the Lord, which tends to downgrade the intellectual life (Noll, 1994).

THOMAS PRINCE

Edwards' older contemporary Thomas Prince (1687–1758), pastor of the Old South church in Boston from 1718 until his death, who became eminent as a preacher, linguist, and scholar, was no mean scientist though his scientific reputation is often tarnished. He published widely—sermons and various historical works, as well as scientific works notably *An Account of the First Aurora Borealis* (Prince, 1717) and his sermon *Earthquakes the Works of God and Tokens of His Just Displeasure*, with an appendix on Franklin's discoveries in electricity (Prince, 1755). Because this sermon was given in the aftermath of the Boston earthquake Prince is often portrayed in a negative light.

This incident is often used to illustrate the conflict of science and religion by contrasting Benjamin Franklin and Thomas Prince on lightning rods. The received version is in 1755 that Prince, a minister in Boston, preached a sermon opposing lightning rods as thwarting the will of God. Finally the good sense of Franklin prevailed and the church was once more seen as the repository of the powers of ignorance. Andrew White made much of it in his *Warfare of Science with Theology*, who presents Thomas Prince in a very bad light (White, 1896, vol. I, pp. 364-6), in his chapter *From "The Prince of the Power of the Air" to Meteorology*, which traces out how the church saw diabolical agency in storms and then after the agency of witches concluded with Franklin's lightning rod. White moved rapidly from Wesley's opposition to witches to opposition to Franklin's lightning rods.

Most readers will know that Franklin flew kites during thunderstorms to test for electricity. It has attracted many adulatory myths, but the important thing is that Franklin, and some European counterparts, carried out the experiments in 1752, but the details do not concern us here.[1] After these experiments which confirmed electrical discharges from lightning. Franklin suggested that lightning rods should be fixed to buildings to protect them from lightning. According to Cohen the first were erected in Philadelphia by August 1752 and over the next few years many were erected in Boston (Cohen, 1990).

On November 18, 1755, a large earthquake (probably 6–6.5 on the Richter scale) struck New England and caused much damage in Boston. The *Boston Weekly News-Letter* reported on November 20:

> Last Tuesday Morning about half an Hour past Four o'Clock, the Weather being serene, the Air clear, the Moon shining very bright, we were surpriz'd with a most terrible Shock of an Earthquake: The conditions were so extreme as to wreck the Houses in this Town to such a degree that the Tops of many Chimnies, and some of them quite down to the Roofs, were thrown down.

Within days Prince preached on the causes of the earthquake, which he published as *Earthquakes the Works of God and Tokens of His Just Displeasure*, with an appendix dealing with Franklin's ideas. Both his discourse and his quarrel with John Winthrop of Harvard operated on two levels—scientific and theological, and it is difficult to disentangle the two. Prince wrote:

> The more *Points* of *Iron* are erected round the *Earth*, to draw the *Electrical Substance* out of the *Air*; the more the *Earth* must needs be charged with it. And therefore it seems worthy of consideration. Whether *Any Part of* the *Earth* being fuller of *this terrible substance*, may not be more exposed to *more shocking earthquakes*. In *Boston* are more erected than anywhere else in *New England*; and *Boston* seems to be more dreadfully shaken. O! there is no getting out of the mighty Hand of *God!*

The last sentence is typical of the Providential beliefs of both Puritanism and Evangelicalism (Prince was both) shared by Mather, Edwards and most of their more orthodox contemporaries. At the same time Prince was well informed scientifically as before the 1750s many had considered lightning and earthquakes to be genetically linked. One of those who had believed that was Franklin himself who in 1737 wrote that:

> that the material cause of thunder, lightning and earthquakes, is one and the same,

In 1752 the antiquary the Rev. William Stukeley echoed Prince's views, writing in the *Philosophical Transactions*:

> From the same Principle I infer, that, if a non-electric cloud discharges its Contents, upon any Part of the Earth, when in a highly electrify'd State, an Earth quake must necessarily follow.

Even after Franklin's experiments of 1752 had confirmed his hypothesis about the electrical nature of the lightning discharge, the theory that earthquakes were caused by lightning was given a modus operandi and thus more worthy of serious consideration.

However Franklin's sentry-box experiments on lightning slowly changed that perception and thus in 1755 Prince was adopting a rejected scientific theory to justify his opposition to lightning rods, and then interwove it with Christian providentialism, which savants like Winthrop and Franklin rejected. Thus on November 26, 1755, Winthrop savaged Prince in his *Lecture on Earthquakes,* which he gave in Harvard College chapel, when he rejected the nexus of lightning and earthquakes and expounded Franklin's researches of the previous few years, which had overturned previous understandings. Jared Eliot may have been right to say, "I think Mr Winthrop has laid Mr Prince flat on back . . . ," but Prince was no religious anti-scientist. Winthrop also challenged Prince's providentialism and asked whether "it is possible, by the help of a few yards of wire, to 'get out of the mighty hand of God.'" To Winthrop it was not only possible but also the right thing to do. In Prince's reply to Winthrop his religious concerns were uppermost as he considered too much attention on the cause of earthquakes may divert people from "Matters of Infinitely Greater Moment," which was to see the earthquake as a warning from God to the people of Boston.

This incident though a small and local one, raised a question for all Christians of orthodox bent, whether evangelical or not, on how far natural events, often involving suffering, could be seen as "the hand of God." Consider the letter Sarah Edwards sent to her sister after the death of their father Jonathan

> What shall I say? A holy and good God has covered us with a dark cloud. Oh that we may kiss the rod [of reproof], . . . The Lord has done it.

Winthrop could not have written such words. Comparison between the Boston earthquake and that in Lisbon on November 1, 1755, is invited as the death toll of over 30,000 from that earthquake made many in Europe to reconsider whether God could be benevolent. Voltaire developed this concern in his *Poème dur le désastre de Lisbonne* written in 1756, which reflected his growing pessimism on life, that it was not "*All for the Best.*" However the Lisbon earthquake made thinkers consider theodicy, but Prince in Boston had questioned whether one should escape from the providential hand of God. Both earthquakes had a seismic effect on belief in God, and their theological aftershocks are still with us.

JOHN WESLEY

John Wesley (1703–1791) is probably the best-known leader of the Evangelical Revival and as the one who paved the way for the foundation of the Methodist church. What is far less known is his interest in the science of his day. The classic work by Andrew White inaccurately charges Wesley and others with a total opposition to science especially to geology. This myth has stuck but White (1896, vol. 1, p. 220) here as elsewhere misquotes and misrepresents Wesley, who needs to be considered in relation to the time he lived. Wesley was fascinated by science (Haas, 1995), but as he died in 1791 made virtually no reference to geology, as the vast age of the earth only became apparent after 1780, and thus Wesley cannot be expected to accept later geological findings.

Wesley was born in an Anglican vicarage at Epworth, went to Oxford University where he founded the Holy Club, whose religious methods earned the epithet *Methodist*. Wesley was an orthodox High Anglican, with a broad theological and intellectual background. He was well acquainted with the science and natural philosophy of his day and their relationship to theology seen through the eyes of Newtonianism and was committed to contemporary scientific medicine. For a time Wesley was very taken with Hutchinsonianism but eventually he concluded that Hutchinson's arguments "have no foundation in Scripture or sound reason" (Noll, 2004, p. 245). One can hardly regard Wesley as a scientist, but he took an interest in science, or rather Natural Philosophy, as it impinged on the Christian faith and in medicine, showing an interest in electrical treatments.

It is an obvious statement, though important, to say that Wesley was not born an evangelical but became one in 1738, largely through the witness of Moravian Pietists. His intellectual outlook was forged in the Oxford of the 1720s with the Oxford Aristotelian logical tradition, the empiricism of Locke and a wide range of writers including Francis Bacon and Robert Boyle. His Christian Faith was very experiential but he also gave qualified endorsement to an empirically based natural theology. Unlike the Calvinists he opposed "systems" and according to Jack Haas due to the difficulty of relating his providentialist theology to mechanistic systems whether Deist or Newtonian. We are thus back to the Prince–Winthrop controversy.

Providence played a large part in Wesley's view of social and natural history, but the understanding of providence changed during the eighteenth century. At Newton's death in 1727 there was a balance between the general providence due to an orderly creation and a special providence, which produced miracles and intervened in nature and the life of individuals. As the century wore on the influence of Deism swung against an interventionist deity. Both Highchurchmen and evangelicals resolved to counter Deism with an apologetic with a greater emphasis on an *immediate* providence. Extraordinary events such as volcanoes, earthquakes, and comets seemed to fit the bill, and possibly opened up the orthodox to a charge of holding to a God-of-the-Gaps. As Haas pointed out "Wesley always had his head cocked for the unusual in nature." His preface of his *Survey of the Wisdom of God* (Wesley, 1763) refers to his wish to "recite both uncommon appearances of nature and uncommon instance of art . . . for surely in these appearances, the wisdom of God is displayed." (This is in marked contrast to William Paley (1804) whose *Natural Theology* perceives the intricate design of God in the most mundane of biological forms. With the desire to look for the unusual and unexplained Wesley seems more of a forerunner of *Intelligent Design* than Paley.) Wesley's providentialism raises many questions of theodicy and in his pamphlet *Serious Thoughts on the Earthquake at Lisbon* (Wesley, 1755) he accepted that the earthquake came from the "Hand of the Almighty, arising to such an effect." But why did the Hand of the Almighty slaughter 30,000 souls?

Wesley was a popular religious writer and produced small tomes for the edification of Christians, rather than serious theological or scientific discourses. His first scientific work *The Desideratum: Or, Electricity made plain and useful* (Wesley, 1759) summarized the work of Franklin, the Newtonian Bishop Hoadly, and other "electricians." Wesley's main interest in electricity lay in its medical use and offered electrotherapy form the 1750s and was a medical pioneer. His major work on natural philosophy *Survey of the Wisdom of God in Creation: A Compendium of Natural Philosophy* began as a two volume work in 1763 and grew to five volumes in 1777. It was a re-packaging of works by early eighteenth-century writers in the physico-theology tradition like Buddeuos, John Ray, Derham, and Mather and was the cheapest treatment of astronomy, mechanics, and most of science to be found in one book.

Apart from his work on electrotherapy Wesley made no contribution to science. His claim to fame was to be a popularizer of science and that was subordinate to his evangelistic and didactic aims. His understanding of science was conventional and Newtonian. He was critical of outlooks, which disagreed with his theology. He regarded Buffon's spontaneous generation as "atheism barefaced," but in 1782 thought Buffon's theory of the earth expounded in volume one of his *Natural History*, which gently hinted at a longer time scale and questioned the Noachian Deluge as "whimsical" but "innocent." It is not known whether he read *Epoques* (Buffon, 1778), which were not translated. In this volume Buffon argued that the earth was at least 74,000 years old. He was drawn to Bonnet's *The Contemplation of Nature* published in 1764, especially "the scale of beings." Wesley used the scale of being to emphasize the unity and connectedness of creation and "the adorable Perfections of the First Cause."

Detailed research is needed to ascertain Wesley's opinion on the age of the earth and from my limited reading I am not convinced he held to Ussher's date of 4004 BC. If he was like many of his educated contemporaries, he could have allowed a few more thousand years, but as he died in 1791 that was too soon to take geology into account. In his *Notes on the Bible*, Wesley wrote that the "inspired penman in this history [Genesis] . . . wrote for the Jews first and calculating his narratives for the infant state of the church, describes things by their outward sensible appearance, and leaves us, by further discoveries of the divine light, to be led into the understandings of the mysteries couched under them" (Wesley, n.d.). This comment is similar to his *Survey* where he wrote that the "Biblical and scientific accounts agreed." Whether Wesley would have said with Thomas Chalmers that the two accounts agreed we do not know, but the later Wesleyan Methodist commentators Joseph Sutcliffe, who also wrote *A Short Introduction to the Study of Geology* (Sutcliffe, 1817) and Adam Clarke writing in the 1820s both held to a six-day creation contrary to most Anglicans. Wesley held to the "accommodation" of scripture, but his comments just cited are too ambiguous to be certain whether he would have welcomed geological findings.

Wesley had "cherished peculiar views on animals" which rejected Descartes' opinion as animals as mere machines. He, along with Bishop Butler, William Whiston, David Hartley, and Matthew Henry thought animal resurrection as possibility (Thomas, 1983, p. 140). In *The Great Deliverance* (Wesley, 1781) he contrasted the present existence of animals with their pre-fall and future state basing his argument on Romans 8:19–22 and that in the words of Paul they would be freed from "the bondage of corruption into a glorious liberty." Later Wesleyans and other evangelicals in the 1840s opposed vivisection and thus continued Wesley's concerns for animals (Desmond), which is also seen in the foundation of the Society for the Prevention of Cruelty to Animals, to which Queen Victoria gave royal approval in 1838 and thus the RSPCA has been the major British animal welfare organization ever since. Some writers like Andrew Linzey have used Wesley's sermons to support animal rights.

Though Wesley's publications contained no original thought, they were important in disseminating scientific knowledge to a wider audience, though presented in an evangelical scheme of natural theology. Wesley's didactic works gave his readers an outline of contemporary scientific thought.

WILLIAM WILLIAMS (PANTYCELYN) AND GENESIS

More amenable to longer ages was the Welsh evangelical hymn writer William Williams or Pantycelyn (1717–1791), who was perhaps the best theologian of the Welsh Revival, if not the best evangelical theologian in Britain. He is known through his one hymn translated from the Welsh *Guide Me O Thou Great Redeemer.* He was one of six children born to a farmer at Cefn-y-coed, near Llandovery in South Wales. Little is known of his childhood except that all his siblings died. As a youth he wished to train for medicine and went to the nonconformist academy at Llwyn-llwyd Academy near Hay-on-Wye where he studied moral philosophy, divinity, mathematics, and the classics. All that was changed in 1738 when he heard Howell Harris preach in the churchyard at Talgarth, near Brecon. Soon after his conversion he felt a call to take holy orders and was made Deacon in the Anglican Church in 1740 and was appointed curate at Llanwrtyd and Abergwesyn, despite being of Independent stock. His vicar, Theophilus Evans, denounced the evangelicals and Williams was refused priest's orders, being found guilty of nineteen offences. Williams could not get priested in the Anglican Church and spent the rest of his life on the fringes of Anglicanism and nonconformity. That did not stop him becoming the greatest Welsh hymn writer and the best theologian from Wales in the eighteenth century, despite of his lack of formal education after leaving Llwynllwyd and unlike most evangelical leaders never went to Oxford or Cambridge.

He wrote prolifically in Welsh, and translations are rare. In 1756 he wrote *Golwg ar Deyrnas Crist* (*A View of Christ's Kingdom*) (Williams, 1756) an epic poem answering the Deists. Pantycelyn's understanding of creation is discussed by Derec Llwyd Morgan (Morgan, 1984), who contrasts him to John Owen, who affirmed Creation in one line, "From nothing was created all 'neath the glorious skies." Pantycelyn is anything but brief. In the notes at the end of the work he gives a long summary of the contemporary state of science, mostly based on Derham's *Astro-theology and Physico-theology* and other works. He had a good grasp of science, far more than one would expect of someone whose education was cut short at the age of twenty. Chapter II of his epic poem is a long poetic account of Creation amplifying Genesis One. Llwyd Morgan points out that "Williams, perhaps taking his cue from Milton, . . . Maintains that the ordered creation was fashioned out of chaos, but that God was also the Creator 'Of Chaos vast and all its turbulence.'" Williams maintains there were two creations: the creation of the basic materials and the creation of the universe with those materials, all of which God accomplished "in one hundred and forty four hours," as it is in Genesis (Morgan, 1984, pp. 224–225). Though the Recreation took 144 hours, Pantycelyn gives no indication how long Chaos had existed. In an almost Miltonic manner Williams weaved contemporary science and physico–theology into the Genesis text, beginning with the creation of chaos. None of Pantycelyn's works were translated in his lifetime but *A View of Christ's Kingdom* was translated in the late nineteenth century. As Pantycelyn's books were widely read by Welsh speakers *Golwg ar Deyrnas Crist* was an important source of scientific knowledge for the Welsh. Though he did not explicitly discuss time, the poem is best read if Williams considered the duration of the earth to be more than 6,000 years. Williams' work is similar to many of his contemporaries and not that different from the writings of Whitehurst. Other evangelicals shared Pantycelyn's biblical interpretation for in 1761 Wesley produced a commentary on Genesis following the common Chaos-Restitution theme, as did the Baptist Calvinist John Gill (1746, revised 1768). Jonathan Edwards in his *Notes on the Bible* wrote in a similar vein and cited Grotius and Aristophanes among other writers. Many poets incorporated Chaos when versifying on Creation or related matters. The ubiquity of Chaos is evidenced by Phillis Wheatley's *Thoughts on the Works of Providence*:

That called creation from eternal night.
'Let there be light,' He said: and from his profound
Old Chaos heard. (Wheatley, 1994, p. 816)

Wheatley was a slave born in Africa who was purchased by John Wheatley of Boston and also wrote a poem on the death of George Whitfield. In a letter to Samson Occom, a Mohegan Presbyterian minister in 1774, she wrote:

> The divine Light is chasing away the thick darkness which broods over the Land of Africa; and the Chaos which has reign'd so long, is converting into beautiful Order. (Wheatley, 1994, p. 826)

The Wheatleys, slave-owners and slave, moved in evangelical circles and are more properly considered in respect of abolitionism, but it sheds light on the whole concept of Chaos and what must be considered a departure from a literal understanding of Genesis. Sadly Phillis died in poverty at the age of 31 in 1784, after the death of the Wheatleys.

However, though these evangelical understandings of Genesis One are similar to those of most Roman Catholic and Protestant exegetes from the eighteenth century, they are very equivocal on the duration since creation. The only unequivocal conclusion is that they did not affirm Ussher's creation entirely within 6 twenty-four-hour days, but it is impossible to conclude what time they did allow, thus undermining the common axiom that all Christians, whether evangelical or not before 1790 were dogmatically literalist and thus opposed to all geological findings. The scene is set for the acceptance or rejection of geological time, which will form the heart of the next chapter.

SUMMARY

None of my four examples of "evangelical scientists" can be considered scientists of any importance. Their approach and outlook was similar to many other clergy of the eighteenth century who took an interest in natural philosophy and antiquarianism. Some merely dabbled and others published works of varying quality. What is not known is the numbers of evangelicals who took such an interest or were members of local societies. Evangelicals did not produce scholars like William Stukeley or Gilbert White, probably because evangelicals did not have the time to pursue such studies, or were simply too few in number.

The interest in these writers is not their originality but the way they had accepted and absorbed eighteenth-century science and then repackaged it for a lay audience. They remained within eighteenth-century physico-theology, because Copernicanism was taken for granted and the implications of vast geological ages were just beyond the horizon. Geological findings were scarcely known until 1778 when John Whitehurst argued for an older earth than Ussher in the first edition of *Inquiry into the Original State and Formation of the Earth.* The evangelical response to Whitehurst is unknown and so far I have found no reference to geological time from evangelicals before 1800.

Towards the end of the eighteenth century some evangelicals showed a tendency to literalism as in the commentaries by the Anglican Thomas Scott and the Baptist Andrew Fuller. Neither makes any reference to any scientific or classical writings and is an *internalist* exposition of the Bible. As a result it is impossible to say whether the writers were opposed to scientific findings or whether it was simply a non-issue as their aim was to explain the Bible for Christian believers. The lack of published research precludes any firm conclusions, and to date I have not found any eighteenth-century evangelical who is critical of geological time. By 1810 or so that had changed. My tentative conclusion is that most evangelicals accepted a youngish age of the earth, perhaps on the lines of a de Luc who was convinced that the earth was tens of thousands years old rather than millions. Had evangelicals been dogmatic on a literal genesis there would have been more controversy over the next few decades.

Numbers of evangelicals did regard scientific studies as a deflection from the real work for Christ and also reckoned that natural philosophy or natural theology had nothing to add to Scripture. John Newton, formerly a slave trader and latterly an anti-slavery activist, believed that Christians might study the natural world, provided "their enquiries are kept within due bounds, and in a proper subservience to things of greater importances." Even so he thought there were few "who have leisure, capacity, or opportunity for these inquiries." This may be due to an exaltation of Scripture with the attendant downplaying of intellectual pursuits, which has long been a feature of popular Evangelicalism.

For the first three-quarters of the century there was no consensus on the duration of time. What the *uneducated* believed no one can say with certainty but the case of Phillis Wheatley should caution against assuming a mere 6,000 years as only the literate have left any evidence. A minority did take the Bible literally and adhere to an Ussher chronology, but many Christians, whether evangelical or not, stretched matters with an indefinite chaos with humanity limited to 6,000 years.

With the benefit of hindsight as we follow it through in the next few chapters, we can see how the future attitudes of evangelicals to science are to be found in embryo in the eighteenth century—a total acceptance of Copernican and Newtonian astronomy and of physical science generally but an ambivalence to the historical sciences like geology and the history of life.

It is difficult to decide whether the lines of William Cowper (1731–1800), an evangelical poet and close friend of John Newton, reflect a concern for geology or not

Some drill and bore
The solid earth, and from the strata there
Extract a register, by which we learn
That he who made it, and reveal'd its date
To Moses, was mistaken in its age.
William Cowper "The Task"

My next chapter will explore that question.

Chapter 4

Evangelicals and Science in the Age of Revolution

INTRODUCTION

The years from 1789 to 1850 were remarkable as an age of revolutions. In 1789, George Washington became President and the French Revolution began. The Industrial Revolution was under way with its attendant expansion of cities and slums. Canals enabled goods to be easily transported, but were soon eclipsed by railways. At sea, sail gave way to steam. The half-century witnessed a vast expansion of science, with evangelicals largely part of the mainstream of science. In chemistry the period begins with the discovery of oxygen and ends on the eve of the formulation of the Periodic Table and the development of organic chemistry. Similar advances occurred in every science. It would be wrong to focus only on scientific discovery, as many lay people were interested in science, whether through popular lectures or books and writings. The interest was at all levels from clerical rock hounds, working-class botanists, upper class ladies, and Mechanics' Institutes. Evangelicals permeated every aspect of society in Britain and America.

CONCERNS OF EVANGELICALS

The conventional picture of science and religion during this period is that there was conflict with the new science of geology, but that does not do justice to the numbers of clerical geologists including evangelicals. The conflict thesis of science and religion tends to color many historians' perspectives and thus it is often assumed that evangelicals *had* to be literalists. The corollary of that is that if a Christian accepted science and particularly geology then they were not evangelical and thus "liberal" in inclination. So Cannon argued forty years ago (Cannon, 1964) in an article entitled *Scientists and Broad churchmen: An early intellectual network.* She posited a network of "liberal" scholars at Cambridge and Oxford, whose liberal perspective enabled them to embrace the implications of geological science. Morrell and Thackray in *Gentlemen of Science* (Morrell and Thackray, 1981, pp. 225–259) argued that liberal Anglicans dominated the fledgling British Association for the Advancement of Science yet overlook the fact that Sedgwick was an evangelical. At both universities there was a broad tradition of intellectual endeavor. Some like Hare at Cambridge and Baden Powell (1796–1860) at Oxford *were* antecedents of the liberal Anglicanism associated with *Essays and Reviews,* but others were not. The Oriel Noetics of Oriel College, Oxford—Edward Copleston (1776–1849), Richard Whateley (1787–1863), and others—were mildly liberal. Sedgwick's Oxford counterpart William Buckland (1784–1856) received much support from J. B. Sumner (1780–1862) and G. S. Faber (1773–1854), leading evangelical theologians of his day and from Bishop Barrington (1734–1826) of Durham, an ultraconservative and evangelical sympathiszer. These examples undermine the simple liberal–evangelical divide, which is based on the supposition that liberals accept the findings of science and evangelicals do not!

THEOLOGICAL ISSUES

During the eighteenth century, science raised no *evangelical* theological questions, but the growth of geological science changed that. The vast ages of geology necessitated the interpretation of Genesis to be clarified, as it was no longer possible to be noncommittal on the age of the earth, or whether animals could die before the Fall. Most churchmen along with evangelicals simply modified the old Chaos-Restitution hermeneutic to allow for geological time and here the ideas of Sumner and Chalmers were those of the wider church (Roberts, 2002, 2007). The flexibility over Genesis 1 did not always extend to the rest of early Genesis, and many still held to a fairly literal view of Genesis 2–11 on the Fall and Flood. Even so, there was much *accommodation* and that the bible's interest in history was moral rather than precise. Most, whether evangelical or not, reckoned that humans appeared in about 4000 BC until the 1830s so that a "traditional" chronology of the Old Testament similar to Ussher's was largely unquestioned.

One key evangelical doctrine is the death of Christ as *Substitutionary Atonement.* Though Substitutionary Atonement was held from the Reformation, it has been largely rejected by non-evangelicals since the mid-nineteenth century, partly because it is considered to be morally repugnant. A strong view on the Atonement, by definition, requires a strong view on sin, as the Atonement reverses the effects of the Fall, but only a minority, notably George Bugg (1769–1851) argued that animal suffering did not exist before the Fall. However in the early nineteenth century most evangelicals, like Sumner, Chalmers, and others, accepted death and suffering before the Fall, yet retained substitutionary atonement.

As Evangelicalism is supernaturalist, it is marked by an acceptance of biblical miracles, which were often seen as breaking the laws of nature, laying up future problems. Some evangelicals posited additional miracles to explain aspects of the Creation and Flood. The Congregationalist minister John Pye Smith (1774–1851) scathingly objected to this "exuberance of miracles." He wrote that he could not envisage animals being "brought into one small spot, from the polar regions, the torrid zone, and . . . Asia, Africa, Europe, Australia and the thousands of islands; . . . without bringing up the idea of miracles more stupendous than any that are recorded in Scripture." And then with bitter irony he wrote, "(t)he great decisive miracle of Christianity, the RESURRECTION of the LORD JESUS—sinks down before it" (Pye Smith, 1839, pp. 116–117).

THE DESIGN ARGUMENT

The recent interest in Intelligent Design has made many revisit the old Design arguments. These had roots in the work of John Ray in the seventeenth century and reached their peak with William Paley (1743–1805), an Anglican archdeacon. His *Natural Theology* published in 1802 set the tone for the Design Argument for sixty years. Paley considered most aspects of the body so as to persuade the reader that must have been designed by God. In a pre-Darwinian age Paley's arguments were appealing. However, evangelicals were ambivalent towards Paley's Design because proving God from nature, that is natural theology, demonstrates the existence of God rather than a Redeemer. Even so, evangelicals used design arguments to demonstrate the existence of God but quickly passed on to the Redeemer, so that all should experience the new birth. When Paley's *Natural Theology* was discussed in the evangelical *Christian Observer*, it was criticized as Paley did not exalt God the Redeemer. In 1817, Wilberforce's spiritual advisor Thomas Gisborne (1758–1846) published the *Testimony of Natural Theology to Christianity*, to correct Paley's deficiencies. Paley had minimized the effects of the Fall and denied that animal suffering was due to Adam's sin. Evangelicals of all stripes, including Sedgwick, had reservations on Design and thought it supporting, rather than conclusive evidence. Yet popular evangelical apologetics appealed to the "proofs of the Creator" from design in nature and astronomy as a prolegomena for the Gospel.

Design arguments peaked with the eight *Bridgwater Treatises* in the 1830s. Most authors were orthodox Anglican clergy but only Thomas Chalmers (1780–1847) was clearly evangelical. Here Chalmers (1833) pursued a *weak* design argument, whereby design was used to evince faith in the Creator rather than prove it. Another *Bridgewater* author was Buckland, who was on the fringe of Evangelicalism. He was one of the greatest exponents of the Design argument as is seen in his *Bridgewater Treatise* on geology (Buckland, 1836) and his discussions and lecture on *Megatherium* (Roberts, 1999). However, as Brooke and Cantor point out, design arguments were more moral compulsion and rhetoric than scientific demonstration (Brooke and Cantor, 1998, pp. 180–181).

OVERVIEW

At Cambridge University several evangelicals became science professors at the turn of the century. Isaac Milner (1750–1820), a close friend of Wilberforce, became the first professor of Natural Philosophy at Cambridge in 1783 and subsequently professor of mathematics and President of Queen's College, from whence he supported Gorham against Sedgwick as geology professor in 1818. Francis Wollaston (1762–1823) was Jacksonian Professor of Chemistry from 1792 to 1813 and was succeeded by William Farish (1759–1837), who held the chair until 1836. Farish survived into the era of Sedgwick, Henslow, and Whewell, but was not as famous. In the late 1820s he was ridiculed when he suggested that trains would soon travel at sixty miles an hour. (The first person to be killed by a train in 1830 was William Huskisson, an evangelical politician.) Another evangelical Thomas Martyn (1735–1825) was Cambridge Professor of Botany from 1762 until his death and was followed by John Henslow. A fair assessment of these scientists must await a future date.

For the next generation, we are bedevilled by definitions of what an evangelical is. Of the Oxbridge scientists Sedgwick was unequivocally evangelical, but many, including the Oxford geologists William Buckland and William Conybeare (1787–1857), and the Cambridge philosopher of science William Whewell (1794–1866) were evangelical sympathizers. The result is that evangelical perspectives on science scarcely differed from other Protestants.

The situation in Scotland was similar as Presbyterians, both evangelical and Moderate, had a strong intellectual side to their faith and there was little to choose in attitudes to science between evangelicals like Thomas Chalmers, Sir David Brewster (1781–1868), John Fleming (1785–1857), and Hugh Miller (1802–1856) and their non-evangelical compatriots.

One aspect of science was the fascination for exploration, as part of every voyage was scientific research. The early nineteenth century saw much exploration on land and sea. Polar exploration was often the most harrowing and the story has been told many times and Fleming and Spufford (Fleming, 1998; Spufford, 1996, passim) offer a few hints at the relationship of polar exploration and Evangelicalism. Contemporary accounts filled the early Victorians with awe at the heroism and tragedy. William Scoresby (1789–1857) prepared the way by serving in the Greenland fishery and occupied himself with scientific questions, including the design of a marine compass. In 1823 at the age of thirty-five he entered the evangelical Queen's College, to prepare for ordination and became an evangelical clergyman. In the 1840s Scoresby wrote for the *Religious Tract Society* on polar exploration.

In the early decades after the Battle of Waterloo, the British were desperate to find the North-West Passage and several futile attempts were made culminating with the evangelical Sir John Franklin's (1786–1847) fateful expedition of 1845. John Ross led the first expedition in 1818 and the following year William Parry (1790–1856) and Franklin led expeditions. During many expeditions they had to overwinter in appalling conditions. Franklin was lieutenant-governor of Van Diemen's Land (Tasmania) from 1837 to 1843 and set sail again for the Northwest Passage in 1845, never to return. The horrors of the voyages were comparable to those of missionaries who went out to West Africa to succumb to tropical diseases. Heat and cold were equally lethal, but both brought forth stories of heroism and moral struggle. Parry and Franklin were evangelicals and letters to their wives reflected this, as Spufford wrote of the Parrys, "Praying together and churchgoing together were important parts of the intimacy of marriage" (Spufford, 1996, p. 97) Their piety was matched by his concern for moral rectitude on board their ships, with Parry seeking to eradicate *rum, bum, and concertina,* or less colloquially drunkenness, buggery, and the lash. Both Parry and Franklin were knighted and Parry promoted to Rear-Admiral.

Perhaps the naval explorer who most impinges on Victorian science is Robert Fitzroy (1805–1865), who captained *The Beagle* during its surveys of South America from 1832 to 1836 and hosted the young Darwin on the world voyage of 1831 to 1836. As Fitzroy is often presented as Darwin's nemesis, both on the voyage and later, it is difficult to consider Fitzroy in his own right. He was an able surveyor often surveying in terrible conditions and was a worthy successor to Captain Cook. After the *Beagle* voyage both Fitzroy (Fitzroy, 1839, 2007) and Darwin (Darwin, 1839) wrote accounts of the voyage, but Fitzroy's has often been eclipsed and is rarely read. Despite popular interpretations Fitzroy and Darwin got on remarkably well during the voyage as they had few quarrels despite being cooped up in such a confined space. Disagreements were more on slavery than science (Darwin, 1983, pp. 41–43) and it seems that Fitzroy adopted *anti-geology* after their return. Few look beyond Darwin's account in his *Autobiography* and overlook the many comments in his correspondence, where it is clear that they got on fairly well, with Fitzroy calling Darwin *Philos* (philosopher) (Burkhardt and Smith, 1985, pp. 506, 509). Religious differences did not appear until Fitzroy's *Narrative* was published in 1839. This was not over the substance of the account but a chapter on geology entitled *A very few remarks with reference to the Deluge* (Fitzroy, 1839, vol. 2, chap. 28; Darwin, 1989, pp. 400–424), in which he notes that though he used to disbelieve "the inspired history written by Moses" he now does not and had adopted the literalist views of the *anti-geologists* discussed below. He seems to have done this after returning from the *Beagle* voyage, much to Darwin's consternation; as he wrote to his sister Caroline:

> You will be amused with Fitzroy's Deluge chapter—Lyell, who was here today, has just read it, & he says it beats all the other nonsense he has ever read on the subject. (Burkhardt and Smith, 1986, p. 236).

It is not clear how long he held such views, and my suspicion is that by 1850 he moved quietly to a typical progressive creationist position, as did many others like William Brande. After a difficult time as governor of New Zealand, he was made Chief of the meteorological department of the Board of Trade in 1854. Here he designed the Fitzroy barometer and began weather forecasts on about 1860, which were not always reliable (Anderson, 2005). He was present at the BAAS in Oxford in 1860 and is alleged by some to have brandished a Bible overhead as a reproof to Huxley. That story is probably mythical. But Fitzroy had no truck with evolution. A few years later Fitzroy committed suicide while shaving. Most of Fitzroy's scientific work was in surveying and meteorology, which did not impinge on geology or evolution.

Much later David Livingstone (1813–1873) went to Africa as a medical missionary-explorer. His story alone would fill several books. The relationship of evangelical faith, exploration, science and the attendant suffering and possible martyrdom can only be hinted at, but I hope that this brief account demonstrates that there was a relationship which needs further exploration to shed light on the wider implications of Evangelicalism, and the possible subjection of science to the missionary nature of Evangelicalism (Sivasundaram, 1995).

Throughout the early nineteenth century, if evangelicals were interested in science—and many were not because of the pressing concerns of evangelism—they took a broad practical interest in it and saw no conflict with astronomy or physics and chemistry. Michael Faraday's (1791–1867) electrical experiments filled people with awe rather than doubt. Natural history caused some problems, but not many asked questions about the origin of species. Even so the publication of *The Vestiges of the Natural History of Creation* (Chambers, 1844) rattled Christian cages both evangelical and not. One science that caused problems to some evangelicals, yet was practiced by many evangelicals, was geology.

GEOLOGY

By 1859 few educated doubted that the earth was millions of years old, but up to 1810 geologists were unsure whether it was millions or tens of thousands. A few clung to 6,000 years. From about 1820, if not earlier, there was a broad consensus on the geological column, and the antiquity of the earth. By 1850 the Geologic Column from the Cambrian to the Pleistocene was firmly in place. Geology developed rapidly after 1770, with different approaches in Scotland (Hutton), England (Smith), Germany (Werner), Switzerland (de Luc and de Saussure), and France (Cuvier, Soulavie, and Brogniart). Popular accounts, which give most of the credit first to Hutton and then to Lyell, are simply inaccurate as any recent history of geology demonstrate (Rudwick, 2004). After 1800 the priority was to determine the history of the earth by the stratigraphical principles developed by William Smith (1769–1839) near Bath and Cuvier and Brogniart in the Paris Basin. By using fossils in conjunction with the Principle of Superposition, geologists were able to put strata in the order of deposition, and thus of relative age.

Smith was a Canal Engineer working near Bath and in 1795 spotted that the same sequence of fossils occurred in two parallel valleys in the same strata. Three local Anglican clergy, Richard Warner (1763–1857), Benjamin Richardson, and Joseph Townsend (1739–1816) encouraged him to publicize his methods. Of these Townsend was an evangelical. After Cambridge and medical studies at Edinburgh he took orders in 1765 and settled at Pewsey Rectory. He was an early evangelical becoming one of the Countess of Huntingdon's preachers along with his brother-in-law Thomas Haweis from 1765 until 1779. After helping to propagate Smith's ideas, he wrote *The Character of Moses established for veracity as an Historian, recording events from the Creation to the Deluge* in 1810. He rejected the vast ages of Hutton and preferred the shorter time scale of de Luc. The importance of de Luc for enabling orthodox Christians to accept geological time is little known, but as Rudwick (2002) has recently shown de Luc was a great influence throughout Europe. Whether Townsend had ever held to an Ussher type chronology is not known, but Torrens has shown that in 1797, well after Smith had worked out his table of strata, he wrote that "of the mineral bodies there is not a doubt that they were all formed at the same time." By 1806 he could write "Fossils . . . have handed down to us the clearest and most unequivocal history of the most ancient inhabitants if the Earth" and that "the time required for the . . . formation into strata . . . would stagger the faith of many" (Phillips/Torrens, 2003, pp. 189–191). During these nine years Smith rejected a short time scale. As for Townsend, by 1810 he reconciled long geological time with Genesis through the Chaos-Restitution interpretation of Genesis (Townsend, 1813), as had Thomas Chalmers.

During the early nineteenth century geology became a historical reconstruction of the world. By 1810 geologists had extended the time scale though they often emphasized Deluges or Catastrophes. Earlier geologists like de Luc and his successors up to about 1830 attempted to make a direct correlation between the Deluge and the uppermost strata, frequently called diluvium. To many, "diluvialism" is both an example of bad science and the church's tyranny over science. Gillispie in his classic work *Genesis and Geology* (Gillispie, 1951) is negative towards Catastrophism, but Stephen Gould is far more positive (Gould, 1987). Taking into account the culture of the eighteenth century, it was almost inevitable that many geologists, particularly religious ones, would begin with one Flood, multiply it, and finally let them drain away. The dominant school of thought in England was Diluvialism or Catastrophism, to which Buckland gave classic expression. In the 1830s Catastrophism waned due to Lyell's Uniformitarianism, and geology was weaned from its scriptural roots. That was a disaster for Christians according to Mortenson of *Answers in Genesis* as argued in his book *The Great Turning Point* (Mortenson, 2004). The stance a geologist took over catastrophism and uniformitarianism did not affect either the progress of geology or to attitudes of geological time in the 20s and 30s.

EVANGELICALS AND GEOLOGY

The story of geology from 1790 until 1850 undermines the simple warfare history of science and religion as it denies a divide between geology and evangelicals. Certainly some evangelicals were opposed to geology, but most evangelicals in Britain and America were enthusiastic for geology and its time scale. Evangelical geological aficionados included university dons at the forefront of geology, parson-naturalists and geologically informed theologians. With geology as the most popular science from about 1815 to 1850, most were active in that time, and several made significant contributions to geology and none more than Sedgwick.

ADAM SEDGWICK: THE EVANGELICAL GEOLOGIST

Adam Sedgwick (1785–1873) of Cambridge University was a leading geologist of his generation. It was he who gave us the Cambrian Period and was heavily involved in the Silurian and Devonian as well. Sedgwick was born in Dent in Yorkshire and graduated in mathematics at Cambridge in 1807. He became a mathematics fellow at Trinity College, and was ordained in 1817. In 1818, without having studied geology, he stood for the Woodwardian Professorship of geology against fellow-evangelical C. G. Gorham of Queen's College, who later achieved notoriety by falling out with the Bishop of Exeter over baptism in 1847 in an event known as the Gorham controversy (Chadwick, 1965, pp. 250–259).

Little is known of Gorham's geology, but he accepted an ancient earth. In the election Sedgwick received 186 votes compared to Gorham's 59. Sedgwick took up geology with a vengeance. He gave lecture courses each semester, and spent every summer for the next forty years in the field throughout England and Wales. In the 20s he worked in the North of England but his most important work was in Wales beginning in 1831. That year he spent from August to October in North Wales with the aim of trying to work out the geological succession below the Old Red Sandstone or Devonian, in strata we now call Cambrian, Ordovician, and Silurian. He began in the Shrewsbury area in August with the young Charles Darwin, who left Sedgwick on 20 August near Bangor to return home for the shooting season, but joined the *Beagle* instead. Sedgwick gave Darwin a superb grounding in geology, which he used to great effect on the *Beagle* voyage (Roberts, 2001). Darwin only read Lyell's *Principles of Geology* while on the *Beagle* and because of his appreciation of Lyell in his *Autobiography* his debt to Sedgwick is often ignored or downplayed (Herbert, 2005).

Sedgwick failed to work down the geological succession, as in North Wales the Devonian is present only in Anglesey. Thus Sedgwick had to start in older rocks and try to link them to the Old Red Sandstone. He covered vast distances on foot and climbed most of the mountains of Snowdonia. In late August 1831 he walked over eighteen miles and climbed 6,000 feet, as well as making many geological notes and collecting many rock specimens. When I rewalked the route with my dog Topper in 1999, it took us over ten hours, without stopping to study the geology, as I had done so on previous occasions. In two months Sedgwick had worked out the basic stratigraphy and structure of a complex area. In the same year Roderick Murchison (1792–1871) traveled through South Wales looking for what lay below the Old Red Sandstone. He was lucky as near Ludlow, the Silurian (using today's terminology) lies conformably beneath the Old Red Sandstone (Devonian). However an evangelical vicar aided him! The Rev. Thomas Lewis (1801–1858), who had learnt geology from Sedgwick, had worked out the succession in his parish of Aymestry. Murchison took the credit and thus the work of one evangelical geologist is hardly known. From 1836 to 1847 several geologists including Sedgwick worked in Devon and Cornwall to elucidate what age the various strata were. Martin Rudwick has comprehensively charted out their work and Sedgwick's vital role (Rudwick, 1985). Though Sedgwick was not the most prolific of writers, he was one of the main geologists who worked out the Lower Paleozoic from the Cambrian to the Devonian. That establishes him as one of the greatest geologists. Yet he was a convinced evangelical.

That evinces several questions, the first being, "What effect did his faith have on his science?" There is no evidence that he had a crisis of faith over science nor any conflict between *geology* and Christianity. Some of his perspectives changed, as until 1831 he was a *Catastrophist* when he largely adopted *Uniformitarianism*. In the early 1820s, like most British geologists, he reckoned the Noachian Deluge to be a worldwide event, which had deposited sediments known as *diluvium*, hence *diluvialists*. (These are now reckoned to be glacial deposits.) In 1825 Sedgwick wrote a paper *On the Origin of alluvial and diluvial formations*, which was a good account of "Ice Age" sediments in Britain (the Ice Age was not recognized in Britain until 1840). He thought these "to demonstrate the reality of a great diluvian catastrophe during a comparatively recent period" and that "It must . . . be rash and unphilosophical to look to the language of revelation for any direct proofs of the truths of physical science." He was close to considering the Ice Age. In 1831 he retracted his adherence to the diluvial theory to the Geological Society of London and later accepted Ice Ages. Some like Pennock argue that Sedgwick and Henslow were adherents of "the detailed hypotheses of catastrophist flood geology" (Pennock, 1999, p. 61; Roberts, 2004) which they allegedly taught Darwin. Thus Henslow and Sedgwick were *supernaturalist*, but, in fact, both were using the Noachian Deluge in a *naturalistic* way. Sedgwick rejected his earlier ideas on scientific grounds, without any effect on his faith. Further a comparison of his geological work before and after his *recantation* in 1831 shows no significant change. However after 1831 he was more wary of attributing things to the direct action of God as he said to the Geological Society in 1831:

> To the supreme Intelligence, indeed, all the complex and mutable combinations we behold, may be the necessary results of some simple law, regulating every material change, and involving within itself the very complications, which we, in our ignorance, regard as interruptions in the continuity of Nature's work. (Sedgwick, 1831, p. 302)

Sedgwick could not accept the evolutionary ideas of Chambers (1802–71) in the *Vestiges of the Natural History of Creation* (1844). At the 1844 British Association meeting in York, Sedgwick savaged the Dean of York, William Cockburn (d. 1858), who tried to prove that all strata had been laid down in Noachian Deluge. The next year, Sedgwick proceeded to maul *Vestiges*, identifying its mistakes on its use of the fossil record, and concluding with a flourish, "I believe some woman is the author . . . partly from the utter ignorance the book displays of all sound physical logic." He was no misogynist as he was the first Cambridge don to allow women to attend lectures.

Sedgwick's real reasons for opposing evolution became apparent when Darwin published *The Origin of Species* in 1859. Sedgwick wrote critical reviews and also a friendly, but critical, letter to Darwin. In his letter Sedgwick criticized Darwin for his scientific shortcomings in his rejection of "the true method of induction" and coming out with speculations as "wild I think as Bishop Wilkin's locomotive that was to sail with us to the Moon." However, Sedgwick's greatest concerns were theological and moral and as he wrote to Darwin, "Tis the crown & glory of organic science that it *does* thro' *final cause*, link material to moral; . . . You have ignored this . . . " (Burkhardt and Smith, 1991, p. 396)

Yet we may ask, "Just how different are Sedgwick and Darwin?" They had much in common. Both accepted vast geological time and that the fossil record demonstrates the progression of life. However Darwin argued for a seamless common descent but Sedgwick held to progressive creations forming new species at regular intervals. Sedgwick considered that Darwin's theory rejected *final causes* that is God which reduced the moral status of humanity. (This is, and always has been, one of the chief theological and moral objections to evolution.) More significant is to compare the last paragraph of *The Origin* with Sedgwick's comments spoken to the geological society in 1831 above. Both sought to explain natural phenomena by natural laws rather than by divine interruptions. A modern YEC may consider both Darwin and Sedgwick to be "uniformitarian and evolutionary." To use the false definitions of *Naturalist* and *Theist* held by Intelligent Designers today, in his geology Sedgwick was *Naturalist* and non–interventionist but *Theist* and interventionist in his interpretation of the Fossil Record. Geological revolutions were *natural* but the creation of species was *supernatural.* This may be an inconsistent approach to the history of the earth and life, but it was commonplace before 1860. The prevalent pre-Darwinian Progressive Creationism was an unstable amalgam of *supernaturalism* and *naturalism.* When we take into account the quotation from Sedgwick in 1831, it is clear that there is only a short step from Progressive Creation to Evolution. Sedgwick resisted that step because of fears of materialism and on moral grounds, but many made that short step, with or without consequences to their faith.

It is easy to criticize Sedgwick for rejecting evolution, but he was seventy-five in 1859. He could remember the Napoleonic Wars and the guillotine. His generation tended to associate evolution with Lamarck and the anti-Christianity of the French Revolution. Thus Sedgwick objected to evolution because of its perceived moral and materialist associations rather than any concerns over geological time and a "literal" Genesis. Sedgwick's objection to evolution was both theological and cultural. Cultural and political considerations are often as important as theological ones in the acceptance of evolution, which was brought out powerfully by Livingstone in his comparison of attitudes of Calvinist Presbyterians to evolution in Scotland, Ulster and the United States discussed in the next chapter (Livingstone, Hart and Noll, 1999, pp. 193–219).

We now ask, "What effect did his science have on his faith?" There is no evidence that questions over evolution shook Sedgwick's faith in any way, but he was not easily ruffled by doubts. Sedgwick was a moderate evangelical, and as a result was pilloried by virulent evangelicals like the Rev. Henry Cole (Mortenson, 1996) who in 1832 wrote a tract *Popular Geology Subversive of Revelation! Being a Letter to Adam Sedgwick* (Cole, 1834). Cole's tract was a most vitriolic attack on Sedgwick. Cole had dismissed Sedgwick's ideas as "palpable evasion," "willing ignorance," "scoffers," "heaven-marked infidelity, presumption and falsehood," and so forth.

A brief study of Sedgwick's letters, writings, and life (Clark and Hughes, 1896) demonstrates that he was not troubled by supposed geological challenges to his evangelical faith. Sedgwick's conservative theology became apparent when he raised serious doubts about Hugh Miller's revelatory day theory of Genesis in a letter to his friend Dean Francis Close (1797–1882) in 1858. Miller rejected Chalmers' *Chaos-Restitution* interpretation of Genesis in his posthumous *The Testimony of the Rocks* (1858) in favor of his concept of *The Mosaic Vision of Creation.* One of the first to popularize Miller's ideas was Dean Francis Close (1797–1882) who gave a lecture to the YMCA in London in 1858 (Close, 1858) and made extensive use of Miller's book. Close was a leading evangelical and was Dean of Carlisle. In the 1820s, while at Cheltenham, he preached on early Genesis, but took Genesis literally and ignored geology (Close, 1826). Within thirty years he moved to the scientifically informed nonliteralism of his YMCA lecture. On reading that, Sedgwick wrote to Close and criticized Hugh Miller for not being faithful to Holy Writ!

Sedgwick had long rejected Chalmers' views on Genesis in that he regarded the Days as indefinite periods. His biography by Clark and Hughes sheds little light on this and his most detailed comment is his letter to Close. Sedgwick thought Miller might do some harm as his overschematic approach was geologically wrong, and concluded "Hugh Miller was a man of *great natural genius,* +in some parts of geology, admirably well informed but it is not always safe to follow him, when he travels beyond his own beat—His "Testimony of the Rocks" is in its way a noble work—it may do much good, but it *may do some harm*—"

He also wrote, "I make no difficulty in the words *Morning + Evening* they are only I think meant to mark the beginning + end of periods or days,—the Mosaic day is assuredly not 24 hours, + if we once admit a prophetic extended meaning of day, our souls are then free, + we are permitted to give any indefinite period, + the word day." But then he wrote, "I do not like the scheme of stretching the Bible, like an elastic band, till we can wrap up our hypotheses in its sacred leaves."

This letter is of great significance as the leading evangelical geologist wrote it to another evangelical. Both had a high view of scripture and both were more than convinced by geological findings. They took Genesis "non-literally" yet Sedgwick, cautious as usual, was reluctant "of stretching the Bible, like an elastic band" and preferred to wait as this "will end in harmony, + true accordance with the word of God."

SEDGWICK'S COLLEAGUES

Sedgwick was the most evangelical of several Anglican clerical geologists; with the not-quite-evangelical Conybeare brothers, William and John J. (1779–1824), and William Buckland at Oxford, and from Cambridge the mineralogist-turned-botanist John Henslow, who did pioneer work on the Precambrian of Anglesey, and Thomas Lewis, who effectively gave us the Silurian. These five and Sedgwick did some of the most important geological work in England and Wales in the two decades from 1820. W.D. Conybeare, Sedgwick, and Lewis were three important geologists elucidating the Cambrian, Silurian, Devonian, and Carboniferous Systems in the 20s and 30s. (At that time the Ordovician was subsumed into the Cambrian and Silurian.) Buckland worked on Mesozoic strata and was a better paleontologist. He was the first to identify the fossilized dung—coprolite—of dinosaurs (which, when polished, made beautiful ear-rings for his lady acquaintances), and achieved notoriety both for that and his enthusiasm for the Deluge.

Coneybeare and Buckland were on the fringes of Anglican Evangelicalism and ended up as Deans of cathedrals, Conybeare at Llandaff (Cardiff) and Buckland at Westminster. Buckland can be considered a "bridge" person between earlier geologists like de Luc who made a direct correlation between the Deluge and geology and the mid-century geologists who ignored the Deluge. Buckland's early works *Vindiciae Geologicae* and *Reliquiae Diluvianae* attempted to harmonize geology with the Deluge, but rejected Diluvialism in his Bridgewater Treatise *Geology and Mineralogy Considered with Reference to Natural Theology* in 1836.

Sedgwick's friend, Lewis was a competent amateur geologist. Henry Lyte (1793–1847) best known as the author of the hymn *Abide with me* was Vicar of Brixham and were it not for his geological interests that hymn would not have been written, as his death was caused by chest problems brought on by spending long hours looking for human fossils in the dank caves in the sea cliffs of Devon. In the 1830s, while vice-principal at St David's College, Lampeter, near Aberystwyth in Wales, Alfred Ollivant (1798–1882) spent much time geologizing in the surrounding Silurian strata collecting specimens for Roderick Murchison and corresponded with Sedgwick. He later became Bishop of Llandaff and in the 1860s became one of the most outspoken critics of the volume of liberal theology *Essays and Reviews.* In the early 1820s one evangelical Oxford undergraduate went to Buckland's geology lectures three years in succession. That was Samuel Wilberforce (1805–1873), son of the anti-slavery campaigner William, who later became a high church bishop and is frequently presented as the episcopal buffoon who was routed by Thomas Huxley over *The Origin of Species.*

SCOTTISH GEOLOGISTS

Evangelicals in the Scottish Presbyterian Church also saw little conflict between geology and Genesis. Although none were as great a geologist as Sedgwick, they did produce the greatest geological popularizer of the nineteenth century—Hugh Miller. Scottish geologists generally were far more sympathetic to Hutton's geology than their English counterparts and this we see in John Fleming and Miller (Livingstone, 1987). The dominant feature of Scottish church life in the early nineteenth century was the Great Disruption of 1843 led by Chalmers. Before 1843 the tensions between the Moderate and evangelical wings of the Presbyterian Church did not affect science. After the Disruption some of the best Presbyterian scientists sided with Chalmers. These included the Sir David Brewster, a physicist, and John Fleming. Hugh Miller could make merry about the Disruption and wrote:

> On the one side we saw Moderate science personified in Dr Anderson of Newburgh—a dabbler in geology, who found a fish in the Old Red Sandstone and described it as a beetle: we saw science not Moderate (i.e. evangelical), on the other side, represented by Sir David Brewster. (Miller, 1870)

To Miller, evangelicals had were better scientists than their liberal counterparts!

Fleming, who had parish appointments before becoming professor of Natural History at Aberdeen and then at the Free Church College in Edinburgh in 1845, almost anticipated Lyell's Uniformitarianism. In the 1820s Fleming made his most significant contributions to geology and opposed the diluvialism and catastrophism of the English School (Fleming, 1826).

The most well known was Miller, a stone mason, self-taught geologist, and man of letters. Miller was brought up in Cromarty and later lived in Edinburgh. He was a great amateur geologist and was held in high regard by many including Louis Agassiz and Sir Alexander Geikie, whom he encouraged to take up geology in the 1850s, which Geikie recorded in his autobiography *A Life's Long Work* (Geikie, 1923). For many years he was the editor of *The Witness* and wrote many books. *The Old Red Sandstone* is an excellent survey of Devonian geology in the 1840s; *Footprints of the Creator* was a best-selling rebuttal of *The Vestiges*. His finest work is his posthumous *The Testimony of the Rocks* (1857), which is a mixture of geology and the theological implications of it, along with an almost satirical chapter on *The Geology of the Anti-Geologists*. Miller died tragically by a self-inflicted gunshot wound, which has given biographers scope for creativity. Flawed or not, Miller was a great advocate of Scottish Evangelicalism and the Free Church. His writings give lie to the common cry that the Kirk was dour, as they are full of intellectual depth and wit.

THEIR AMERICAN COUNTERPARTS

In the early nineteenth century American science was beginning to thrive and produced two important geologists. The first was Benjamin Silliman (1779–1864) who was the first professor of chemistry and natural history at Yale. He had studied at the University of Pennsylvania and Edinburgh. He founded of the *American Journal of Science and the Arts* (known as Silliman's Journal) and became the first president of the American Association for the Advancement of Science. He did more for American science by teaching and inspiration than by research. Timothy Dwight, the evangelical president of Yale, had appointed Silliman in 1801 because he wished to appoint a man who saw science as an ally of faith. Alongside his science Silliman presented a "concordist" approach to Genesis and geology and emphasized the parallels between the order in Genesis and that found by geologists. Silliman drew up his concordance in 1833 which showed plants to be created before animals as one finds in Genesis with plants created on day three and animals on days four and five, which reflected the geological understanding of the day as pre-Carboniferous strata were hardly known until the early 1830s. Clearly the discovery of Paleozoic fossils undermined Silliman's concordist system (Davis, in Miller, 2003). But it served for a time.

One student of Silliman who continued his harmony of geology and Genesis was Edward Hitchcock (1793–1864) from Massachusetts. An attack of mumps in his twenty-first year left him with lifelong ill-health, and led him return to his father's congregationalism. From 1815 to 1819 he was preceptor of Deerfield Academy and then entered Yale to train for the ministry, when he attended Silliman's geology lectures. From 1821 to 1825 he was a pastor in Conway, Massachusetts, but was dismissed through ill-health. After further study with Silliman he became professor of chemistry and natural history at Amherst College. He was appointed geologist for Massachusetts and Vermont and in 1833 produced his *Report on the Geology, Mineralogy, Botany, and Zoology of Massachusetts.* Shortly afterwards he reported the first dinosaur tracks, which he mistook for birds' tracks. He published *Elementary geology* in 1840 and *The Religion of Geology* in 1851, in which he expounded the relation of geology and Genesis. He adopted the "Chaos-Restitution" interpretation and argued forcibly for an ancient earth. He criticised those who did not as "groaning under the burden of Milton's mythology," a not-so-veiled criticism of the (baneful) influence *Paradise Lost* has had on popular understanding of Genesis.

There are close parallels between these two American scientists and their British counterparts, especially over harmonizing geology and Genesis. However they received criticism from the Old Testament scholar Moses Stuart (1780–1852) of Andover Seminary during the 20s and 30s, because they had violated "the laws of exegesis in order to accommodate a geological theory." Stuart was not prepared to let science influence his interpretation, but Hitchcock's colleague writing in Silliman's Journal charged Stuart with re-interpreting Moses' statements about the firmament in Genesis 1 vs 6–8. Nothing has changed today!

EVANGELICALS, GEOLOGY, AND GENESIS LEADING TO DARWIN

By the 1820s most evangelicals adopted the Chaos-Restitution Theory of Chalmers and Sumner, seemingly removing any conflict between Genesis and Geology. However, this statement needs qualification, as the only people who can be studied are those who wrote either in books or magazines. The pages of the *Christian Observer* the leading Anglican evangelical journal give an insight into Anglican Evangelical attitudes to geology. The editor, S. C. Wilks (1789–1872) tried to avoid controversy, but ensured the anti-geologists were always answered, relying on W. D. Conybeare for geological guidance. From 1827 the division between Bugg and Faber dominated several volumes, and the correspondence became acrimonious with articles such as "On the Infidel Tendency of Certain Scientific Speculations" (vol. 34, 1834, pp. 199–207) and then "Replies to a Layman on Geology" (pp. 306–316), written by Conybeare. For twenty years, Wilks, the editor, attempted to guide his readers into accepting geology without totally censoring the anti-geologists. A modern-day parallel can be seen in *Christianity Today.*

Before the 1820s dissenting and Methodist evangelicals were more likely to interpret the Bible literally. By the reign of Victoria their leading scholars had accepted geological findings. Most notable was the Congregationalist John Pye Smith, a Biblical scholar, who published *The Relation between the Holy Scriptures and Some Parts of Geological Science* in 1839, originally given as the Congregational Lecture in 1838. Smith adopted a novel exegesis of Genesis 1, by arguing that God had recreated a small portion of the earth in six days and put Adam and Eve there. The rest of the planet had been there for millennia, thus accommodating geological ages. Pye Smith gave an excellent resume of geology, and criticized the "anti-geologists." In 1837 another Congregationalist George Redford (1785–1860) also wrote on the relation of geology and Genesis in *The Holy Scriptures Verified*, grappling with the issues in a muddled way, more or less accepting the Gap Theory and for his geology looking to Fairholme as well as Buckland. His muddled arguments indicate that he was not a dogmatic literalist.

Though there was a considerable diversity among evangelicals of all denominations, the majority were supportive of geology. Only a vocal minority considered geology to be infidel. Commentators frequently adopted a non-literal approach to Genesis, most notably the Free Kirk Robert Candlish (1806–1873).

THE HARMONY OF GENESIS AND GEOLOGY

During the early nineteenth century there were numerous harmonies of Geology and Genesis. Though some were "anti-geologies," the majority accepted geology and propounded their harmonies with variable geological competence. The most widely sold and competent was Buckland's Bridgewater Treatise. By the 1850s the vast majority of educated Christians accepted geology, the enthusiasm for "anti-geology" had waned, evincing the astronomer Rev. Robert Main's (1808–1878) comment in the conservative *Replies to Essays and Reviews* (Wilberforce, 1862) edited by Samuel Wilberforce: "No educated Christian accepts 4004B.C. as the date of creation." That was true then, but not today!

The Chaos-Restitution theory was the most widely held reconciliation of Genesis and geology until mid-century, which Hugh Miller challenged in *Footprints of the Creator* (Miller, 1847, p. 332), his anti-evolutionary critique of the *Vestiges* and in *The Testimony of the Rocks* (Miller, 1858). Within a few years Gilbert Rorison was arguing for a totally pictorial exegesis of Genesis in Wilberforce's *Answers to Essays and Reviews* (Wilberforce, 1861, pp. 281–286) and the Chaos-Restitution interpretation began to go out of fashion. Archdeacon John Pratt of Calcutta was one of the last serious writers to expound it. After that it was taken up by nascent fundamentalists in the late 19th century, and enshrined in the *Schofield Reference Bible*, while the Day-Age interpretation gained ground among the more "intellectual" conservatives, notably by J. W. Dawson.

Most typical of the 1850s are the volumes by Pratt, Hitchcock, and Miller. John Pratt (d1871) was Archdeacon of Calcutta, and in the midst of his missionary work wrote on the isostasy of the Himalayas. An evangelical, he published *Scripture and Science not at variance* in 1856 and revised it in 1871, when he still held the Gap Theory and rejected Rorison's exegesis in *Replies to Essays and Reviews*. From its title the American Edward Hitchcock's *The Religion of Geology* (1853) sounds unpromising. Hitchcock was no mean geologist, and was aggressive in justifying geology to those like the Hebraist Moses Stuart. Hitchcock saw the problem as being caused by too literal a reading of *Paradise Lost* and that "the theologians having so mixed up the ideas of Milton with those derived from inspiration," thus giving rise to Bishop John Colenso's (1814–1882) complaint "The truth is that we literally groan, even in the present day, under the burden of Milton's mythology" (Colenso 1863, vol. iv, p. 148). Though Colenso was notorious for his views on biblical criticism in the 1860s, his approach to Genesis One was similar to that of evangelicals and he quoted extensively from them.

Pride of place must go to Hugh Miller's *The Testimony of the Rocks*, which has already been mentioned.

These three represent the moderate, scholarly evangelical. Others were less moderate, as is shown by George Eliot's essay on the immoderate evangelical—John Cumming (1807–1881). He wrote at least twice on science, first a lecture given at Exeter Hall in 1851 and then his peculiar *Church before the Flood* (Cumming, 1854). Eliot's criticisms of Cumming are fair (Eliot, 1973), but a dismissal of Cumming will miss an essential point. Cumming was very conservative, yet, *accommodates* the whole of geology into the first two verses of Genesis. Joseph Baylee (1808–1883), Principal of the Anglican theological college, St. Aidan's Birkenhead, was also an ultraconservative, who wrote on geology and Genesis and allowed geology to sit alongside his almost literal Genesis (Baylee, 1857). This acceptance of geology is easily lost in a cursory reading as Baylee claims to be literalist and it demonstrates the need to study Victorian (or any) writing on their terms and not with spectacles provided by the twenty-first century. The exegesis may not be convincing, but it shows that ultraconservatives accepted geology.

SCRIPTURAL AND ANTI-GEOLOGISTS

It may seem that geology presented little challenge to evangelical faith in the nineteenth century. That was not so for some Christians, who argued that geology undermined the truth of Genesis and are frequently termed

Scriptural Geologists, which can cause confusion as geologists like Sedgwick and Buckland held both geology and a conservative interpretation of Genesis. So they could rightly be termed Scriptural Geologists. Millhauser (1954) to a limited extent and Mortenson (2004) define Scriptural Geologists as those who claimed that all geology was within the confines of Six Days and a Flood. To Mortenson "scriptural" means a literal hermeneutic. Therefore, all who reject literalism have *compromised* their biblical faith, be they Sedgwick or not. Throughout his work, Mortenson made strident criticisms of conventional geology and claimed that several "Scriptural Geologists" were competent geologists. However his failings are twofold. First he misunderstands conventional nineteenth-century geologists and wrongly claims that their geology is based on *assumptions* of the vast age of the earth, and secondly he fails to discern that his "competent" scriptural geologists misunderstood geology. Hence, his work has value in giving biographical details, but not in the assessment of the scriptural geologists.

Against Mortenson, John Lynch takes a wider view of "scriptural geology" in his introduction to *Creationism and Scriptural Geology, 1817–1857* (Lynch, 2002), a seven volume series reprinting some of eight "scriptural" geologists. Five were hostile to geology and three supportive. Thomas Chalmers, John Pye Smith, and Hugh Miller did not regard geology as infidel. The hostile five were John Mellor Brown an Anglican clergyman, Granville Penn (1761–1844), grandson of the founder of Pennsylvania and an Anglican layman, George Young (1777–1848), a Presbyterian minister and two laymen George Fairholme (1789–1846) and John Murray (1786?–1851), a chemist. Only two had good field skills in geology, Miller who needs no introduction and George Young, who did some competent work around Whitby in the 1820s. By choosing both "pro-" and "anti–geologists" Lynch undermined the polarized historiography, which usually surrounds "Genesis and Geology" discussions. All eight writers can be rightly described as *Creationists and Scriptural Geologists*, as they understood geology from the perspective of Creation and Scripture. Chalmers, Pye Smith, and Miller were respected evangelicals whose works indicate the change in biblical interpretation over that half century from Chalmer's semi–literalist Gap Theory to Miller's poetic vision. However, in a recent lecture to the *Evangelical Theological Society* Mortenson (2001) stresses the theological compromises of Chalmers, Pye Smith, and Miller, who had "succumbed" to the Enlightenment.

There is less correlation of evangelical fervor and opposition to geology from 1817 to 1857 than today. I am aware that most historians, whether Millhauser or Mortenson refer to flood geologists as *Scriptural Geologists* but I prefer the term *anti-geologist* used by Miller in *The Testimony of the Rocks* in his chapter *The Geology of the Anti-Geologists* (Miller, 1857). Miller as an evangelical was not going to let others claim the term scriptural.

Anti-geologist is theologically neutral and focuses on attitudes to geology, not scripture.

The flowering of "anti-geologists" came as a deluge in the mid-20s and annoyed Uniformitarian and Catastrophist alike. Their cry was that geologists were mistaken and ungodly. Some had good scientific credentials outside geology, like Brande of the Royal Institution, John Murray and Andrew Ure (1778–1857) of Glasgow, others were evangelicals for example Bugg, Nolan, Cole, Best, Mellor Brown, and Young and some were traditionalist clergy for example Vernon Harcourt (brother of the cofounder of the British Association), Dean Cockburn of York, and Edward Nares. Despite their variety the anti-geologists had a common theme; the earth was a few thousand years old being created in six 24-hour days and the strata were laid down in the Noachian Deluge. Many emphasized that there was no death or suffering before the Fall (Genesis 3) and thus no animals had lived for more than a few hours before Adam. This was to retain the centrality of the Atonement, as death is the curse of sin. (Most orthodox Christians, e.g., Sumner, Chalmers, and Bishop Samuel Wilberforce did not reckon that animal death before the Fall affected the Atonement.)

The importance of the "anti-geologists" can be overstated as they attracted much attention, particularly in retrospect. The "anti-geologists" were attacked most vigorously by other Christians, as was Ure's *A New System of Geology* (1829), which was scathingly reviewed anonymously in the *British Critic* of 1828. Lyell identified the reviewer, "A bishop, Buckland ascertained (we suppose Sumner), gave Ure a dressing in the British Critic and Theological Review! They see at last the mischief and scandal brought on them by Mosaic systems." Sumner was an evangelical.

There appear to have been no *anti-geologists* in America until the Lord brothers writing in the 1850s in reaction to Hitchcock's *The Religion of Geology*. Eleazer Lord responded with *The Epoch of Creation: The Scripture Doctrine Contrasted with the Geology Theory* (1851) and his brother David *Geognosy, or the Facts and Principles of Geology against Theories* (1855). Both criticized geologists and their Christian apologists like Pye Smith, Hitchcock, and Miller and argued for a belief in a six-day creation. By the time of the Civil War American Scriptural Geology had almost gone, soon to be resurrected by Ellen White (Stilling, 1999).

THE EVANGELICAL ANTI-GEOLOGISTS, 1817–1845

Many anti-geologists were evangelical clergy and laity. The first work, which challenged geology was Thomas Gisbourne's *The Testimony of Natural Theology to Christianity* in 1818. Gisbourne was a friend of William Wilberforce and the last patient to be treated by Erasmus Darwin in 1802. The book was eirenic, but objected to geology, because the existence of death in the animal world implicit in the existence of prehistoric life before Adam contradicts the view in the opening lines of *Paradise Lost*

> Of man's first disobedience, and the fruit
> Of that forbidden tree, whose mortal taste
> Brought death into the world, and all our woe,

The storm broke in the 1820s in the *Christian Observer*, creating internecine warfare among evangelicals, and was paralleled by the publication of aggressive critiques of geology. It began with reviews of G. S. Faber's *A Treatise of the Three Dispensations* in 1823, which was classic theology on the "dispensations" of Abraham, Moses, and Jesus Christ, but the third chapter *Respecting the length of the Six Demi-Urgic Days* caused the problem. Here Faber summarized geological findings under the guidance of Buckland. Bugg took great objection. Several years later Bugg wrote to the *Christian Observer* criticizing the editor S. C. Wilks for taking the "side of modern geologists" and listed the five difficulties of the bible versus geology, which were

(1) Geology claims that death was there before Adam sinned,
(2) Geology denies the Six Days of Creation
(3) "Scriptural Creation" is handed over to Geology.
(4) Prevents missionary work among the Hindoos.
(5) Removes the basis of the Sabbath. (Bugg, 1828, p. 329)

A few years after Faber's work, Bugg published his magnum opus *Scriptural Geology* in two volumes, which was an answer to Buckland. Bugg claimed that "whatever is contrary to that Bible must be false." He started from the premise that the Mosaic narrative gives the general order of the strata with *one* physical revolution on the third day and that "Christian Geologists are bound in honor and conscience to agree." What follows is a variety of theological argument, a rejection of contemporary geology, and a reinstatement of the Deluge as the source of all strata. Bugg's motivation was theological as he was unable to accept animal death before the Fall.

Frederick Nolan (1784–1864) was a notable Oxford divine of his day, whose career parallels that of Faber. Both were leading evangelical theologians publishing prodigiously on the similar subjects of evangelical beliefs, polemics against the Oxford Movement and millenarianism. The pair made forays into geological science, Nolan rejecting it and Faber welcoming geological findings. In 1832 Nolan was elected to the Royal Society and in 1833 he gave the Bampton Lectures entitled *The Analogy of Revelation and Science established* (Nolan, 1834). Nolan argued that the findings of geologists were mistaken and the earth really was a few thousand years old. Buckland's anger was undisguised as his wife Mary wrote to William Whewell on May 12, 1833,

> we have had the Bampton Lecturer holding forth in St Mary's against all modern science,... Denouncing all who assert that the world was not made in 6 days as obstinate unbelievers, etc." (Morrell and Thackray, 1984)

Though Nolan's *Bamptons* were soon eclipsed by Keble's Assize Sermon on July 11, 1833, which marked the start of the Oxford Movement, Nolan's lectures highlighted a rumbling problem within the churches. At that time geology was the science of the day with its strange extinct beasts, its vast time scale, with the present day "towering o'er the wrecks of time." Geology had captured the imagination of the British public.

There were other evangelicals who took up cudgels against geology during those two decades. To obtain an exhaustive list would involve the detailed scanning of journals and libraries. In his research Mortenson identified about thirty and I have found another dozen or so. In mere numbers this is a fraction of Christians who wrote positively of geology. They had passed the peak of their activity in about 1840 and thereafter dwindled, though still making some impact in the 1850s. There are a variety of reasons for their decline. A major factor was simply increasing age; younger evangelicals were more open to geology, following on first from Chalmers and Faber, then Pye Smith and Miller, and then those like Birks and Pratt. Faber and Birks were strong millenarians, which indicate the lack of correlation of *anti-geology* with millenarian views, as is the case in the twenty-first century and the Seventh Day Adventists.

SCIENTIFIC ANTI-GEOLOGY

This may seem to be an oxymoron, but numbers of anti-geologists argued that their geology was more scientific than conventional geology.

A frequent contributor to the *Christian Observer* during the 1820s and 1830s was George Fairholme (1789–1846), who signed himself as "A Layman on Scriptural Geology." Fairholme was Scottish born and had no university education. According to Mortenson his denomination is unknown, nor are his evangelical convictions. As well as contributing to the *Christian Observer* and the *Philosophical Magazine*, Fairholme wrote on the *General View of the Geology of Scripture* (1833) and the *Mosaic Deluge* (1837). The preface of the latter discusses the theological results and skepticism caused by geology and especially the rejection of a universal deluge, "there cannot be conceived a principle more pregnant with mischief to the simple reception of scripture." All emphasis is put on the universality of the Deluge—"if false.... then has our Blessed Saviour himself aided in promoting the belief of that falsehood, by.... alluding both to the fact and the universality of its destructive consequences to mankind" (p. 61). Fairholme made much of stems of tall plants, which intersect several strata (*Polystrate fossils*).

In the *General View of the Geology of Scripture* (1833) Fairholme gave the air of geological competence, enhanced by citing geological works. His geology does not bear comparison with geological writers of his day, whether Buckland, Sedgwick, or amateurs like Pye Smith. Though he claimed to carry out geological fieldwork, there is no evidence that he did more than ramble though the countryside. His lack of geological competence is best seen in his discussion of the relationship of coal to chalk. (In the Geologic Column coal is found in the Upper Carboniferous or Pennsylvanian strata and chalk in the Upper Cretaceous.) Fairholme wrote

> the chalk formation is placed far above that of coal, apparently from no better reason, than that chalk usually presents an elevation on the upper surface, while coal must be looked for at various depths below the level of the ground. (Fairholme, 1833, p. 243).

He had previously discussed this (op cit pp. 207–210) and concluded, having misunderstood an article in the *Edinburgh Encyclopaedia*, that

> Nothing can be clearer than this account; and it appears certain, that, as in the case of the Paris Basin, this lime-stone formed the bed of the antediluvian sea, on which the diluvial deposits of coal, clay, ironstone, and free-stone, were alternately laid at the same period. (Fairholme, 1833, p. 209)

It is clear that Fairholme regards Carboniferous Limestone and the Cretaceous chalk as the same formation, and wrote on coal fields that

> they lie among sandstones, . . . , but have, in no instance, been found below chalk, which is one of the best defined secondary formations immediately preceding the Deluge, . . .

Thus the Cretaceous strata are pre-Flood and the Coal Measures were deposited during the Flood!

To a geologist today and in 1833 that is risible! When Fairholme penned this, it had been known for decades that Chalk *always, always* overlie the Coal Measures with a vast thickness of strata in between. In 1799, William Smith drew up a list of strata from the coal measures to the chalk, which was extended in the table accompanying his geological map of 1815 (Phillips, 1844/2003). Cuvier and Brogniart, who had worked extensively in the Paris Basin, gave the same succession. Thus by the standards of his day, Fairholme was talking nonsense as he was when he wrote

> But during the awful event [the Deluge] we are now considering, all animated nature ceased to exist, and consequently, the floating bodies of the dead bodies must have been bouyed up until the *bladders burst*, by the force of the increasing air contained within them. (Fairholme, 1833, p. 257)

It is impossible to agree with Mortenson's assessment, "By early nineteenth century standards, George Fairholme was quite competent to critically analyze old-earth geological theories"(Mortenson, 2004, p. 130). Though Fairholme took it upon himself to criticize geology, he did so from sheer ignorance, as is evidenced by his claim that Chalk always underlies Coal. Fairholme, like all anti-geologists, attempted from his armchair to find fault with geology, which he regarded as infidel, but his "scientific" objections were a total misunderstanding of geology. No wonder they were rounded on by Sedgwick, who in *A Discourse on the Studies of the University* (1834/1969) wrote that the anti-geologists

> have committed the folly and the sin of dogmatizing on matters they have not personally examined. (p. 106)

and regarded some as "beyond all hope of rational argument." Then, as now, the advantage of writing ridiculous works is that the refuting of them is beyond the wit of rational people.

PHILIP GOSSE

The last of the scientific literalists was Philip Gosse (1804–1881), an unusual Victorian naturalist, whose life is frequently seen through the eyes of his son Edmund in the book *Father and Son* (Gosse, 1907, 1949). Gosse was originally a member of the Brethren. His most (in)famous book was *Omphalos* (1857), which was an attempt to get out of the logical and scientific impasse of the 1850s, when Progressive Creationism reigned supreme. There was a twofold theme to *Omphalos*. First he holds that Genesis One has to be taken to mean 144 hours of creation. Secondly this he reconciles with "science" by his principle of "Prochronism," and criticizes geologists because "they have not allowed for the Law of Prochronism in Creation" (p. vi). Expounding that Law he stresses that the course of nature is a circle; for example the life cycle of a moth.

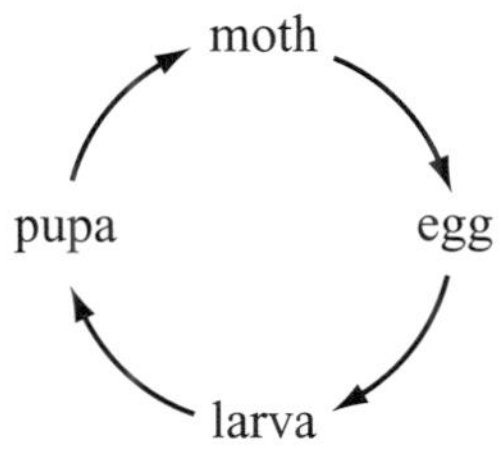

Creation is God suddenly breaking into the circle, thus Adam had a navel (omphalos) but had no history though his navel gave the "appearance" of history (pp. 123–124). Thus "The strata of the surface of the earth (with fossils) may possibly belong to the prochronic development of the mighty plan of the life history of this world." He asks "who will dare say that such a suggestion is a self-evident absurdity?" The reaction to Gosse was predictable and severe. To Charles Kingsley, Gosse made God a liar; as to create a world with such an appearance of antiquity a mere six millennia ago would be dishonest. However, there is an impeccable logic to Gosse, and *Omphalos* is the only consistent "anti-geology." His picture of breaking into the circle of life is logically no worse than Progressive Creationists, who held that at regular intervals in the words of Micron (Paradise Lost, vii line, p. 463) "the grassy clods now calfed, now half appeared" not only "the tawny lion" but also all mammals, dinosaurs, reptiles, and amphibians at regular intervals throughout the history of the world. In a sense Progressive Creationists were having their cake and eating it, by adopting a naturalistic geology, yet on life forms were as supernaturally interventionist, that is, creationist in the strict sense as any anti-geologist. Seen in this light, the paleontology in the 1850s was crying out for Darwin to remove the implicit absurdities.

Gosse's thesis is logically irrefutable, and unintentionally exposes the fatal flaw in pre-Darwinian Progressive Creationism. That flaw is to accept "Natural Law" for the astronomical and geological development, but to insist on intervention or miracle for the cause of life. Darwin realized this inconsistency both in early notebooks and in later work. Perhaps the only consistent alternative to a thoroughgoing evolution is Gosse's Prochronism, which raises even more theological problems than does evolution.

ASSESSMENT

The anti-geologists are often given an importance they never had in their day. Unlike today's YECs they made little impact, and Victorian evangelicals were far more successful than today's in persuading evangelicals that geology did not nullify an evangelical faith. By 1855 few evangelicals either in Britain or America held such views. George Bugg stated in his "*Scriptural Geology*" of 1826 "Whatever is contrary to that Bible must be false"(Bugg, 1826, vol. I p. 16), meaning "Whatever is contrary to a literal interpretation of the Bible." Most anti-geologists believed the literal view of the Bible was the correct interpretation, and Bugg condemned any non-literalists, however evangelical they were. The anti-geologists were (and are) not alone in that. To many people, whether today or last century, Christian Orthodoxy means Literalism, and thus opposition to Geology and Evolution, an outlook YECs have revived. However though Literalism is a recurrent phenomenon within the churches, it is questionable whether it ever was the traditional and orthodox view.

As was presented above their scientific arguments are worthless by the standards of the 1830s, as they were no more than armchair geologists who familiarized themselves with geological literature or those competent in other fields who think that they can pontificate on a different subject.

It is difficult to ascertain how these controversies affected the average layperson. Some insight can be gained from the diaries of a Shropshire lady, Louise Charlotte Kenyon (d. 1860s) of Pradoe, near Oswestry. Her diaries from 1822 to 1836 are fascinating to read alongside Darwin's *Correspondence* for the same period. Her physician was Dr. Robert Darwin and in 1833 her daughter Charlotte married the Rev. John Hill, who had been previously engaged to Fanny Mostyn Owen, who had sent a series of love letters to Charles Darwin in 1828. I must not digress on love stories of that era! In the 20s Louise attended chemistry lectures in Oswestry and after her daughter's wedding made a close study of geology, quoting the *Christian Observer* "Geology is one of the most interesting subjects that can occupy the mind of man." She read first Penn and then Ure, followed by "Mr Murray's *Truths of revealed religion* in which he proves by geology the truth of the Mosaic account of creation." However over the next decades she either organized or gave scientific talks at Pradoe Church. However geology worried her, and she asked her daughter-in-law to help by writing to the Rev. J. Cornish who replied on November 6, 1856, seeking to wean her away from "anti-geology." The letter indicates just how real the problem of geology was for some Christians, and continues to be today.

OTHER SCIENCES

There was a tremendous amount of scientific activity in the early nineteenth century, but much was uncontroversial. There were no theological challenges from atomic theory or electricity even for the most evangelical. Beyond detailed studies of Brewster and Faraday, there has been little research on evangelicals and science, though there is greater interest today. As a result few evangelical physicists and chemists have been identified, apart from Wollaston and Farish mentioned earlier. It is probable that a good proportion of university science teachers were evangelical whether in English, Scottish, or American universities, paralleling those active in geology.

Sir David Brewster led a kaleidoscopic life at the centre of science and religion in Scotland. He went to Edinburgh University at the age of twelve and never graduated but received many honors for his science, F. R. S. in 1818 and a knighthood in 1832. From 1811 to 1818 he researched polarized light and invented the kaleidoscope. Brewster never obtained a University chair but that may be due to his support of the evangelical John Leslie (1766–1832) for the Chair of Mathematics at Edinburgh in 1805, which reflected the rift between Moderate and evangelical Presbyterians. In 1838 he was made principal of the United Colleges of St Leonard and St. Salvator at St Andrews, a position he nearly lost due to his support of the Great Disruption of 1843. He became Principal of the University of Edinburgh in 1859 and the following year his young second wife fainted during the Huxley-Wilberforce "debate." Brewster's interests were very wide-ranging through science to natural theology to Evangelicalism. As a Scot he preferred the Natural Theology of Chalmers to English schemes of biblical reconciliation. This along with understandings of Laplace's Nebular Hypothesis and the plurality of worlds drew him into controversy with William Whewell in the 1830s. However after the publication of *Vestiges* in 1844, he rejected the Nebular Hypothesis as he saw that leading to evolution. From then on he was unwilling to substitute natural law for individual acts of the Creator. As events turned out Brewster was correct to see that Creation by natural law led inexorably to evolution, which he opposed again in 1860 as Darwin expounded it.

One of the greatest scientists of this period was Michael Faraday (1791–1867), who during his time at the Royal Institution from 1813 until his death made tremendous advances in electricity. Much of his early research was on chemistry. His discovery of electromagnetic rotations in 1821 proved to be the principle of electric motors and dynamos and in 1831 he detected electromagnetic induction. He continued his electrical research into the 1850s and became one of the most highly regarded scientists of his day. Religiously he belonged to a tiny Christian group, the Sandemanians. His father joined that church in 1791 and Michael in 1821, becoming a deacon in 1832 and an elder in 1840. The Sandemanians were founded in the early eighteenth century by breaking away from the Presbyterians. They were biblically based but eschewed evangelism and may not be strictly evangelical.

Faraday kept his religious beliefs to himself and it is difficult to say much beyond that he saw his science as an expression of his faith in the Creator God, who had made an ordered universe. He saw no conflict between his physical science and his faith, but there has long been a question whether his almost Biblicist faith clashed with geology. One of his colleagues, William Brande, rejected long geological ages in the 1820s but later considered himself a Lyellian. During the 20s Faraday studied the geology of the Isle of Wight while on holiday, and that, and his friendship with many geologists, indicate that he was no anti-geologist. Faraday's religious views and their relationship to his science remain elusive, despite the amount of research on every facet of his life.

EVANGELICALS AND POPULAR SCIENCE

During this half-century, one of the popular entertainments for the aspiring classes were scientific lectures, whether Humphry Davy in the opening decade or Faraday in mid-century at the Royal Institution. As well as these star performers in the metropolis, science lecturers were popular in the provinces as well and ranged in competence from the skilled to the mediocre.

One skilled amateur lecturer was Thomas Dick (1774–1857), a Scot who trained for the ministry but due to an indiscretion was not ordained. Though scarcely a scientist, he was competent in astronomy and other sciences and influenced the young David Livingstone. Dick blended his science and Evangelicalism together to make a plausible and saleable mix for both lecture hall and the market for popular tracts and books. He wrote nine books and many shorter pieces, and saw no conflict between his faith and science whether astronomy or geology. He is a good example of how a popular evangelical reconciled his faith and science (Astore, 2001).

After the formation of the Young Men's Christian Association in 1844, part of that organization's evangelical ministry was to present edifying lectures at Exeter Hall in London. These lectures were published as volumes of twelve lectures apiece. The content of the lectures varied but were designed to build up the Christian character of the young men of London. As these were by leading evangelicals of the 1850s they give a window into how evangelicals considered science. In 1851, the year of the Great Exhibition in the Crystal Palace, the Rev. John Cumming, a Scottish minister discussed earlier gave a lecture entitled *God in Science.* In his lecture Cumming regaled his audience with anti-popery before telling them that God designed the natural world. He spoke favorably of geology and cited Hugh Miller and Hitchcock in support. Dean Francis Close of Carlisle Cathedral gave the other lecture in January 1858, the year of Darwin and Wallace's joint publication on Natural Selection. These two examples are significant as Close and Cumming were conservative rather than liberal evangelicals yet had no problem with geology. For the vast majority of evangelicals, or least those with some education, science, whether chemistry, astronomy or geology, held no terrors for them, though the specter of evolution loomed over them to demote the pinnacle of God's creation to an ape.

The use of science in evangelical apologetic was common and the chief publisher of tracts, the *Religious Tract Society* (founded in 1799), ensured that part of their voluminous list of works contained many on science. The name was almost a misnomer as many of the tracts were booklets or books rather than tracts. In *Science and Salvation* (Fyfe, 2004) Aileen Fyfe has studied the industry of evangelical works on science in depth. Throughout the book she emphasized that science was little problem to most evangelicals and wrote, "Evangelicals in the mid 19th century, however, did not consider the sciences, in themselves, to constitute a danger to the faith and salvation of the nation" (Fyfe, 2004, p. 3). Though, of course, the publication of *Vestiges* in 1843 caused problems.

One reason why the R. T. S. moved into publishing volumes with scientific content was that by 1840s there was a plethora of cheap literature available and much neither wholesome nor Christian. As a result the R. T. S. determined to provide "scientific literature" which had a "Christian tone" as Thomas Arnold would express it. It was not always easy to get the balance between Christian content and tone and retain potential readers. During the 1840s the R. T. S. had several competitors in providing cheap serious nonfiction as in 1846 Henry Bohn began to publish cheap editions of about 500 pages at 3/6 (60cents) in *Bohn's Standard Library*. They covered all subjects from classics to natural history and theology. The R. T. S. was aiming at a slightly lower market and published scientific books along with religious and theological works. Their choice of scientific authors varied from the competent to hack writers. Significantly they did not employ any of the anti-geologists.

As part of their monthly volumes they reprinted some of Scoresby's dated works on the Arctic and whales. Thomas Dick was a regular writer as were Thomas Milner and William Martin, who had worked for seven years at the Zoological Museum in London. Milner had been a Congregationalist minister before turning full-time to writing. His most popular work was *The Gallery of Nature* published in 1846. As Fyfe pointed out, "By the time Milner was writing, an ancient and changing earth was hardly contentious among practising geologists, evangelicals or not" (Fyfe, 2004, pp. 239) Milner argued, almost apologetically that if the earth was more than 6,000 years old then there was no significant consequence for the Christian faith. This may seem like special pleading, but many of his readers would have never have questioned a simple reading of Genesis, which they would have learnt in Sunday school.

JUST BEFORE "THE ORIGIN"

In 1859 evangelicals had every reason to be self-confident; in England there was a scientifically competent evangelical Archbishop, John Sumner, and thriving evangelical churches of all persuasions, in Scotland the evangelical Free Church was thriving, and in America evangelicals were dominant. Most educated evangelicals completely accepted the findings of all sciences and probably did not anticipate the storms ahead.

Chapter 5

Post-Darwinian Evangelicals

The second half of the nineteenth century saw the expansion of the European empires and the development of the American West. None were glorious. New technologies were used in the American Civil War, the telegraph, railways, and the first submarine. One of the technological marvels was the laying of the Atlantic telegraph, which gave fame to the physicist William Thomson, who was knighted for his endeavors and later raised to the peerage as Lord Kelvin. The first trans-Atlantic telegraph messages were sent in 1858, but the telegraph was not fully operational until 1865. The advances in science were immense from Clerk-Maxwell's electromagnetic wave theory to Mendeleef's Periodic Table, but it is difficult to escape the shadow of Darwin, which can obscure the wider picture.

During the late nineteenth century, Christians, including evangelicals, universally accepted technology and the physical sciences. Some scientists were evangelical, but few were of the first rank. With few exceptions, evangelicals also accepted the whole of geological and astronomical science, and young earth ideas were virtually extinct by 1860—at least among the educated. The scientific issues which created a problem for the churches were evolution and the antiquity of man. In the 1860s most evangelicals opposed evolution, but many gradually accepted them, only to be rejected by most American evangelicals at the end of the First World War.

THEOLOGICAL ISSUES

In 1860, evangelicals in both Britain and America were at their zenith and then gradually declined until the 1920s for many reasons. First, the religious zeitgeist moved away from Evangelicalism, partly due to the rise of Anglo-Catholicism in the Church of England, and on both sides of the Atlantic Roman Catholicism was growing in numbers and respectability. Secondly, the rise of biblical Criticism undermined traditional Protestant belief in the Bible (Rogerson, 1984; Neill, 1964; Noll, 1986). The roots of biblical Criticism are usually traced back to F. C. Baur and associates in Germany in the 1830s. From their particular philosophical perspectives, they approached the Bible with historical skepticism and a rejection of the supernatural, which cut to the root of the traditional understanding of the Bible. An evangelical would have to give up his Evangelicalism if he adopted such an approach. The novelist George Eliot is one example, but others simply moved to a more liberal faith. Some biblical critics like Westcott, Hort and Lightfoot on the New Testament and Delitzsch on the Old took more moderate positions, which were more acceptable to evangelicals. However the most conservative evangelicals remained with the near-literalist expositions of the early nineteenth century of Thomas Scott and Adam Clarke. In one sense Biblical Criticism was not new, as there was a long tradition of conservative criticism going back to Erasmus and Calvin, as well as more skeptical criticism in the eighteenth century.

Biblical critics raised questions on the authorship and reliability of the Bible. Moses was no longer considered to be the author of the Pentateuch (Genesis, Exodus, Numbers, Leviticus, and Deuteronomy) and early Genesis was seen as myth or legend. The discovery of the Epic of Gilgamesh in 1872 by George Smith and other ancient near eastern epics on the creation and flood shed new light on the origins of the biblical accounts of creation and flood (Young, 1995). Most critics thought that the bible was derivative and that Patriarchal longevity was mythical. By the early 1860s radical biblical criticism had made few inroads into the British and American churches but the publication of *Essays and Reviews* (Jowett, 1860) and Colenso's volumes caused a conservative backlash in Britain, which reverberated across the Atlantic. The result was polarization between liberal and conservative and paved the way for the modernist-fundamentalist split. It was furthered by the increasing emphasis on Inerrancy and Dispensationalism, which encouraged a default literal hermeneutic and encouraged the hemorrhage of intellectual evangelicals. As a result the theological questions faced by evangelicals in the late nineteenth century were more about biblical criticism than science.

CULTURAL CHANGES

Many scientists in the early nineteenth century were clergymen, but they declined in number after mid-century and those like Hitchcock and

Sedgwick were not replaced by the rising generation. Thus by 1900 there were few clergy scientists like G. F. Wright and Thomas Bonney (1833–1923). There were many reasons for this decline:

1. The more demanding calls on clergy gave less time to pursue science as a sideline, resulting in clergy knowing far less about science.
2. It was difficult for amateurs to contribute to science, due to its increasing complexity. As universities offered sciences as degree subjects, those studying arts or theology would know less science than their forbears half a century earlier.
3. Many involved in science aligned themselves with the rising agnosticism and later scientific humanism, distancing churches from science.
4. There was a rising awareness of science as a profession and of ministry as a vocation in a narrower sense.
5. There was an increasing *perception* that science and religion were incompatible, culminating in the Draper–White conflict understanding of science and religion.

SCIENTISTS DECLARE ON *ESSAYS AND REVIEWS*

The biggest theological controversy to strike the English church in the 1860s was not *The Origin of Species*, but the publication of *Essays and Reviews* in 1860 (Jowett, 1860) and Colenso's work on the Pentateuch. Though these were about biblical criticism and the restatement of theology, questions raised by science were never far away. In *Essays and Reviews*, Goodwin criticized the harmonizing of geology and Genesis by Buckland and Miller; Baden Powell denied miracles; Benjamin Jowett (1817–1893) wrote on biblical interpretation and used the famous, but unoriginal, phrase "that the bible should be interpreted like any other book" (Jowett, 1860). Two clergy, Rowland Williams (1817–1870) and H. B. Wilson (1803–1888) were charged with heresy. Before that, in February 1861 Sumner, the evangelical Archbishop of Canterbury, and twenty-five other bishops wrote to the authors expressing concern as the could not "understand how their opinions can be held consistently with an honest subscription to the formularies of our Church [The 39 Articles of 1559]." Of these bishops at least six were evangelical and none opposed geological science (see Chadwick 1970, vol. II, pp. 75–96 for an overview). During the next month, leading scientists intended to send a memorial to Frederick Temple (1821–1902), one of the essayists and later Archbishop of Canterbury (1899–1902), to express their support of the essayists for "their enquiries conducted in a spirit so earnest and reverential." It is not clear whether the letter was ever sent but it was published without the signatories (Burkhardt, 1994, pp. 416–419). These seem to include Darwin and Lyell but not Huxley. Though it was never sent, it does demonstrate the unease afforded to conservative theology by more liberally minded scientists and thinkers.

Several years later in 1865 a manifesto (*Declaration of Students of the Natural and Physical Scientists*) was drawn up for scientists to declare their belief in the truth of the Bible and that some "researches into scientific truth are perverted by some . . . for casting doubt upon the truth and authenticity of the Holy Scriptures." Some 600 scientists and medical men signed the statement. Many of these are unknown medics and scientists, others are clergy or missionaries, but several were prominent scientists: J. H. Balfour, Sir David Brewster, James Joule, Sir Henry Rawlinson, Adam Sedgwick, Phillip Gosse, and fifty-seven Fellows of the Royal Society. Some evangelical scientists did not sign: H. B. Tristram, Charles Babbington, and John Salter. The proportion of evangelicals is unknown. This manifesto was aimed at the rising tide of agnostic scientists like Darwin and Huxley (Moore, 1979, pp. 83, 363). However, it is significant that many scientists, whether Christian or not, signed neither manifesto.

EARLY RESPONSES TO DARWIN

The Origin of Species was not the seismic shock to Christians as often claimed. It cannot be emphasized enough that very few Christians *who expressed themselves in print* believed in a six-day creation after the mid-1850s Thus, the theological questions were not to do with the interpretation of the Bible, but whether the chanciness of evolution destroyed Design and whether "animal ancestry" reduced humans to mere beasts without moral sense. That was the essence of the response to Darwin in 1860.

Bishop Wilberforce has come to be considered the most notorious opponent of Darwin, and despite the careful scholarship of Brooke (Brooke, 2001), James (2005), and others, Huxley's fanciful account is still repeated as gospel. Though by 1860 Wilberforce had moved from evangelicalism to highchurchmanship, his theology was as conservative. Shortly before the famous meeting with Huxley in Oxford in June 1860, Wilberforce published a long review of *The Origin* in the *Quarterly Review* ([Wilberforce], 1860). It is a careful review, expressing the concern of most educated Christians, including scientists. His review, though anti-evolutionary, deals mostly with scientific objections and is competent, well-written and amusing, referring to "our unsuspected cousinship with mushrooms." Darwin told Hooker it was "uncommonly clever" and advised his son William to read it (Burkhardt, 1993, pp. 293, 305). Only in the last few pages of a forty-page review does Wilberforce give theological objections to Darwin. These centered on design and the moral status of humanity.

Some of the main opponents of Darwin were geologists and physicists, whereas Christians were initially divided. Despite popular opinion,

Christians varied in their response from total acceptance, (Baden Powell, Frederick Temple), qualified acceptance (Asa Gray, Charles Kingsley), through ambivalence (John Henslow) to rejection (Sedgwick), *all within an Old Earth framework*, which is far removed from "the furore caused by Charles Darwin's theory of evolution which so traumatically disturbed those who up to that time had read the story of the seven-day creation as literally and historically factual, . . ." as Peter Carnley, former Archbishop of Australia, described the situation. Since Jim Moore's ground-breaking book *The Post-Darwinian Controversies* (Moore, 1979) there have been many good studies, both detailed and wide-ranging so that this statement in inexcusable. To repeat old ideas that the church reacted vigorously against Darwin because of concerns over a six-day creation is to fly in the face of the best scholarship. Despite all expectations evangelicals responded similarly to other Christians, except more opposed evolution.

The first evangelical to accept evolution, or rather natural selection, was a Durham vicar, H. B. Tristram (1822–1906). He was a competent ornithologist, who made a detailed study of the larks in the Holy Land. After reading Darwin's 1858 paper while preparing his own article, he incorporated some of Darwin's ideas, hence was the first to use Darwin's theory. He was present at the Oxford Museum and was convinced by Wilberforce's *scientific* arguments against Darwin! A few years later, he became re-convinced that Darwin was right and remained an "evolutionist" for the rest of his life. His family reported him as saying, "When life evolved, or rather God created . . ." Tristram is a good example to consider both from the scientific and religious angle as he shows that acceptance of evolution in 1860 was not straightforward.

Another evangelical scientist present at the Oxford debate was Charles Babbington (1806–1895), a botanist who collected beetles with Darwin at Cambridge in 1828 and described insects collected on the Beagle voyage. He followed Henslow as Professor of Botany at Cambridge in 1861. In 1884 Babbington was the opening chairman at the commissioning of the "Cambridge Seven," as they were "sent" to China as missionaries. One could not be more evangelical than that! Yet in 1860, as he listened to the discussions on natural selection in Oxford, Babbington was convinced that "descent with modification" had occurred. This is partly because it is often difficult to classify plants into distinct species, as with the British blackberry. In his memoirs of Babbington, the New Testament scholar J. B. Mayor, discussed the incident and commented that Babbington would have wished to side with Wilberforce *theologically* but with Hooker and Huxley *scientifically*. However, Babbington does not appear to have written on the subject.

The most illuminating response came from the paleontologist John Salter (1820–1869). In the 1840s he worked with Adam Sedgwick on the geology of North Wales. As evangelicals they read their Polyglot Bibles together on Sundays. In 1854 he became paleontologist to the Geological Survey on a salary of £250. While with the Survey, Salter carried out his most important work, including superb paleontology for William Ramsey's geological memoir of North Wales. The evidence on whether he accepted evolution is contradictory and needs further study. In 1860 to the fascination of Darwin he presented Devonian and Carboniferous spirifers [brachiopods] "arranged (not at my request, for I thought he was dead against me) after my [Darwin's] diagram in the 'Origin'" (Darwin to Murray, 28 April 1860, Burkhardt, 1993, 176), showing how a group of fossils fitted in with Darwin's branching tree picture of evolution. Unfortunately there is no other documentation. Salter's later life was marred by mental illness, and after differences with Huxley he resigned from the Survey in 1863. In 1869 he jumped off the Margate–London steamer, after giving his son the gold watch Sedgwick presented to him.

The most well-known evangelical response to Darwin was by the American botanist Asa Gray (1810–1888) who was Professor of Botany at Harvard from 1842. Gray was the foremost American botanist of his day who had met Darwin in England in 1851 and corresponded with him (Dupree, 1988) In 1857 Darwin sent him a précis of his theory and a copy of *The Origin of Species* in 1859. Gray facilitated the publication of an American edition and reviewed Darwin favorably in the *American Journal of Science* and *Atlantic Monthly* interpreting Darwin theistically. Darwin and Gray discussed the "theology" of *The Origin* in letters (Roberts, 1997). At the end of *Variation under domestication* Darwin made it clear he could not accept Gray's theology that God had directed evolution and wrote, "However much we may wish it, we can hardly follow Asa Gray in his belief, 'that variation has been led along certain beneficial lines' . . ." (Darwin, 1868, vol. II, p. 428) Gray described himself as "one who is scientifically, . . . , a Darwinian, philosophically a convinced theist, and religiously an acceptor of the 'creed commonly called the Nicene' as the exponent of the Christian faith." He retained his Nicene orthodoxy despite being at Unitarian Harvard. Gray's faith was eirenic and reckoned Darwin "is about as far from being an atheist as I am." He wrote, "I don't believe the Bible teaches science" and that evolution "can be held theistically or atheistically." He was equally eirenic when he reviewed Hodge's *What Is Darwinism?* in *The Nation* in 1874 as he challenged Hodge's answer, "It is atheism." Gray became acquainted with G. F. Wright in 1874, who persuaded him to publish *Darwiniana,* which included both his 1860 articles and important later writings. In 1881 Gray gave lectures to Yale Divinity School published as *Natural Science and Religion,* which caused no stir because of the accommodation between evolution and Protestantism. That accommodation survived largely intact until the Scopes decade, by when many evangelicals had come to reject evolution. Though Gray is usually considered an evangelical, Noll and Livingstone aptly describe him as "a fairly conservative, relatively traditional Christian."

The geologist James Dwight Dana (1813–1895) went on the Wilkes' expedition of 1838–1842, when he made important observations on coral atolls and volcanic islands, complementing those of Charles Darwin. He was appointed as Silliman Professor of Natural History at Yale in 1856. He wrote two standard works on mineralogy, *Manual of Mineralogy* and *The System of Mineralogy*, and, finally, *The Manual of Geology* in 1863. Dana was a devout Congregationalist with a high view of the authority of scripture and design in the natural world. In 1856, the year he was appointed to Yale, he was involved in a controversy with Tayler Lewis, who had published *The Six Days of Creation* in 1855. Like many theologians before and since, Lewis argued that scientific theories were tentative, so thus science should have no bearing on biblical interpretation. He wrote that, "We can get on very well without geology." Lewis here was arguing for a total separation of theology and science, almost anticipating Karl Barth a century later, rather than repudiating science, in contrast to Moses Stuart. At a stroke, Lewis had dismissed all schemes of "reconciling" geology and Genesis. During 1856 and 1857 Dana published a four-part response to Lewis in *Bibliotheca Sacra,* then a congregational journal. Dana had taken his scheme of "reconciliation" from Guyot, which was a form of Day-Age harmonizing.

It is easy to see the dispute as between a scientist and a biblicist, but it was deeper than that. Dana, following a tradition which went back two centuries, accepted the authority of the Bible, but wished to interpret the Bible according to the light of knowledge drawn from extra-biblical sources, in this case geology. He maintained the geology had shed new light on Genesis. In contrast, Lewis argued that biblical interpretation should be independent of the passing whims of science. However Lewis cannot be considered a "biblical creationist" as he regarded science as dealing with secondary causes and not subject to a construct based on Genesis. Dana, too, followed his geology independently of Genesis and his concordism was more impressionistic than pre-Raphaelite. Both had effectively severed the link of geology and Genesis, which had informed the work of harmonizers like William Buckland.

A counter-factual history would be Dana's response to his complimentary copy of *The Origin of Species* in 1859, if he had not been struck down by mental illness, which incapacitated him until 1863. Dana was a catastrophist whose progression of life consisted of replacements of life forms rather than transformation. Dana did not reply to Darwin until December

4, 1862, when he had still not read *The Origin*. Two months later, after sending Darwin a copy of his *Manual of Geology*, he responded with a measured rejection of Darwin on geological grounds. These were the absence of transitional forms and that the fossil record does not show the smooth transition needed for Darwin's theory. Over the next ten years Dana changed his mind. In the second edition of the *Manual of Geology* in 1874, he said that the evolution of life had proceeded by natural means with only a few supernatural interventions. He did not go "the whole orang," but had shifted from his previous catastrophist progressionism. There is no evidence that he had a crisis of faith.

Along with other physicists like Tait and Kelvin, Sir David Brewster was also unhappy with Darwin. His wife was present at the Oxford debate of 1860, when she notoriously fainted. Brewster responded in an article entitled *The facts and fancies of Mr Darwin* in *Home Words* (Brewster, 1862). Neither Darwin nor Huxley thought that Brewster had given any sound objections. Two years previously in 1860, his fellow Free Churchman, James Duns (1820–1909) who succeeded Fleming as Professor of Natural Philosophy at New College, Edinburgh, wrote a critical review in the *North British Review*. In a letter to Asa Gray, Darwin unjustly dismissed Duns as "a dabbler in natural history." However within a decade most Scottish evangelicals had accepted evolution.

These examples of competent evangelical scientists show that there was no fixed response to Darwin, except that they accepted the vast age of the earth. All gave considered responses, often moving to cautious acceptance. Though William Whewell was on the fringes of Evangelicalism, his cautious but slightly negative response reflects the stance of most religious scientists.

However most evangelical theologians were less liable to accept evolution. Of all the works of evangelicals from the 1860s I have read, almost all accept the basic correctness of geology. In Britain there were a handful of dissenters. After Phillip Gosse wrote *Omphalos* in 1857(Gosse, 1857), he did not venture again into print. B. W. Newton (1807–1899), like Gosse a disenchanted former Brethren wrote a tract *Remarks on "Mosaic Cosmogony" and Genesis II.5* (Newton, 1882) in response to Goodwin's chapter in *Essays and Reviews*. As a former Oxford classics don, Newton was a competent linguist, but he would not allow the slightest deviation from a twenty-four-hour day. In America the Southern Presbyterian theologian R. L. Dabney (1820–1898), who served first as a chaplain to the Confederate army and then chief of staff to "Stonewall" Jackson, wrote on *Geology and the Bible* in the *Southern Presbyterian Review* for 1861, in which he criticized both the Gap Theory and Day-Age interpretation. He favored a literal view and thought geology theory was too provisional to take into account. Like Gosse he favored a "mature creation theory." Later in 1873 in the same journal he was embroiled in controversy with James Woodrow, uncle of Woodrow Wilson, who was dismissed from his professorship at Columbia Theological Seminary in 1886. Though these two may have had a considerable following from evangelicals, most educated evangelicals eschewed literalism.

In 1861 on the occasion of the British Association meeting at Manchester, the Rev. Hugh McNiele (1795–1879), an evangelical Vicar in Manchester, gave a sermon in which he upheld the findings of geologists, but not Darwin. That was the general opinion of most of his fellow clergy. In 1856, the Archdeacon of Calcutta, John Pratt, published the first edition of *Scripture and Science not at Variance* (Pratt, 1856, 1871) which underwent many revisions until 1871 to take into account Darwin and *Essays and Reviews*. As much as Pratt accepted geology, he could not accept Darwin and Natural Selection because that is based on chance and thus contrary to design. In the last edition, which Pratt wrote after reading *The Descent of Man*, published earlier that year, which he considered to be "overflowing with facts" and "the untempered mortar of free speculation." Briefly, he rejected an evolutionary origin of man for religious reasons.

Another Anglican opponent of Darwin was Thomas Birks (1810–1882), who followed F. D. Maurice (1805–1872) as Professor of Moral Theology at Cambridge in 1872. Birks was a prolific scholar in the evangelical tradition. Though no scientist, he was well-informed on geology, which held no threats to him. He was educated at Cambridge and served his curacy under Edward Bickersteth. That acquired him the vicar's daughter as a wife and a commitment to Premillenialism. He edited Paley's works and wrote widely on theological subjects, with *The Bible and Modern Thought* (Birks, 1862) as a response to *Essays and Reviews* in 1861. The work is a wide-ranging and learned response to the Essayists dealing with revelation, the historicity of the Bible, miracles, and the inspiration and interpretation of scripture, and adopts a mild Biblical Criticism. One chapter (XIV) is on *The Bible and Modern Science* and is an examination of Goodwin's *Essay on Mosaic Cosmogony*. Birks rejected Goodwin's mythological approach and had doubts about both Rorison's poetic interpretation and Miller's optical presentation, regarding the Gap Theory of Chalmers as the "true relation of Genesis and Geology." James Moore (1979, p. 201) describes Birks as "Britain's foremost anti-Darwinian" and rather unjustly refers to his "exceedingly illiberal" books on evolution; *The Scripture Doctrine of Creation* (1872) and *Modern Physical Fatalism and the Doctrine of Evolution* (1876). He argued that Special Creation is mandated by scripture and "a priori reasoning." He also echoed Sedgwick's challenge to Darwin for rejecting inductive and deductive inference in science, re-iterating the objections to Darwin in the 1860s. Birks was one of the last of the evangelical Anglicans to oppose Darwin as later evangelicals like Handley Moule (1841–1920) and Dean Wace (1836–1924) were far more accepting of evolution.

In America, one of the most significant conservative theologians was Charles Hodge (1797–1878) of Princeton Theological Seminary, who with Warfield was the architect of Biblical Inerrancy (Noll and Livingstone, 2003). Hodge stood in the tradition of the seventeenth-century Calvinist scholastics, and still used the seventeenth-century Francis Turretin's theological works as textbooks. In 1816 he went to Princeton Seminary, where he remained all his life except for two years studying in Europe from 1826. He maintained his Calvinism with an eirenic tenacity, evidenced by his opposition to Thornwell's proposal in 1845 to declare Roman Catholic baptism invalid. On slavery he was a moderate abolitionist. However, he is remembered his many theological writings. Hodge wrote little on science until his last decade. In 1862 he referred to Darwin's work as unconvincing, as he thought it absurd to think that all life could derive from a few life forms.

Hodge's main foray into Darwinism was in 1874, when many Christians had accepted evolution, in *What is Darwinism?* (Hodge, 1874), which attracts dismissive accounts presenting Hodge as a literalist. Hodge considered whether Darwin's view of natural selection by chance would allow any design or Teleology, and whether ultimately Darwin had any room for God. He focused on the question of Teleology and design, since Hodge considered Darwinism to be based on randomness and chance. To Hodge, Darwinism is not the same as Evolution, and though he does not accept Evolution, he is not fundamentally opposed; "If God made them, it makes no difference, how he made them: whether at once or by a process of evolution" (Hodge, 1874, p. 95). His book is a sustained attempt to show that "the answer to our question, What is Darwinism? It is atheism." as Hodge concludes his book. However, Hodge immediately added the qualification that; "It does not mean that Mr Darwin himself and all who adopt his views are atheists, but it means his theory is atheistic" (Hodge, 1874, p. 156).

The essence of Hodge's argument is Darwin's theory of natural selection operates by chance and thus has no place for God. Thus Darwinism that is, natural selection, is by definition atheistic. Hodge accepted a limited natural selection but also inconsistently insisted on the fixity of species (Hodge, 1874, pp. 143–149). Hodge defined Darwin's use of natural in natural selection more narrowly than did Darwin, writing that it "is a selection made by natural laws, working without intention or design ... In using the expression natural selection, Mr Darwin intends to exclude design or final causes" (Hodge, 1874, p. 85). Gray reviewed Hodge in *The*

Nation in May 1874, describing Hodge as ultra-orthodox. His conclusion is that "Excellent as the present volume is . . . and . . . it shows that Darwinism may bear an atheistic as well as a theistic interpretation, . . . it will not contribute much to the reconcilement of science and religion" (Gray, 1963, p. 230). Gray quoted Charles Kingsley "We know of old that God was so wise that he could make all things; but, behold, he is so much wiser than even that, that he can make all things make themselves" (Gray, 1963, p. 232).

Gray sent a copy of his review to Darwin on June 16, 1874, "You will see what uphill work I have in making a theist of you." Darwin replied, "The more I reflect on this subject, the more perplexed I grow" (Hodge, 1874, p. 33). Gray wrote *Evolutionary Teleology* as the last chapter of *Darwiniana* (Gray, 1963, p. xxi.) and argued that Darwin had reintroduced purpose and design into the organic world. He regarded Paley's idea of design to be "confusion of thought," as it was too mechanistic and "an idea which has been set up as the orthodox doctrine, but which to St Augustine and other learned fathers would have savoured of heterodoxy" (Gray, 1963, p. 294). However this would hardly have convinced either Darwin or Hodge.

The works of Pratt, Birks and Hodge, when considered against their social milieu, are fine conservative works, written by theologians with a good grasp of science. They are devoid of reactionary dogmatism. In fact, some YECs today would consider them "theological compromisers," and often seem to airbrush out Hodge and Warfield. Their presentation of Darwin's evolutionary writings is scrupulously fair. Their theological opposition to Darwin was based on several points; the balance of design and chance, the uniqueness of humanity (compromised by animal descent) and the moral and spiritual status of humanity. However as time went on, more evangelicals accepted evolution, but often insisted on creative acts for the first life, conscious beings and humanity.

In Britain the ambivalence of the Victoria Institute (VI) is illuminating. It was founded in 1865 to counteract the liberal and materialist tendencies of *The Origin of Species* and *Essays and Reviews*. Despite its evangelical principles, it was broader than some of their founders may have wished. For over a century the *Victoria Institute* was influential in British evangelical circles on science and religion, but has now been eclipsed by *Christians in Science*. The VI organized regular conferences and published *The Journal of the Victoria Institute* (now *Science and Christian Belief*). Despite Philip Gosse being a founder, the journal always had a broad perspective. The first volume in 1866 contained several articles supporting the common geology/Genesis consensus and in the volume for 1867 George Warington wrote an affirmative article *On the Credibility of Darwinism*. The following year he wrote on *The Biblical Cosmogony* and, followed Rorison, in that as Genesis did not deal with science and thus there was no need for harmony between geology and Genesis.

LATER RESPONSES TO DARWIN

By the mid-seventies, many within all Protestant churches had accepted evolution and the evangelicals were only slightly behind. When Frederick Temple gave his Bampton Lectures in 1884 (Temple, 1884), evolution had ceased to be controversial, though it is difficult to say how far it was accepted at a popular level. In the United States the mantle of Calvinist orthodoxy passed from Charles Hodge to his son Archibald and to B. B. Warfield. A. A. Hodge grudgingly accepted evolution and Warfield regarded himself as a "Darwinian of purest water" and even claimed that Calvin allowed for evolution.

David Livingstone in *Darwin's Forgotten Defenders* (Livingstone, 1987) and other studies has dealt with Evangelical attitudes to evolution and concludes that from the mid-1860s until the First World War most evangelicals were evolutionists and the rest were old earthers. Livingstone has also compared the attitudes to evolution of three groups of orthodox Calvinist Presbyterians in the 1880s (Livingstone, 1999). They had similar theologies and had close links. Those from the north of Ireland were antievolutionary, but old earthers; the second from Princeton, including Warfield and Hodge, were mostly evolutionists, and the Scots betwixt and between. The reasons are fairly simple; Asa Gray and James Dana advised the Americans, and the Ulstermen reacted against John Tyndall's British Association 1874 lecture in Belfast. Among Scottish Presbyterians, who formed the third group, only James Duns of the Free Church writing before 1882 rejected evolution but Principal Rainy of New College declared himself an evolutionist in his inaugural lecture of 1874. In 1894 the Free Church James Iverach wholeheartedly embraced evolution, but a little cautiously for man, as did James Orr in his classic work *The Christian View of God and the World* (Orr, 1897). Henry Drummond in *Natural Law in the Spiritual World* (Drummond, 1883) waxed lyrical about Christianity *and* Evolution. There was a fourth group of Presbyterians from the Confederate states. Their leading theologians, R. L. Dabney and J. H. Thornwell were both antievolutionary and anti-geology regarding geological methods as circular reasoning and thus fallacious. They also used their theology, and their understanding of Genesis, to support slavery, in marked contrast to abolitionists Hodge and Gray.

As the Victorian era continued more and more educated Christians came to terms with evolution. Among Anglican Evangelicals Bishop J. C. Ryle (1816–1900) of Liverpool directed his clergy to the Bucklands and

Sedgwicks for their grasp of science in the 1880s. Even in the 1890s his writings show no concession to evolution, but he seems to take geology for granted. E. A. Litton in *An Introduction to Dogmatic Theology* (1882) regarded the antiquity of man as of no consequence but rejected evolution, and Bishop Moule of Durham writing in 1889 accepted Genesis (along with Revelation) as poetic, and accepted evolution, but drew the line at the creation of man (Moule, 1889). In the 1860s his father the Rev. Henry Moule, poet and inventor of the dry-earth closet, only just accepted geological time. Some evangelical clergy continued with the semi-literalism of Baylee and Cuming as did William Brown Galloway (1811–1903), who wrote such turgid volumes as *Science and Geology in relation to the universal deluge* (Galloway, 1872), which only just allow for geological time, and Samuel Kinns (Kinns, 1883).

To summarize for the late nineteenth century, biblical literalists among mainstream evangelicals are almost non-existent, though a few contributed to the *Fundamentals*. Evangelicals mostly either followed some kind of Progressive Creationism, for example the geologist J. W. Dawson, Bishop J. C. Ryle, or Evolution as confined to the non-human sphere, notably James Orr and Bishop Handley Moule. Some evangelicals adopted evolution from monad to man, like B. B. Warfield. Non-evangelicals accepted evolution, mostly as a non-Darwinian guided evolution. As Darwinism went into eclipse it was replaced by "guided" evolution, in which direction or orthogenesis operated. Thus evolution was more susceptible to a theistic interpretation, as a guiding hand was apparent, and may explain why evolution was more acceptable a century ago than it is today. As Bowler points out in *The Non-Darwinian Revolution* (Bowler, 1988), evolution was rapidly accepted in the 1860s but Darwinism was not. This non-Darwinian evolution is typified by the (dinosaur) paleontologists Edward Drinker Cope and Alpheus Hyatt. Cope accepted that the pattern of evolution was "conceived by the Creator, according to a plan of His own." His private life was not quite Christian and he died of syphilis in 1897 at the age of 47.

Evangelicals were ambivalent toward evolution and there was no firm consensus in the latter half of the nineteenth century. Through few evangelicals rejected geological time (restricted by Lord Kelvin to less than 100 my) numbers never accepted evolution and others insisted on divine interventions for the formation of life, consciousness and humanity, which some saw as God's creative acts in Genesis 1 verses 12, 21, and 27. This ambivalence is best seen in the life and career of George Frederick Wright, who began his scientific career as a Christian Darwinist under Asa Gray, and gradually came to reject Darwinism and, perhaps, facilitated the metamorphism of Evangelicalism into fundamentalism in the early twentieth century. To him we will shortly turn.

THE ANTIQUITY OF EARTH AND HUMANITY

Since the discovery of Deep Time in the eighteenth century, no geologist could give dates for the age of the earth. Throughout the early nineteenth century geologists tried culminating with the Rev. Samuel Haughton's estimate for the base of the Cambrian as 1,800 my in 1860. The work of the physicist William Thomson (later Lord Kelvin) in the early 60s soon reduced the maximum age of the earth to 100 my and then to 24 my. Most geologists, including Charles Darwin, capitulated to these estimates though it mitigated against evolution. Kelvin's dates were widely accepted until the first attempts of radiometric age-dating in 1905.

Of more concern to evangelicals were arguments for the antiquity of humanity put forward from about 1860. Before then evidence was too scanty to give any firm date. So long as humans had only existed for only some 10,000 years, one could adopt a chronology similar to Ussher's. Lyell in *The Antiquity of Man* (Lyell, 1863) concluded that humans first appeared 100,000 years ago, which was unacceptable to most evangelicals, as it challenged any semblance of history in Genesis 4–11. Van Riper (1993) groups responses from 1855 to 1880 as Lyellian (100,000 years), Prestwichian (20,000 years), and traditional (Ussherian?) (6000–8000 years). The first group included Darwin, Huxley, Lubbock, and Wallace, and the last were confined to religious publications. The Canadian geologist, J. W. Dawson (1820–1899) continually argued for a traditional date and at his death was the only leading scientist not to accept evolution. From 1860 Dawson published many books reconciling geology and Christianity, with titles like *Archaia* (Dawson, 1860) and *The Story of the Earth* (Dawson, 1874). These exude geological competence, but he always favored Kelvin's shorter timescales (Burchfield, 1976) and a low human antiquity, which was music to evangelicals as they could retain the traditional chronology, which Schofield put in his *Reference Bible* of 1909.

Another evangelical geologist, George F. Wright (1838–1921), a Congregationalist minister who was encouraged to take up geology by Asa Gray, was persuaded against his earlier Darwinian views by considerations of geological time. From his early years he took an intermediate position between Lyell and Darwin on one hand and the heirs of Ussher on the other. In the 1870s and 1880s Wright was Darwinian as expounded in *Studies in Science and Religion* (Wright, 1882), but retained the special creation of humans, as did Wallace. After he returned to Oberlin College, Ohio, in 1881 he began an intellectual drift to the right but continued his fieldwork on glacial geology. He began to question evolution partly because of the materialism of Spenser and Huxley. He also was worried by the Higher Criticism of C. A. Briggs. When he first heard Briggs in 1891 he was convinced that Moses did not write the Pentateuch. On reflection he reacted against Brigg's liberalism and became more conservative and began to associate evolution with higher criticism, a position still put forward today.

In 1892 he published *Man and the Glacial Period* (Wright, 1892) which is a useful compendium on the state of glacial studies. Wright hoped for a favorable response. That was not granted him, as he insisted that there had been only one period of glaciation, and rejected the findings of recent glaciologists, who had unraveled a series of Ice Ages rather than one as was originally thought in 1840. Wright concurred with Joseph Prestwich, that the one Ice Age had lasted 25,000 years, but by the 1890s few geologists accepted that and Wright was taken to task by the geologists Chamberlin and McGee, the latter calling him as "a betinseled charlatan." Dana regarded McGee's dismissal was "a disgrace to American Science," but Numbers is correct to state that Wright's "theological convictions had undoubtedly colored his scientific conclusions" (Numbers, 2006, p 44). In *The Origin and Antiquity of Man* (Wright, 1913) he reiterated his case and refused to accept "Man's origin by purely naturalistic agencies." He argued that the earth was less than 100 million years old and that life, that is the base of the Cambrian, began some 24 million years ago. These conclusions, drawn from Kelvin, allowed him to accept a short 25,000 year Ice Age. Even so, he followed Flinders Petrie's dating of the first Egyptian dynasty at 4777 BC. On geological time Wright was restrictive. He objected to Lyell's "unlimited" geological time with the base of the Cambrian 500 million years ago (close to today's 550 million years). He commended Darwin for downsizing his almost limitless time in 1859 to some 100 my and favored Walcott of Burgess Shale fame and a Presbyterian for allowing only 27.5 my. It is difficult to be certain why Wright changed from a thoroughgoing evolutionist to a skeptic who took a limited view of geological time, which hardly gave time for evolution. Numbers (2006, pp. 33–50) gives some pointers.

Time, they were a-changing! A few years before in 1905 the English physicist John William Strutt, later Lord Rayleigh (1842–1919) began to apply radioactivity to date rocks and showed that a mineral containing radium was 2 billion years old because of its helium content. In the same year Bertrand Boltwood suggested that Lead may be the end product of the decay of uranium and calculated the ages of forty-three minerals from 400 to 2,200 my. The radiometric dating game had begun and by 1913 Arthur Holmes (1890–1964) in *The Age of the Earth* reckoned the base of the Cambrian to be 600 my and the age of the earth to be 1.6 by. Geologists would never again talk of less than billions. The immediate effect was to render untenable any suggestion that humans had been around for less than 50,000 years. The loose agreement with "biblical chronology" which

Dawson and Wright claimed was consigned to history. From then on the choice was, either to accept billions of years for the age of the earth *and* a 100,000 years or more for humans or to accept that humans are recent that is less than 20,000 years and to REJECT all radiometric age dating. As we shall see that first occurred in the 1930s and became a major thrust of YEC after 1961.

SCIENCE AND LOSS OF EVANGELICAL FAITH

One of the enduring themes of Victorian religion is the loss of faith of many erstwhile evangelicals. This phenomenon is often termed "Honest Doubt," as those who doubted strove to be extremely honest in all their thinking, a virtue learnt from their evangelical homes. Murphy in a classic paper (Murphy, 1955) argued that the cause of doubt was moral rather than intellectual, with a repugnance at doctrines such as eternal punishment, the savagery of substitution atonement, and the idea of punishment for the misdoings of Adam. Murphy's ideas were taken up by Josef Altholz in his suggestive but flawed article *The Warfare of Conscience with Theology* (Altholz, 1988), which has allusions to Andrew White's famous book, and repeats some of its inaccuracies. Elsewhere Frank Turner argues persuasively that no scientist in the nineteenth century gave up his faith because of science, but he is driven by the assumption that an orthodox theological belief stifled proper scientific research. He also assumes that the Broad Church was more open to science than its orthodox counterparts.

The number of intellectual and literary Victorians brought up as evangelicals is legion. Some drifted off into a form of unbelief and others to a liberal or catholic brand of faith. Probably one of the most famous accounts of the change between two generations is Edmund's Gosse's *Father and Son* (Gosse, 1907, 1949), which was first published in 1907. Gosse reacted against the claustrophobic faith of his father into a literary agnosticism. It is a moving book, but is inaccurate in his portrayals of Victorian science and religion. As a result it gives a distorted picture and encourages the perception that Evangelicalism is hostile to science. It is also significant that it came from a time when many were writing of the *conflict* of science and religion.

The great novelist Mary Ann Evans, or George Eliot (1819–1880), was brought up as an evangelical and in her early twenties rejected that faith and was drawn to the radical ideas of Charles Bray. Her novels *Scenes of Clerical Life, Adam Bede* (which Darwin read while writing *The Origin of Species*), and *Middlemarch,* are full of sympathetic portrayals of evangelicals. In 1846 she translated D. F. Strauss's *Leben Jesus* and in 1855 wrote a scathing essay on the follies of a Presbyterian minister in *Evangelical Teaching: Dr Cumming* for the October *Westminster Review.* Though she and George Lewes were well informed scientifically and on good personal relationships with Darwin and his circle there is no evidence that science had any part in her rejection of Christianity. Many biographers of Eliot *assume* it, as does Frederick Karl in his biography, as he claims that Eliot's grasp of geology would have aided her loss of faith. In about 1840 she read Pye Smith's *Relation between the Holy Scripture and some parts of Geological Science,* which could hardly have assisted in what Karl refers to as "the geological assault on the religious explanation of the earth's origin" (Karl, 1995, p. 46) or that "geology was the giant that could topple the church" (Karl, 1995, p. 129). Whatever the reasons were for Mary Ann Evans loss of faith, science was not one of them.

John Colenso (1814–1883) was the most notorious Anglican bishop of the Victorian era, as he was charged with heresy and acquitted after the publication of *The Pentateuch and the Book of Joshua* (1862–1879). Since then Colenso has been a hero to liberals! He was born in Cornwall and by 1830 was helping the local evangelical vicar. He went to Cambridge in 1832 and in 1836 graduated as the second best mathematics student. He was ordained in 1839 and published several arithmetic textbooks. So far he was an exemplary evangelical but in 1842 he met Sarah Frances Bunyon, whom he married in 1846. Like many others, both Frances and John rejected their Evangelicalism and molded their faith around the theology of S. T. Coleridge and F. D. Maurice. By now Colenso was a vicar and developed a great interest in missionary work. In 1853 he became Bishop of Natal and published a commentary on *St Paul's Epistle to the Romans* in 1861, which rejected substitutionary atonement and incurred the wrath of Bishop Gray of Cape Town, his superior. The following year he began to publish *The Pentateuch and Book of Joshua critically reviewed.* Much of these volumes are on critical issues of the Mosaic authorship of the Pentateuch and the vast numbers of the Israelites in the book of Numbers. Colenso claimed that he had been forced to face these questions for the first time by discussions with local Zulus, as when he was translating the story of the Flood, he was asked "Is all this true? . . . that all the beasts and birds, and creeping things . . . came thus and entered into the ark with Noah?"

Colenso wrote that he had learnt geology since leaving England in 1853 and that

> I now for certain, on geological grounds, a fact of which I had only misgivings before, viz. that a *Universal Deluge,* . . . , could not possibly have taken place . . . (Colenso, 1862, vol. 1 pp. vii–viii)

In Part 4 on the first eleven chapters of Genesis, Colenso dealt with the problems of the animals on the ark and wrote with great wit on the problem of animals traveling to the ark prior to embarkation.

How then could these snails, and worms, and snakes, and lizards, of all kinds, have found their way to the Ark, across vast countries, mountains, seas, and rivers, from the distant localities in which they lived? or how could they have *returned* to them?'

'Could, then, the *sloth* and *armadillo,* from the tropical regions of South America, have marched up to the Icy North, . . . , after many years of painful wandering, . . . , have been received into the Ark? (Colenso, 1863, vol. IV, p. 191)

Now if a sloth can travel at 0.068 mph it would have taken many years to travel some 10,000 miles!!!! (These types of questions are still being raised.) This is great fun, and Colenso was clearly enjoying himself writing against the backdrop of the Drakensburg Mountains, but it was not new. Colenso must have known this, as he relied on books by evangelical writers, Pye Smith, Miller, Hitchcock, and Pratt (Colenso, 1863, vol. IV, pp. 191–193, 197, 200, 203), some of whom were writing a quarter of a century before. His major episcopal opponent Samuel Wilberforce was no mean geologist and reacted badly to personal attacks in Colenso's writings. Colenso was looking at Genesis in the same way as Goodwin did in his chapter in *Essays and Reviews*. They both challenged the traditional approach of seeking to reconcile or incorporate Genesis into all human thought on the origin of things, which had gone back through Buckland to Grotius and Mersenne in the early seventeenth century.

The upshot of the publication of his commentary of Romans and his volumes on the Pentateuch was a major controversy, which resulted in Colenso being indicted for heresy. The primary cause was over his rejection of substitutionary atonement and then over critical questions of the authorship of the Pentateuch, rather than comments over the occupants of the Ark. His gamboling with Noah's menagerie probably only served to antagonize conservatives like Samuel Wilberforce and other Bishops who correctly dismissed his geological arguments as trite, puerile, and repetitions of long-refuted objections. Despite the fact that Colenso is hailed as a pioneering scholar who took science into account for his understanding of Genesis, when his sources are considered it is difficult not to conclude that his stories of Zulus disbelieving the Flood are creative. For a contrary viewpoint see Rogerson (1984, pp. 220–237) and Parsons (1997).

One of the greatest Victorian men of Letters was Sir Leslie Stephen (1832–1904), best known as the editor of the British *Dictionary of National Biography* and father of the novelist Virginia Woolf. He came from an evangelical dynasty; his grandfather James Stephen (1758–1832) was a member of the Clapham Sect with William Wilberforce, and his father Sir James Stephen (1789–1859) was colonial under-secretary and prepared the successful bill abolishing the slave trade in 1833. He was educated at Eton and Cambridge and after becoming twentieth wrangler in 1854, a fellow of Trinity Hall and ordained in 1855. His great passion in life was mountaineering, made several first ascents in the Alps and wrote the *Playground of Europe.* His main work was the *History of English Thought in the Eighteenth Century* published in 1867 and he wrote widely. It seems that he began life as an evangelical but gradually abandoned his beliefs. After adopting agnosticism he resigned his orders in 1875. We see in his family a microcosm of growing doubt. His grandfather was an evangelical of firm convictions, his father was softening at the edges, Leslie, himself, moved from a childhood Evangelicalism to a moral Victorian agnosticism and his daughter Virginia was part of the Bloomsbury group.

Leslie spoke about his loss of faith in no uncertain terms. Writing in 1903 he claimed that in 1862 he began to have doubts about the literal truth of Noah's Flood and as a result he no longer took services in the college chapel. Owen Chadwick dismissed his account in no uncertain terms calling it "exaggerated" and simply pointing out that forty years previously Canon Buckland in Oxford had scorned such literalism, as had Sedgwick in Cambridge. However the historians of science Frank Turner and von Arx take Stephen at face value (Turner and von Arx, 1982). Stephen's account is untenable as no fellow at Cambridge or Oxford could possibly have thought that a universal flood was orthodox belief in 1862 as one is hard pressed to find any books in the 1860s, which argue for a universal flood. That, along with Stephen's great intellect, means that we need to take his account as one of *perception* rather than *reality*, as his memory of the 1860s is at best exaggerated.

The stories Colenso and Stephen about their change of faith are as memorable as they are inaccurate. Both are masterpieces of creative writing and have passed into the canon of de-conversion stories. The inaccuracy of their accounts becomes very clear, when we consider the evangelical writings of Miller and Pratt of the 1850s, which are almost cavalier in their insistence of a local flood and an ancient earth. However, these two accounts illustrate the problem of *perception* affecting an individual as well as on a wider sphere. But these accounts and the experience of many indicate how evangelicals were losing many able young men and women.

There was not only the considerable hemorrhaging of evangelicals, either to Liberal Protestantism, Anglo-Catholicism or to unbelief, but some groups across the Atlantic were hardening their arteries. There were two groups, in particular; the Germanic Lutherans of the Mid-West and the Seventh Day Adventists, who though, strictly speaking, were outside Victorian Evangelicalism, have become important in retrospect as they produced most early twentieth-century YECs and thus made a major impact on evangelical understandings of science in the twentieth and twenty-first century. In the late nineteenth-century geocentric or Young Earth writings are almost limited to these two groups.

Whether many adherents of evangelical churches in Britain and America were also young earth and literalist it is impossible to say, as there seems to have been no research on the subject. Further, the *hoi polloi* of evangelicals leave nothing in print. However, as many lay evangelicals, however well-educated, tend to adopt a default literal hermeneutic today it is reasonable to conclude that many would unreflectively opt for literalism and all that entails. Even in Britain this is still evident today as many church members during the last century, evangelical or not, assume that literalism is the orthodox Christian understanding. I have observed this among my parishioners in my ministry and have often led them away from an uncritical literalism. It is what they *learnt* in church and Sunday school, even if they were not *taught* it. The *perception* has a greater impact than the *reality.*

LUTHERAN GEOCENTRICITIES

Whereas no other Protestants questioned the Copernican system in the nineteenth century, some conservative Lutherans in the mid-west rejected heliocentricity in favor of geocentricity. To go back to the sixteenth century, Martin Luther (1483–1546), possibly dismissed Copernicus's theory of a heliocentric universe, and perhaps that fact explains why modern geocentric beliefs seem to have been more common among Lutherans. Exactly how strongly Luther objected to Copernicanism is difficult to ascertain as his comments are to be found only in his *Table Talk*, which was collected together by his followers and may not be reliable. At the end of the sixteenth century both Tycho Brahe (a geostatist rather than a geocentrist) who rejected Copernicanism and Johannes Kepler were Lutherans, but even so Lutherans were more hostile to Copernicanism than Calvinists.

As a result of German immigration to the mid-west strong Lutheran communities formed especially in Missouri and Wisconsin. One of the leaders was German-born Carl Ferdinand Wilhelm Walther (1811–1887) who came to St. Louis in 1838 to be a pastor. He also founded a seminary, a publishing house, and a theological journal and was president of the German Evangelical Lutheran Synod of Missouri, Ohio, and other states (forerunner of the Lutheran Church Missouri Synod), which dates from 1847 and a leading figure in the Evangelical Lutheran Synodical Conference of North America, which was formed in 1872. (Noll, 1992, p. 216) These bodies were largely self-contained until large-scale immigration from Germany ceased in the 1930s, and they kept aloof from the indigenous American evangelicals. Even in the 1930s they were reluctant to pray with other Protestants and more recently would not allow Southern Baptists and others to receive communion with them. Walther disparaged Copernican astronomy in the pages of the synod's official publication, *Der*

Lutheraner. The theologian F. A. Pieper (1852–1931) also rejected Copernicanism, as did August Graebner (1849–1904), who insisted that if the Copernican system disagreed with the Bible, "the heliocentric system must fall" (Numbers, 1991, p. 106). Several other works were published like J. C. W. Lindemann's (d1879) *Astronomishe Unterredung zwischen einem Liebhaber der Astronomie und meheren berühmten Astronomen der Neuzeit, worin deutliche Auskunft gegeben wird über die Untrüglichkeit des Kopernikanischen Sonnen-Systems* published in1873. Hence most geocentric works published in America between 1870 and 1920 were written by members (mainly clergy) of the forerunners of the LCMS and that geocentricity was widely taught within the synod. These were discussed by Friedrich E. Pasche's 1906 book *Bibel und Astronomie*, which also discussed about sixty passages from the Bible that indicate an unmoving earth and/or a moving sun. In 1915 he also wrote in English, *Fifty Reasons:Copernicus or the Bible. Philosophy and vain deceit, or true science? Which Is Right?* and showed that there were fifty reasons why Copernicanism was wrong!

During the twentieth century these Germanic Lutherans slowly gave up geocentricity, but not their insistence on a young earth. which is to be seen in the numbers of Lutheran flood geologists in the inter-war years and the division of the Lutheran Church into the Lutheran Church Missouri Synod (LCMS) and the Evangelical Lutheran Church of America (ECLA) in the 1970s, which was over inerrancy and biblical literalism.

THE RETURN OF BIBLICAL LITERALISM

Following the decline of the Scriptural/anti-geologists after the 1840s, there were very few evangelicals who ventured into print arguing for a literal six-day creation. I emphasize "venturing into print" as I am sure that there were considerable numbers within popular Evangelicalism, both pastors and worshippers. A fruitful line may be to survey popular religious magazines of the period, particularly those of the separatist evangelical groups. My tentative impression is that many evangelical church members did not think deeply about the subject and adopted a default literalism, but were not dogmatically YEC. Many were disturbed by Darwinism and the perception that humans are only monkeys.

The support for a literal six-day creation declined sharply after the mid-1850s. There were a few pamphlets issued in response to Hugh Miller's *The Testimony of the Rocks* after 1858, but they had little influence in dissuading even the most evangelical from accepting some kind of harmony of Genesis and geology. Some, like those of Baylee and Cumming, were based on a naïve and apparently literal exegesis, but they did not question geological ages. After the publication of *The Origin of Species* the writings of B. W. Newton and R. L. Dabney seem to be almost unique and even such a thorough bibliography as Tom McIver's (McIver, 1992) only records a few more from this period.

As far as I can ascertain, the only groups, who overtly supported a young earth approach after the mid-60s were the Seventh Day Adventists and some Lutherans. At that time they had minimal importance as they were outside mainstream Evangelicalism. However despite their marginal importance, they became better known through the writings of McCready Price in the Scopes era, whose work is the basis of the modern YEC movement. Throughout the nineteenth century there were a proliferation of Adventist sects in America and Britain. These had their origin in the Millenarian tradition, which flourished from the late eighteenth century and still thrive among evangelicals today. Many Millenarians were to be found in mainstream churches and in Britain the Anglican Church was well represented with writers like G. S. Faber, F. Nolan, E. Bickersteth and Thomas Birks, who have been discussed earlier. Some Millenarians founded their own groups outside the churches, and extended their Millenarianism to predicting the end of the world. There were eccentrics like the illiterate John Wroe who thought that the Second Coming would take place in Ashton-under-Lyme in Lancashire. Within a year he was hounded out of town, perhaps because he had taken ten virgins to comfort him.

An American counterpart was William Miller (1782–1849), the most famous of all Millenarians. He was converted in 1816 and became drawn to prophecy initially through British Millenarianism. For two decades he did little but from 1843 became active and attracted a following of some 50,000. Slowly Miller's followers drifted away from their churches and began to form their own denomination, which by October 22, 1844, was nearly complete. That was the day, which the Millerites had fixed as the Second Advent and after their hopes were not realized the Millerites declined and referred to that date as the *Great Disappointment*. From the ashes a small apocalyptic sect, the Seventh Day Adventist, smoldered on. Following the Great Disappointment numbers of Millerites took stock and concluded that the Second Coming was imminent, but would be delayed until the world had been warned to observe the Sabbath on Saturday—the seventh day of the week.

In the early 1840s the young Ellen Harmon became involved with a small Adventist chapel in Portland, Maine and in August 1846 married an Adventist preacher John White. Soon after that Ellen White began writing her many books, which were often culled from other sources and rewritten by friends. According to Ronald Numbers the Seventh Day Adventist Church was formally organized by John and Ellen White and Joseph Bates in 1863. Like the Millerites, the Seventh Day Adventists expected the imminent Second Coming and worshipped on Saturday because of the Fourth Commandment, which also refers to Creation in six days. Hence they insisted on a literal interpretation of Genesis. (Today, this is still a crux of the YEC arguments.) In her *Spiritual Gifts: Important Facts of Faith* (White, 1864) White wrote that to follow infidel geologists with vast indefinite periods "strikes directly at the foundations of the Sabbath of the fourth commandment" (cited Numbers, 1992, p. 74) and repeated these arguments in *Patriarchs and Prophets* (White, 1890, p. 97).

This work gives the biblical story from Creation to King David, written in popular style. She insisted on a literal creation week and in chapter IX *The Literal Week* voices her concern at geology. She favored the Deluge being the source of strata rather than long ages and regarded geology as "one of Satan's devices to lead the people to accept the fables of infidelity . . ." (White, 1890, p. 99). She waxed lyrical on the Flood, "The mountains once so beautiful in the perfect symmetry, had now become broken and irregular. . . . And upon countries that were not inhabited, and those where there had been the least crime, the curse rested more lightly." Her understanding of coal is unusual; "At this time immense forests were buried. These have since been changed to coal, . . . The coal and oil frequently ignite . . . Thus rocks are heated . . . and volcanic eruptions follow" (White, 1890, p. 94). So far, the source of White's geology is not known, Stilling wisely suggests that it may come from the Lord brothers. However Ellen White had less grasp of geology than any of the British anti-geologists mentioned in the last chapter and clearly had not read them. Thus to the Adventists, geology with its long ages *had* to be rejected or else their *raison d'être* to change the Sabbath would be nullified and so White's visionary writings passed into the official doctrine of the Seventh Day Adventist Church.

In 1863 Ellen White developed health as a new aspect to her ministry, as one Sabbath day White had a vision of the relationship of physical health to spirituality and the importance of good diet, fresh air, and exercise. She also had other concerns as she thought that the practice of masturbation would result in "imbecility, dwarfed forms, crippled limbs, misshapen heads, and deformity of every description." The result was the SDA's work in medical missions, which are now worldwide. In 1866 the Whites opened a Health Reform Institute at Battle Creek, Michigan, which majored on hydrotherapy, which was the rage in the nineteenth century. The Whites realized they needed a medical director and in 1875 Dr. John Kellogg (1852–1943), a younger member of the church whom the Whites financed through Bellevue Medical College in New York City, was appointed.

Kellogg, who we remember mostly at our breakfast tables through his invention of cornflakes in 1897 as well as peanut butter, proved to be an interesting appointee. He changed the focus at Battle Creek to medical and surgical procedures and Numbers reckons that without his influence Adventist medical work would have ceased. Kellogg made a considerable contribution to medicine and to foods. While teaching physics at Battle Creek College, in 1879 Kellogg wrote a small work *Harmony of Science and the Bible* (Kellogg, 1879), which ran counter to Ellen White's teaching as Kellogg reckoned that a main factor in causing conflict between science and religion was the "Holding of the Bible as unimpeachable authority on *all* subjects" and argued for the separation of science and religion. Even though Kellogg still believed in a special creation, his way of thinking was not amenable to Ellen White and other leaders so eventually Kellogg and the Adventists went their separate ways. Kellogg later became an evolutionist and a eugenicist and founded the Race Betterment Center in 1914.

However other Adventist writers, notably Uriah Smith and Alonzo Jones, published in the *Review and Herald*, supporting Ellen White's visions and refuting the geologists, whom they noted were often in contradiction to each other. These writers were putting in place the distinctively Adventist approach to science and religion, which was developed in the next century by George McCready Price, who attended Battle Creek College from 1891 to 1893. Whatever the validity of their arguments, it is clear that they could do no other because of their insistence that the Fourth Commandment on the Sabbath necessitated a six-day creation. Very few evangelicals accepted such a literalist approach. Their ideas were restricted to sidelines of American society until into the latter half of the Twentieth Century. I venture to suggest that were it not for Ellen White and the Germanic Lutherans YEC would not have come to the fore today.

CONCLUSION

A short chapter can only give the barest outline of evangelical attitudes to science in a half-century which saw the publication of the *Origin of Species* and the rise of so many ideas contrary to an evangelical worldview. My selectivity has been to both describe the past and to explain future developments. For the vast majority of *educated* evangelicals science placed no threat to their faith. No mention has been made of the advances in physics and chemistry, or even astronomy, as they caused no controversy for any Christian, whether evangelical or not.

It must be written in capital letters that *most evangelicals had no objection to geology and thus did not insist on a six day creation*, but many were concerned at the possibility of evolution and an existence of humanity for more than Ussher's allotted 6,000 years. In the eyes of many YECs today Victorian evangelicals were hopelessly compromised theologically, but this had no effect on their evangelical zeal, nor their devotion to the bible. Despite their acceptance of so much science, which is now anathema to the YEC, evangelicals were full of confidence in their faith in Jesus Christ, and on the whole were unperturbed by the findings of geologists and biologists. However there was a gap between the clergy and the educated laymen with many rank and file Christians, who regarded science with suspicion. This was shared by many in the revivalist churches, and became apparent after WWI. However, a turn of the century commentator would have said that there were no future battles to be fought and how wrong they were!

Chapter 6

Evangelicals in the Shadow of Scopes

SCIENTIFIC DEVELOPMENTS

In about 1900, Lord Kelvin thought that all major scientific discoveries had been made. In 1905, Einstein put forward his theory of "Special Relativity," and in the 1920s cosmologists, like Fr. Le Maitre, were determining the age of the universe. The discovery of radioactivity in 1895 resulted in two benefits—the use in radiometric age dating from 1905 and the medical use of X-rays. The unlimited energy of radioactivity resulted both in nuclear power and the atomic bomb. From the 1930s as the neo-Darwinian synthesis was forged by Mayr, Dobhansky, Huxley, and others. That put paid to orthogenetic and teleological evolution, which had reigned since 1880. In the inter-war years many British and Americans favored eugenics resulting in the sterilization of mentally-retarded people. Eugenics was rejected after the Second World War.

Work in paleoanthropology grew apace with important discoveries throughout the world. Notable was Raymond Dart in South Africa, but most notorious were Piltdown Man, a hoax which developed a life of its own and lives on as part of the YEC arsenal against evolution, and Nebraska man. Space precludes any mention of advances in physics and chemistry, but this was the time when biochemistry became a science in its own right. Medical advances culminated in the discovery and utilization of Penicillin. Technology bounded ahead, with the aeroplane and the rise of motorized transport shrinking the world. Electronic communications came into being with both radio and television, and an evangelical, Ambrose Fleming, invented the thermionic valve.

THE FUNDAMENTALS AND POSTWAR DECLINE

As the twentieth century dawned evangelicals were dominant but began to decline soon after the publication of the *Fundamentals* (*Fundamentals*, 1910–1915) in 1910. In the first decade evangelicals could claim considerable scientific respectability as were some of the writers on science in the *Fundamentals.* These twelve booklets were written to buttress the fundamentals of Christianity and reflect a broad Evangelicalism. Prof. James Orr (1844–1913) of the United Free Church College, Glasgow, wrote two of the most important scientific articles. In Volume IV he contributed *Science and Christian Faith,* a brief but well-argued article. Orr began by criticizing those, like Draper and White, who pit science against Christianity. He mentioned scientists who were Christian or sympathetic. After a short discussion of miracles he then briefly described the relation of Christianity to the sciences of astronomy and geology.

His last section was on *Evolution and Man* (*Fundamentals,* 1910–1915, vol. iv, pp. 102–103), which he saw as "the point in which 'conflict' between Scripture and science is most frequently urged." Yet Orr had no concerns as "[a] newer evolution has arisen which breaks with Darwin on the three points." These were Chance, the insufficiency of Natural selection and the understanding that speciation could be rapid rather than slow. Orr wrote that evolution was a new name for creation, except that "the creative power now works from *within.*" However in Vol VIII, Henry Beach from Colorado contributed an essay entitled *Decadence of Darwinism* and G. F. Wright wrote on *The Passing of Evolution,* in which he continued his theme that limited geological time precluded Darwinian evolution and paved the way for anti-evolutionism. These are some of the scientific articles in *The Fundamentals.* Some articles represent careful conservative scholarship, others are harbingers of 1920s anti-evolutionism.

As the First World War loomed, confidence among evangelicals was high but after the end of the war their influence declined rapidly in both the United States and Britain. Many evangelical leaders died within a few years of 1918 and by 1930 evangelicals were considered a spent force. One result was evangelical disengagement from science, culture, and the intellectual world. During the interwar years evangelicals were at a low ebb in intellectual attainment and influence, but, with historical hindsight, all future developments were there in embryo.

It is easy to dismiss all evangelicals as fundamentalist, but there is a problem of defining both Fundamentalism and Evangelicalism, which may often be seen to be synonymous. The British avoided the term fundamentalist, but Americans often did not. This era was also the heyday of liberals both in United Kingdom and United States, who took control of most mainline denominations. In America many evangelicals simply left and found their own churches, but British evangelicals remained as a beleaguered rump.

THE INTERWAR YEARS IN THE UNITED STATES: A WIDER PERSPECTIVE

After the First World War American evangelicals were left on the sidelines, and often formed their own "pure" evangelical churches. Further doctrinal or personal disagreements caused more splits resulting in the bewildering variety of churches. There were several other results. First, there was often a rejection of the old intellectual conservatism epitomized by the Princeton theologians and the Baptist Augustus Strong (1836–1921). In its place, many adopted the Dispensationalism of Cyrus Scofield (1843–1921), editor of the *Scofield Bible* and rejected any hint of modernism. This meant that many evangelical clergy moved from an intellectually based conservatism to one with a literalist hermeneutic. The popularity of Dispensationalism encouraged this as all the prophetic books of the Bible were interpreted literally. This was applied to early Genesis as well, though the Gap Theory espoused by Dispensationalists did allow for all geological time. However Ussher's chronology for the creation of humans in 4004 BC and the Old Testament was retained. Thus geology was more than acceptable but evolution and the existence of early humans before 6,000 years ago was a great no-no. Most of the fundamentalist leaders of the inter-war years followed this position with the great Fundamentalist preachers like William Bell Riley (1861–1947) and John Roach Straton (1875–1929) being strident in their Pre-Millennialism and anti-evolutionism, but often inconsistent over geology, sometimes looking to McCready Price (1870–1963). Another vibrant strand in American evangelicalism was the Holiness Movement, which derived from the nineteenth-century Methodist churches. Their approach to faith was very experiential and emphasized that "Orthodoxy of opinion was secondary to the Orthodoxy of life." Though Holiness leaders were skeptical of evolution, they did not take part in the anti-evolution crusades of the 1920s (Numbers, 1998, pp. 111–136). Despite the force of their rhetoric and the number of anti-evolution books and tracts, the anti-evolutionist movement failed to attract any mainstream scientists. The only biologist with any postgraduate qualifications was S. James Bole (1875–1956), a member of the Holiness movement, who taught at various colleges including Wheaton College.

The success of the fundamentalist anti-evolution campaign was limited to their own members as many, whether Christian or not, were repelled by what they perceived as anti-intellectualism. Into this popular anti-evolutionism stumbled a schoolteacher, John Scopes (1901–70) whose trial, when shorn of myth, brings out the strengths and weaknesses of the pre-Depression evangelicals.

THE SCOPES TRIAL, 1925

The Scopes trial sums up evangelicals and science at this time.[1] The *myth* of the Scopes trial has taken on a life of its own and has misinformed opinion ever since. In the United States science for evangelicals was dominated by the Scopes trial, and the abiding images of those decades are provided by *Inherit the Wind*. This portrays American evangelicals as having a hillbilly faith based on anti-intellectual literalism. Edward Larson in his Pulitzer Prize winning book *Summer of the Gods* (Larson, 1997) corrects much of that, but old story still rules. The events that led up to the trial are the most bizarre in the history of science and religion.

As we saw in Chapter 3, the warfare model of the relationship of science and Christianity dominated the twentieth century. In a sense the warfare model both fuelled the events of the Scopes trial and their interpretation as it had become the received wisdom of many secular Americans. The attorney Clarence Darrow (1857–1938) knew the books of Draper and White by heart, and the defense co-counsel at Dayton, Arthur Garfield Heys said, "Of all the books I have read for this trial, the 'Warfare between Science and Religion [*sic*],' by Prof. White, is, to my mind, one of the most interesting and readable." With attitudes like this, it is not surprising that fundamentalists initiated the anti-evolution crusade and objected to the new Darwinian biology textbooks, such as Hunter's *Civic Biology* used in Tennessee. Further in the 20s the eugenics movement was at its height and many eugenicists were evolutionists—R. A. Fisher, Leonard Darwin, and H. F. Osborn. By 1935, thirty-five states had passed laws to compel the sterilization of the eugenically unfit. As Hunter wrote in his *Civic Biology*, "If such people were lower animals, we would probably kill them off to prevent them from spreading."

Christian anti-evolutionists like the "Great Commoner" William Jennings Bryan (1860–1925) and Billy Sunday (1862–1935) denounced eugenics as inspired by evolution. Bryan called it brutal and at Dayton argued it was a reason not to teach evolution. Billy Sunday bracketed eugenics and evolution in his 1925 Memphis crusade. The Modernist theologians Shailer Mathews (1863–1941) and H. E. Fosdick (1878–1969) both supported eugenics. From our post-Nazi perspective it is difficult not to grant the moral high ground to the Fundamentalists.[2] It also gave reasons to reject Modernism. Ironically in 1939 when most had rejected eugenics William J. Tinkle (1892–1981) "was still advocating selective human breeding in his creationist textbook, *Fundamentals of Biology*" (Numbers, 1992, p. 223).

Anti-evolutionism was only part of fundamentalist militancy as their main target was theological Modernism, which swept through every denomination. As a result conservatives formed a loose coalition to combat this threat to orthodoxy. Some stressed the German roots of higher criticism and attributed a "survival of the fittest" mentality to German militarism. These were combined into the distinctive Fundamentalism of the 1920s and 1930s and the formation of the World's Christian Fundamentals Association (WCFA) in 1919. With William Jennings Bryan's opposition to the war and anti-evolutionism, this led to Dayton.

The alliance of Bryan and Fundamentalists like Riley does not demonstrate that they were in total agreement. Riley was a dispensationalist, but Bryan believed in the power of reform to make life better. Bryan had a thirty-five year career in public life, becoming a Democratic Congressman in 1890. With his oratory he became known as the Great Commoner and secured three presidential nominations. After supporting Woodrow Wilson in his presidential campaign of 1912, Bryan became secretary of state and resigned from office after America's entry into World War One. Yet he was a progressive reformer and supported both prohibition and female suffrage. As his biographer Lawrence Levine commented, "In Williams Jennings Bryan reform and reaction lived happily, if somewhat incongruously, side by side." His anti-evolutionism came from his Christian convictions but he was no six-day creationist. He was willing to accept evolution for the animal kingdom but not for man. He was very much in line with earlier Christians, like James Orr.

Into this heady political and religious mix the Scopes trial was born. Matters began late in 1921 when Kentucky's Baptist State Board of Missions passed a resolution asking for a law against teaching evolution. Bryan heard about it early the next year and adopted it. The campaign spread quickly, with John Roach Straton (1875–1929) advocating anti-evolution in New York, Norris in Texas and T. T. Martin throughout the south who with William Bell Riley was offering to debate evolutionists, and providing the support of the WCFA. Three years later these four were the most prominent ministers supporting Scopes' prosecution.

In January 1925 Senator Shelton introduced a bill in the Tennessee Legislature. The next day John W. Butler put forward similar legislation in the House of Representatives, demanding a $500 fine for a public school teacher teaching "any theory that denies the story of the Divine Creation of man as taught in the Bible . . ." The House passed it by seventy-one votes to five. The public was caught off guard and opponents began to work on the Senate and wrote letters to the press. In February Billy Sunday returned to Memphis for a second crusade. On March 21, the Senate passed the Butler bill by twenty-four to six and it was sent to the Governor to sign. Despite protests from evolutionists and liberal churchmen it was made law in Tennessee. The American Civil Liberties Union saw the bill as contrary to civil liberties and offered legal help to any schoolteacher challenging the law.

What happened next is slapstick comedy. Dr. George Rappleyea, a mine manager, who attended a modernist Methodist church, read in the *Chattanooga Times* on May 4 of the ACLU's offer of legal help. The most credible version of legend says he hurried to Robinson's drugstore and suggested getting publicity for Dayton. With seven others, including several attorneys, he obtained support from the ACLU. They then called in the high school's science teacher and football coach, John Scopes and Rappleyea asked him if he had been teaching from Hunter's *Civic Biology*. When he admitted his felony Robinson told him, "Then you've been violating the law" and then asked, "John, would you be willing to stand for a test case?" The die was cast.

Scopes was not a radical and taught physics, math, and football rather than biology. Like his father he was agnostic. He preferred sport to politics and occasionally attended Dayton's Methodist church. The following day, a front-page article in the *Banner* carried the story how George Rappleyea was prosecuting a teacher for violating state law. Anyone reading that the prosecution was acting for the ACLU would have known it was not an ordinary criminal case. Many Tennesseans did not appreciate Dayton's publicity stunt.

The preliminary hearing took place on May 9 for action in August. The prosecutors included two local attorneys Sue (a man) and Herbert Hicks along with Bryan, though he had not pursued law for thirty years. According to Larson, this changed the whole issue from a narrow constitutional test to one where evolution as well as Scopes was on trial. The ACLU's hopes for a test case were dashed again when Clarence Darrow offered to duel Bryan. Darrow, who was then sixty-eight, is best described as an atheistic pugilist of considerable notoriety, which had increased after his successful defense of the Leopold-Loeb case, when Darrow saved two from death by appealing to psychological determinism. The historian Will Herberg described him as 'the last of the 'village atheists' on a national scale'. The humanist Edwin Mimms from Vanderbilt University wryly commented, "When Clarence Darrow is put forth as the champion of the forces of enlightenment to fight the battle for scientific knowledge, one feels almost persuaded to become a Fundamentalist." The ACLU tried to displace Darrow as defense lawyer, but Scopes wanted him.

The trial began on July 10, with five hundred visitors from the media. With America's finest journalists present, including H. L. Mencken, the trial became a media event and dominated the national newspapers for a week. Judge Raulston arrived at 8.30 A.M. with a Bible and statute book and as temperatures were set to top 100 degrees he allowed attorneys to dispense with coats and ties. He was followed by the defense, Darrow, Malone and Neal, and then Scopes and Rappleyea. At nine o'clock Bryan entered with the other prosecutors to great applause. The court opened with an "interminable" prayer punctuated by amens. Shortly before noon a thousand people left the stifling courtroom, to find four steers being barbecued. After lunch the jury was selected and the veniremen chosen were fundamentalist inclined, which Mencken did not regard as impartial. After that court was adjourned for the weekend and most visitors headed to the Great Smoky Mountains to escape the heat, while Bryan preached at Dayton's Southern Methodist church.

On Monday the court was refilled to capacity and business began. The defense challenged the constitutionality of the antievolution statute to quash the indictment. Neal and Hays began the defense so that Darrow could close dramatically. Hays compared the statute to a law against Copernicanism claiming that "Evolution is as much a scientific fact as the Copernican theory. McKenzie and Stewart took up the prosecution. Then Darrow took the floor and argued that the antievolution law was illegal as it established a particular religious viewpoint in public schools. Darrow's speech was electric and Mencken wrote that "It was not designed for reading but for hearing." Responses to the speech were varied, some hissed (morons to Mencken) and others applauded. Court resumed next morning only to be adjourned due to power failure, which prevented Judge Raulston from preparing his ruling on the motion to quash the indictment.

Wednesday was the hottest day and during lunchtime Scopes went swimming in a mountain pool with two of the prosecutors, Wallace Haggard and William Bryan Jr. and returned late. That afternoon the defense's first witness, the zoologist Maynard M. Metcalf, was called and Darrow prevented Scopes from taking the stand, because he was not, in fact, a biology teacher and that would collapse the whole trial. Metcalf was an Oberlin graduate and also taught a college-age Sunday-school class. Darrow persuaded Metcalf to explain evolution. The next day William Bryan Jr. opened for the state. Hays followed, to be answered by Bryan Sr., who rose to the occasion with an hour-long attack on teaching evolution, followed by Malone with an appeal for freedom. Stewart was last and put the case for statutory interpretation rather than testimony for or against evolution.

The next day the court met for an hour before closing for the weekend. Monday was crunch time and every seat was filled by 8:30 A.M. when proceedings began with prayer aimed at the defense. Hays read out the statements of the witnesses for the defense, eight scientists, three of whom sought to reconcile evolution with Genesis, as did the four religious witnesses, including Shailer Mathews. Then Hays summoned Bryan who stated, "They came here to try revealed religion. I have come here to defend it, and they can ask me any questions they please." As the *Nashville Banner* reported, "Then began an examination which has few, if any, parallels in court history. In reality, it was a debate between Darrow and Bryan on Biblical history, on agnosticism and belief in revealed religion." They jousted over Jonah and the whale and the long day of Joshua. When it came to Genesis 1, Bryan demonstrated his acceptance of the Day-Age interpretation, resulting in the following exchange:

Have you any idea of the length of these periods?

No; I don't

Do you think that the sun was made on the fourth day?

Yes.

And they had evening and morning without the sun?

I am simply saying it is a period.

They had evening and morning for four periods without the sun, do you think?

I believe in creation as there told, and if I am not able to explain it I will accept it.

This gave the defense what they wanted in that as Hays said, "Bryan had conceded that he interpreted the Bible." Scopes reported in his autobiography that, "The Biblical literalists . . . were . . . disappointed that Bryan gave ground" (Larson, 1997, p. 189) This part of Bryan's testimony was altered in *Inherit the Wind*. Soon afterwards Raulston adjourned for the day and Darrow's supporters were jubilant. Next day the jury was sent out after Darrow had suggested that the judge should instruct them to find the defendant guilty. They did and recommended a $100 fine. The following Sunday afternoon Bryan died in his sleep.

Then the Scopes trial took on a life of its own. Soon the Scopes legend began to grow, beginning with the publication of F. L. Allen's *Only Yesterday: An Informal History of the Nineteen-Twenties* in 1931 and culminating with the release of the film version of *Inherit the Wind* in 1960. As Larson wrote, "Far more than what happened in Dayton, these two works shaped how later generations would come to think of the Scopes trial." Allen intended to give a racy account of the Roaring Twenties, but altered what happened at the trial, incorrectly stating, "Bryan affirmed his belief that the world was created in 4004 BC." By doing this Allen wrongly persuaded future generations that not only was Bryan a six-day creationist but that this was a central tenet of fundamentalism. As Larson points out Allen made many distortions, which became the Scopes legend. Larson then names other writers who adopted Allen's account such as the historians Furniss and Hofstadter.

Hofstadter drew parallels with the Scopes trial and McCarthyism in his *Anti-intellectualism in American life*, which was what the liberal left wanted to hear. This association with McCarthyism inspired Jerome Lawrence and Robert E. Lee's play, *Inherit the Wind*, in which anti-evolutionism was not the danger but McCarthyism, as they felt that McCarthyism paralleled some aspects of the Scopes trial. When I first saw the film some thirty years ago I thought it was a docudrama of the Scopes trial. Whatever the intentions of the playwrights, this is how most viewers perceive it as it molded their understanding of 1920s anti-evolutionism.

The characters in *Inherit the Wind* have their counterparts in the Scopes trial, but much was changed for artistic and political reasons. Prominent in the film is the fundamentalist mob singing about the old-time religion. Bryan was metamorphosed into Brady, a mindless reactionary demagogue who told Drummond (Darrow) that creation took place "on the 23rd of October in the Year 4004 BC at—uh, at 9 A.M.!" Brady collapsed and died at the end of his closing speech. At the time critics savaged the play and the movie and by 1967 Joseph Wood Knutch could say, "Most people who have any notions about the trial get them from the play, *Inherit the Wind*, or from the movie."

The response today of YEC leaders to the Scopes trial exposes the ambiguity of the participants. The Bryan of *Inherit the Wind* would be more to their liking. Henry Morris wrote, "Probably the most serious mistake made by Bryan on the stand was to insist repeatedly that he had implicit confidence in the infallibility of Scripture, but then to hedge on the geological questions, relying on the day/age theory. George McCready Price had warned him against this very thing. Darrow, of course, made the most of it, ridiculing the idea of people claiming to believe the Bible was inspired when its meaning was so flexible that one could make it say whatever he wished!" (Morris, 1984b, p. 66). Writing in 1942, Price complained that Bryan had "conceded the entire geological arguments to evolutionists, with the pitiful results now known to all the world" (Numbers, 1992, p. 99).

The Australian, Carl Weiland, reviewing Larson's book also criticized Bryan for accepting geological time. He wrote "In fact, it may surprise many readers to know that the 'Great Commoner,' as the populist Bryan was affectionately known, would have felt perfectly comfortable with any of today's 'intelligent design' theorists and long-age creationists. In a pinch, he would have been able to cope with some form of theistic evolution, it seems, so long as Adam's soul remained divinely created.... And of course, it is well-known that in the witness box, the wily Darrow showed up the inconsistencies in Bryan's acceptance of millions of years in the face of the Bible's clear statements on six days. Not to mention that Bryan, not having a clear stand or understanding on the historicity of Genesis, had no coherent response to the question of Cain's wife, either. The message this gave people was quite clear—if even this great 'champion' stumbled in the face of 'science,' Christians had no answers, and the Bible could not be trusted."[3]

It is almost impossible to consider the Scopes trial dispassionately as it is hard to separate myth from history. Many see it as the precursor of the recent YEC debates and education bills. In one sense it is, but there are vital differences. In 1925 the contentious issue was the teaching of evolution, but not geology. Today the intention is to reject both evolution and geology. Perhaps the real victor of the Scopes trial was George McCready Price, who is the grandfather of modern creationism, to whom we now turn.

GEORGE MCCREADY PRICE

George McCready Price (1870–1963) is one of the few people, who are more important in retrospect than in their lifetimes. Price was ridiculed by scientists, never more than partially accepted by Fundamentalists, and respected only within limited circles. His importance is that Henry Morris drew upon his writings for his Young Earth Geology in *The Genesis Flood*. Price was born in New Brunswick in 1870 and after he lost his father at twelve, his mother joined the Seventh Day Adventist church. There Price was introduced to the writings of Ellen White. At seventeen Price married a woman twelve years his senior and then sold White's works in the Maritime Provinces of Canada. From 1891 he spent two years at the Adventist Battle Creek College in Michigan and did a one-year teacher-training course in New Brunswick where he took elementary courses in science—his only science education.

He went into teaching and devoured evolution books concluding that "all turned on its view of geology, and that if geology were true, the rest would seem to be more or less reasonable." He almost accepted evolution but could not harmonize it with Genesis as interpreted by White. Finally he accepted White's revelations in *Patriarchs and Prophets* and concluded that the deluge had buried the strata and attendant fossils. After reading many geological tomes he claimed to find a flaw in "evolutionary geology." That fatal flaw in geology was that, Price claimed, geologists dated rocks by their fossils and the fossils by their position in the geological column, and is thus a circular argument. He first argued this in *Outlines of Modern Christianity* (Price, 1902) and developed the theme over the next half century. Like Wise and Austin a century later, he claimed that water eroded out the Grand Canyon before the sediments had completely hardened, thus allowing its rapid formation.

His second book *Illogical Geology* (Price, 1906) gave an alternative explanation for overthrusting, whereby "older" strata are thrust over "younger" strata. He cited Cambrian strata lying on top of Cretaceous strata in Alberta, and soon after publishing his book discovered the Lewis overthrust in Montana, which covered several thousand square miles. He then pronounced a new geological principle, the *Law of Conformable Stratigraphical Sequence* "Any kind of fossiliferous rock may occur conformably on any kind of fossiliferous rock, old or young." The conclusion of the discovery was clear to Price and that is that the deposition of the strata was catastrophic rather than Uniformitarian and so "flood geology" explained "beautifully every major problem in the supposed conflict between modern science and modern Christianity."

In 1923, he published his magnum opus *The New Geology* (Price, 1923), a massive 736-page volume, which appears like a college geology text from the 20s. It was full of unauthorized geological photographs from respectable sources. It was no wonder that Schuchert described this as a "geological nightmare" as if it were true, then geology would be impossible. His new geology which was so opposed to standard Uniformitarian geology he called the "new catastrophism" to distinguish it from the old catastrophism of Cuvier and Buckland a century earlier.

These books by Price were a new approach to challenging evolution. It was a result of his conviction that geology was 90 percent of evolution, and thus if you could undermine geology, evolution would go. Price was correct as if the geological timescale were wrong then there would be no time for evolution. As a result Price spent little time on biological questions. In the mid-1920s he was at the height of his powers and *Science* regarded him as "the principle scientific authority of the Fundamentalists." He also published in periodicals as diverse as *Moody Monthly* and the *Catholic World*.

During the 1930s public interest began to wane and in the 1940s former students like Harold Clark devised their own theories of flood geology. Divisions like this hindered the advance of Deluge Geology, which had to wait another twenty years. Even so for a half a century from 1902, Price published innumerable books on the theme of his Deluge Geology and attempted to form various deluge societies. The greatest recognition that Price received was to be included as a prime pseudoscientist in Martin Gardner's *Fads and Fallacies in the Name of Science* (Gardner, 1957), which described him unprophetically as "the last and greatest of the antievolutionists." In 1950 Lawrence Kulp commented that Price's ideas had "grown and infiltrated the greater portion of fundamental Christianity in America." A few years later Bernard Ramm noted the "staggering" influence of Price and said that Flood Geology was "the backbone of much Fundamentalist thought.' His ghost is still with us.

THE INFLUENCE OF MCCREADY PRICE AND DELUGE GEOLOGY

Price's greatest influence was within Adventist and Missouri Lutheran circles. Bryan had referred to him at the Scopes Trial. Price did not convince many American evangelicals but he made many doubt geological dating. Many fundamentalist writers were adamant that evolution was wrong and led to immorality, even though they incorporated geological ages by adopting either the Gap Theory or the Day Age interpretation. Several theological writers during the interwar years expressed either agreement with or sympathy for Price's deluge geology. Despite his allegiance to geological time, Jennings Bryan valued Price's work and Rimmer tried a pick'n'mix with Price and Scofield.

Price's books were used in some fundamentalist colleges. *Evolutionary Geology* (Price, 1926), a shorter version of *The New Geology*, was used as a geology textbook at Wheaton College in the 1920s and this was mentioned at the Scopes trial. However it was discarded in the 30s when Paul Wright began to teach geology. From that time Wheaton College was clearly old earth. However in the 1960s copies of Price's *Evolutionary Geology* were re-bound (Prof. S. Moshier kindly gave me a copy and explained its history) and used for a different purpose. Rather than to teach students what geology they should believe, they were used as an exercise in critical thinking as the students were encouraged to look for the flaws in Price's work. That stopped by 1970 and an exercise like that would be dynamite today as many students entering Wheaton are YECs, as I found when I taught geology in the Black Hills for Wheaton in 2001.

Support for McCready Price came from surprising quarters, and from my Anglican perspective none is more surprising than W. H. Griffith Thomas (1861–1924). As I discussed in the previous chapter late nineteenth-century evangelical Anglicans unanimously accepted geology and the majority accepted evolution with minor reservations. During the first half of the twentieth century Liberal Catholics came to dominate Anglicanism, with Modernists coming a close second. As a result most Anglicans not only accepted evolution but held Genesis to be mythological as in the report *Doctrine in the Church of England* (1937). There was not one evangelical on that commission which was chaired by William Temple (1881–1944). He was Archbishop of Canterbury from 1942 to 1944. Evangelicals declined in numbers and in England split into Liberal and Conservative Evangelicals. The former were closer to Modernists in many ways and the latter deliberately eschewed the "excesses" of American Fundamentalism. They tended to follow Handley Moule (1841–1920) and, to a lesser extent, J. C. Ryle (1816–1900) whose tradition was carried on in the early decades by Dean Henry Wace of Canterbury, Bishop Edward Knox (1847–1934) of Manchester and Griffith Thomas.

Griffith Thomas rejected evolution through the influence of McCready Price. Thomas was a significant evangelical theological writer of his time and recommended for Anglican evangelical clergy until the 1970s. After being the principal of an Anglican seminary in Oxford, Wycliff Hall, he went to Toronto in 1910. He was a leader in the World Conferences on Christian Fundamentals and moved to join the newly formed Dallas Theological Seminary in 1924, but died suddenly before classes began. In his two widely-read books *The Catholic Faith* and *The Principles of Theology; An Introduction to the Thirty-Nine articles*, Thomas argued for a qualified acceptance of evolution. *The Catholic Faith* was published before Thomas left England, and then revised in 1920. Thomas wrote, "Evolution as a law of nature is undoubtedly true to an extent, and it may yet prove to be a very great extent, . . . at present it can only be called a working hypothesis . . ." *The Principles of Theology* was not completed before his death and was edited by Dyson Hague (1857–1935) and published posthumously in 1930. Here, when speaking about Creation, Thomas was critical about the association of evolution with "the philosophy of materialism" and then wrote that, "time has shown that the Darwinian theory is not necessarily to be identified with the general doctrine of evolution." He concluded after giving similar sentiments as quoted from *The Catholic Faith* that, "If we regard Evolution as modal it is not only not anti-theistic, but in many respects gives a far deeper, richer and fuller conception of the Divine working than the older theories" (Thomas, 1943, pp. 10–11). Thomas had based his arguments on recent writers such as James Orr (see Chapter 4) and other protestant apologists such as McCosh, Stokes, Salmon and Webb. It is difficult to ascertain when Thomas wrote this, but from his references it possibly pre-dates his leaving England.

Then came the change. In 1917 Griffith Thomas met McCready Price at a meeting for fundamentalist leaders in Colorado Springs and adopted his ideas. The next year he wrote a booklet *What About Evolution?* and later an article for *Bibliotheca Sacra* entitled "Evolution and the Supernatural"(Thomas, 1922). In contrast to previous writings Thomas adopted a wavering tone and was skeptical of an evolution of man. At the end he raised several questions about evolution, but his penultimate paragraph makes it clear that he rejected geology. Citing Price (Q.E.D. p. 125f) he rejected the whole geological succession stating, "This is now known to be a mistake." It was "this mistake" which formed the major theme in Price's books and in the writings of Theodore Graebner, who maintained that the use of fossils as age markers is based on circular reasoning from the assumption of evolution, foreshadowing Morris's arguments (Morris and Whitcomb, 1961, pp. 132–135).

Thomas's new views made no impact in Britain, and to my knowledge no other British Anglicans showed the slightest sympathy to Price. In February 1924 Price gave a lecture to the Victoria Institute (VI) in London to a frosty welcome. A few anti-evolutionists like Douglas Dewar cited Price favorably, but that was more in passing. Price's "New Geology" slowly made inroads into American Fundamentalism and created the conditions for the rapid acceptance of Morris's YEC in the 1960s.

Price's New Geology found more acceptance with the Dutch Calvinists and when Valentine Hepp, Bavinck's successor as professor of theology at the Free University, Amsterdam, gave the Stone Lectures at Princeton Seminary in 1930, his lecture on *Calvinism and Geology* (Hepp, 1930, pp. 183–223) acknowledges his debt to Price. Hepp was far more skeptical of geology than Bavinck, who accepted the geological timescale. It is ironical that the previous year Gresham Machen (1881-1937) had left Princeton in protest of its nascent liberalism to set up the conservative Westminster Theological Seminary and that Machen himself accepted geology (Marsden, 1991, pp. 182–201). This vignette highlights the complex nature of American Christianity in the 1920s.

As the older generation of Germanic Lutherans passed on theologians in the twentieth century began to accept Copernicanism but made no concessions to geology or evolution. Wisconsin-born Theodore Graebner (1876–1950), the professor of philosophy at Concordia Seminary at St. Louis, was more tolerant of heliocentrism than his father, but not of geology nor evolution. While at Luther College, Prof. Tinglestad convinced him that evolutionists were "reasoning in a circle" by dating rocks from their fossil content. This he argued in several works beginning with *Evolution; an investigation and a criticism* in 1921 and finally in *God and the Cosmos* in 1932, which was in three parts, *Atheism, Materialism and Evolutionism.*

Graebner was the first of several Lutherans associated with Price's new geology. Another was Byron Nelson (1893–1972), grandfather of Paul Nelson, a contemporary YEC exponent of Intelligent Design (ID). Nelson reveled in Price's *New Geology* in 1923 and wrote a master's thesis at Princeton Theological Seminary of the Genesis "kinds," which was published in 1927 as *After its Kind* (Nelson, 1927), which closely followed Price's geology. This was followed by *The Deluge Story in stone* an account of the history of flood geology.

From Adventist and Lutheran flood geologists along with an assortment of others especially Dudley Whitney (1883–1964) was born the *Religion and Science Association* in 1935. This only lasted a few years as it succumbed to dissent over whether Price's deluge geology was preferable to the orthodox geology, legitimized by the Gap Theory favored by Allan Higley (1871–1955), who taught chemistry and geology at Wheaton and had a Ph.D. in organic chemistry from Chicago. Higley's disagreements with members of the *Religion and Science Association* were nothing compared to those he had at Wheaton, from whence he left in disgrace in 1939. Numbers chronicles several other societies formed in the following years.

It is not easy to assess the influence of Price and *Deluge Geology* on American Fundamentalism in the early twentieth century. In 1915, its influence had not even begun and almost all evangelicals accommodated geology, helped by the Scofield Bible. But after 1918 Deluge Geology entered into Fundamentalism like a leaven. Many considered it, some adopted it and others like Harry Rimmer seemed to pick'n'mix between the Gap Theory and Deluge Geology and passed on their ambivalence to their hearers.

Suffice it to say that in 1915, very few of the nascent fundamentalists adopted deluge geology (today's YEC). By 1930 it was possible to claim that the Scopes trial was about a literal Genesis. As was demonstrated above that is only partly true, and could only gain credence in the 30s if enough Fundamentalists were literalists. That there were is evidenced by the activities of "deluge geologists" and Ramm's claim that Price's geology provided "the backbone of much of Fundamentalist thought about geology, creation, and the flood" (Ramm, 1955, p. 125). Many mid-century Fundamentalists were like Rimmer and held an unstable amalgam of deluge and conventional geology.

In 1910 when the *Fundamentals* were published the vast majority of American evangelicals accepted geological time and accommodated this to Genesis by either the Gap Theory or the Day Age Theory. The small minority who did not, mostly read Genesis without reference to science. Price had unsettled that certainty and through his books created doubt. Though few agreed with Price, many spoke highly of him and respected him as a scientist. Price had not gained many converts but had sown confusion in the evangelical camp and this is nowhere seen better than in Harry Rimmer.

HARRY RIMMER

Harry Rimmer (1890–1952) was the most prominent scientific apologist for fundamentalism in the 30s and 40s. He was a Presbyterian minister and considered himself a scientist. Before he accepted a pastorate in Duluth, Minnesota in 1934 he spent much of his time as an itinerant evangelist and often spoke against evolution. He came from a humble background being born to British immigrants in California. During a grim childhood he was expelled in the third grade and worked from the age of fourteen for railroad, lumber, and mining camps. Despite poor education he was an avid reader. After a short spell in the army he entered Hahnemann Medical College, a homeopathic school but dropped out after conversion.

In 1913 he probably went to San Francisco Bible College for few months and began preaching from his beloved Scofield Reference Bible. About 1920 he became an itinerant speaker for the YMCA and became interested in the relationship of religion and science. Despite no training Rimmer presented himself as a scientist and in March 1921 formed the Research Science Bureau, which by the 1930s had 3,000 members. The primary purpose was to raise funds for Rimmer's ministry and to hold the copyright for his publications. As well as many anti-evolution pamphlets and books he printed sermons on the Virgin Birth, Joshua's Long Day, and Jonah and the Whale.

Rimmer's antievolution pamphlets have been collected and published by Edward Davis. Tracing out their publication history is difficult as they were often later reprinted in another form with little change (Davis, 1995, pp. xxxi–xxxiv). The great strength of Rimmer's writings is their popular style combined with an authoritative tone in discussing science. The weakness is his cavalier attitude to science, as when he states that Piltdown Man was found in a cave in Sussex, whereas it was "found" in recent gravel. Rimmer was implacably opposed to evolution as tracts like *Monkeyshines, Fakes, Fables, Facts concerning evolution* indicate. But before we dismiss him out of hand his tract *Embryology and the Recapitulation Theory* makes telling criticisms of "Ontogeny Recapitulates Phylogeny" as did Stephen Gould forty years later. His geology was inconsistent as though he rejects "evolutionary geology," he did not quite give his support to Price's *New Geology*, which described as "[t]he most remarkable and up-to-date book of Geology extant today" (Davis, 1995, p. 369). Ultimately Rimmer could not decide whether the earth was young or not, nor the length of the Genesis days. He alternated inconsistently between Scofield's Gap Theory and a literal view, drawing on McCready Price. However, that did not affect his position as the leading fundamentalist scientist.

After 1935 Rimmer collected his anti-evolution pamphlets and repackaged them as books—*The Harmony of Science and Scripture* (Rimmer, 1936) and *Modern Science and the Genesis Record* (Rimmer, 1937), but published little after that. As Davis summed up Rimmer, "What distinguished him from most other leading fundamentalists, however was his claim to have scientific expertise, . . . By his example Rimmer offered fundamentalist Christians support in the literal truth of biblical statements about nature, . . ." (Davis, 1995, p. xxi). He convinced many at the time, but before he died his star was fading. In 1947 Van Kampen Press did not republish *The Theory of evolution and the Facts of Science* on the advice of members of the American Scientific Affiliation (ASA). By the 50s he was almost forgotten with the help of the ASA and Bernard Ramm.

THE INTERWAR YEARS IN BRITAIN

British evangelicals went into decline after 1900, and were neither fundamentalist nor anti-evolutionist. Despite the evolutionary agnosticism of H. G. Wells, who co-authored *The Science of Life* in 1929 with Julian Huxley and works like Bertrand Russell's *Religion and Science*, which was a popular rehash of the conflict thesis, British evangelicals had little antagonism to science or to evolution in particular. This may be because British evangelicals "stressed personal piety over doctrinal purity" according to Numbers. Also many evangelical leaders were Oxbridge educated and thus had a broad education and some knowledge of science. Undergraduates at Oxbridge were not isolated from the whole gamut of thought from the Classics, the arts and the sciences as could a student at an American evangelical college. However, Christian scientific apologetics had passed to those of a more liberal bent like William Temple, Charles Raven, and the modernist Bishop Barnes. Much may be because the stressed on a subjective faith along the lines of the Holiness and Keswick Movements rather than a cosmic, or intellectual faith. Questions of science were sidelined, or regarded as long settled.

The VI had lost its Victorian vigor, but continued with meetings and journals. In those years archeology, especially of Mesopotamia, was of great interest. Questions of Genesis One had long been settled but archaeology promised historical support for parts of the Old Testament questioned by radical critics. While in Britain in 1924, McCready Price gave a lecture to the VI, which was published in the *Journal of the Transactions of the Victoria Institute (JTVI)*, entitled *Geology and its Relation to Scripture Revelation* (JTVI, 1924, vol. LVI, pp. 97–123). The response was mixed but Price simply faded from the British scene. In Britain Price could cite only two divines fighting evolution, one Presbyterian and one Wesleyan. Against this I have picked up much anecdotal evidence from British adults who attended Sunday-school classes from the interwar years where youngsters were taught or caught the idea that the Genesis days were twenty-four hours.[4] This was not the deliberate policy of the churches and their ministers, but is probably the result of poorly trained Sunday-school teachers doing what they thought right. At times the teaching was abysmal as when a young Sunday school teacher taught Snow White instead of the Bible!

The inter-war years were a lean time for evangelical writing, and little touched on science. The emphasis was very much on a personal faith with the focus on the death of Christ and the need for personal repentance and faith. If science was mentioned there was a bland acceptance of geology and some kind of evolution, but this was passed over quickly as in Basil Atkinson's commentary on Genesis.

There has been little research into British evangelical scientists of the period, except for Ronald Numbers in his study *The Creationists*, who only considered those who could be termed Creationist (Numbers, 1992, pp. 140–157). Evangelicals were at an all time low, and the paucity can be seen in the membership list for the VI for 1924. Out of 500 about five were scientists of some sort, with a clutch of medics and the rest in the forces or from the clergy.

The most prominent evangelical scientist was Sir Ambrose Fleming (1849-1945) who was professor of electrical engineering at University College, London and invented the thermionic valve in 1904. His achievements in the physics of telecommunications were immense and in his later life became an evangelical apologist, having become president of the VI in 1927. Fleming is claimed to be a Creationist scientist by Henry Morris (Morris, 1982, p. 102) and others but he was not Young Earth, though he insisted on the special creation of the soul and accepted divinely guided development rather than evolution. He did not consider himself an evolutionist as to him as there were "discontinuities in nature" and "all such processes require guidance." He first argued this in an article in JTVI in 1928 and later in several books. These were sufficient to elicit responses from Sir Arthur Keith, who was an ardent supporter of Piltdown Man. Keith stressed that Fleming accepted evolution, but not for man. By 1935 Fleming was more strident in opposing evolution and in response Keith (1935) wrote *Darwinism and its Critics.*

Fleming was the most prominent inter-war anti-evolutionist but not the most strident. That honor must be shared by Bernard Acworth (1885–1963), Douglas Dewar (1875–1957), and Lt. Col. Lewis Merson Davies (1882–1955). Dewar graduated from Cambridge in natural science and entered the Indian Civil Service. While in India, Dewar became an authority on Indian birds and returned to England in 1924 and began to question evolution on moral grounds and his field work on birds. And so a series of anti-evolutionary books followed: *Difficulties of the Evolution* (Dewar, 1931) and *More Difficulties of the Evolution Theory* (Dewar, 1938) which was a reply to Arthur Keith's *Evolution and its Modern Critics* and other works. Dewar became more skeptical both of biological evolution and geological time and was possibly the first to doubt radiometric age-dating.

Merson Davies served with the Indian Army in the Northwest Frontier and there became interested in geology. On returning to Scotland he obtained both a Ph.D. and a D.Sc. from Edinburgh in geology and specialized in foraminifera. However despite his opposition to evolution and religious modernism and his adherence to "a literal Bible" and "a literal second coming," he had no truck with Price's deluge geology though he respected Price. His main work was *The Bible and Modern Science* (Merson Davies, 1934). In 1932 anti-evolutionism took a new turn when Bernard

Acworth and others founded the *Evolution Protest Movement*, with Ambrose Fleming as president. The main protagonists were Dewar and Davies. The EPM did not accept Deluge Geology and opposed evolution from an old earth position and took on such evolutionary worthies as the paleontologist Arthur Morley Davies and the atheist Joseph McCabe, a former Roman Catholic priest. They failed win over C. S. Lewis despite his minor doubts about evolution. After the Second World War the EPM almost faded from public view, until it became YEC and was renamed the *Creation Science Movement* in 1980.

With so few evangelical scientists, only two need mention. The first was a doctor Arthur Rendle Short (1880–1953), who published several books from an evangelical evolutionary perspective in the middle decades. However his son, John Rendle Short, also a medical professor, claims that he developed doubts about evolution toward the end of his life. The other was Robert. E. D. Clark (1906–1984), who gained a Ph.D. in organic chemistry and spent most of his career as senior lecturer in Chemistry at the Cambridgeshire College of Arts and Technology. Clark doubted evolution, but rejected flood geology. Even so he accepted P. J. Wiseman's visionary interpretation of Genesis and thus accepted an ancient earth and even the common ancestry of man and ape. As typical of his era his books were apologetic rather than ethical and had such titles as *Scientific Rationalism and Christian Faith* (Clark, 1945) and *Darwin, before and after* (Clark, 1948). For many years he was active in the VI. Though Clark was influential in his day, he was eclipsed by the later leaders of the Research Scientists Christian Fellowship (RSCF) like Donald Mackay.

JONAH AND THE WHALE AND OTHER FISHY FUNDAMENTALIST TALES

Jonah and the Whale has long been a large fundamentalist red herring! Others are the Long Day of Joshua and the Dial of Ahaz. These are perennial favorites of fundamentalists and are still used as a litmus test to test whether someone believes in total inerrancy. Rimmer dealt with each at length as did many others. They have not lost their popularity as an internet search will show.

Joshua's Long Day refers to the battle between the Israelites and Amorites in Joshua chapter ten, where it says in verse 13, "The sun stayed in the midst of heaven, and did not hasten to go down for a whole day." To take this literally, one must conclude that on that occasion the day was longer than twenty-four hours and thus it has been part of the fundamentalist apologetic to give scientific explanation of it. Harry Rimmer made much of it in *The Harmony of Science and Scripture* and argued that it is common knowledge among astronomers that a full day is missing and that in 1890

Prof. Pickering of Harvard traced it back to the time of Joshua. Recent versions allege that NASA computers support this but even *Answers in Genesis* (or rather *Creation Ministries International* of Australia) regards this as an urban legend!

The most popular is Jonah and the Whale and Rimmer's type of apologetic is still flourishing. Here, rather than considering Jonah as an allegory, the story is taken as history. The focus of the story is not the repentance of the people of Nineveh, but survival in a whale. Fortunately a contemporary story was doing the rounds. An English seaman called Bartley claimed to have been swallowed by a whale off the Falkland Islands in 1891, and after the whale was caught, was released when fellow whalers cut open the stomach and he emerged with rather bleached flesh. In almost every version there are differences and contradictions, which are exposed with great skill and humor by Ted Davis (Davis, 1992). Davis followed up all possible references and found that all leads led to a fog of inconsistency and confusion, resulting in the worst kind of apologetic.

The persistence of these arguments is a result of seeing the Bible as totally inerrant both in what it says about God but also about history and cosmology. This is compounded by a desire to interpret the Bible literally without interpretation.

THE FORMATION OF THE AMERICAN SCIENTIFIC AFFILIATION AND NEW ENGAGEMENT WITH SCIENCE

The foremost American evangelical scientific organization, the American Scientific Affiliation (ASA), celebrated its diamond jubilee in 2002. It has long been in the exposed middle ground as it is condemned by YECs for being too "liberal." David Hart is right to speak of *The Fundamentalist Origins of the ASA* (Hart, 1991). First, the principle founders Moon, Houghton and Everest all came from a prewar fundamentalist stable in that they were not attached to mainstream churches and were largely dispensationalist and inerrantist. They were hardly liberal, mostly "gappists" with a nod to six-day creationism, anti-evolution to the core yet had an openness, which soon landed the ASA in controversy and within two decades schism with Morris and the YECs. They were "escaping from fundamentalism" but would not have admitted it. Of the three Will Houghton (1887–1947) was president of Moody Bible Institute, Irwin Moon (1907–1986) "a self-taught amateur scientist" and a former pastor, who had left in 1937 for his itinerant ministry on "Sermons form Science." Moon rejected a chance to study physics at Yale to go to the Moody Bible Institute. Later, through him, Moody Institute produced its "Fact and Faith" Science Films. F Alton Everest (1909–2005) was a professor of engineering at Oregon State University, who had done early development work on television.

The birth of the ASA could not have been less promising. Like Jennings Bryan, Houghton and Moon were worried about the moral and religious implications of a science without God. Despite that they enlisted the help of evangelical scientists and the nascent ASA leadership decided not to make anti-evolution its modus-vivendi. As a result the ASA after 1942 was open to all science, resulting in opposition from groups like the Deluge Society and, from 1946, Henry Morris. This openness resulted in friction from the beginning with the YECs feeling that all they opposed in "evolution" was being accepted and proclaimed by fellow evangelicals. Thus the formation of the Creation Research Society in 1963 was not simply the opposition of the ASA to *The Genesis Flood* but a culmination of problems going back to 1942.

The early ASA did not have a sophisticated philosophy of science or theology as they claimed allegiance to an "unerring" Bible and "the real facts of science." The very expression "the real facts of science" gives philosophers of science apoplexy, as it reflects the naïve view that science discovers "facts" and ignores theory. Soon the ASA were quietly opposing the republication of Rimmer's *Theory of Evolution and the Facts of Science* in 1947. At about this time Russell Mixter (1906–2007), Professor of Biology at Wheaton College, and the up-coming geochemist J. Laurence Kulp (1921–2006), a Wheaton graduate, were recruited to the ASA and with the philosopher-theologian Bernard Ramm gave much direction to the ASA for several decades.

Kulp, who had Ph.Ds in both chemistry and geology, took on Flood Geology in the late 40s, after several founder members of the ASA, including Dudley Whitney (1883–1964) attempted to align the ASA with Flood Geology. Peter Stoner (1888–1980) would have nothing to do with it and in a letter to Everest likened Flood Geology to a mathematician suggesting that $2 \times 3 = 7$. (Stoner wrote several books on scientific apologetics during the following decades.) In 1947 a group met in Stoner's house to discuss "The age of the Earth by Radioactive Methods." This was published as *A Symposium on "The Age of the Earth,"* which included papers by Ramm and Kulp, and a contrary paper by Walter Lammerts, one of the future founders of the Creation Research Society, which questioned radiometric age dating.

Everest encouraged Kulp to challenge the legacy of Price. Kulp obliged in 1948 with a paper on the "Antiquity of Hominoid Fossils" arguing that such creatures had walked this earth for hundreds of thousands of years. He was opposed by Edwin Monsma, biology professor at Calvin College, who objected to Kulp's Uniformitarianism and that "wholesale death and destruction took place before the fall." After that Kulp savaged Flood Geology in a paper given in 1949 entitled "Flood Geology," to the chagrin of Henry Morris. The disagreement went beyond the ASA and in the 50s

InterVarsity's HIS magazine became a forum to debate Flood Geology. Everest's hopes to silence Flood Geology failed and the publication of *The Genesis Flood* in 1961 gave it new life.

Meanwhile Russell Mixter was leading the way in a new "creationist interpretation" of evolution, with the ASA's second monograph *Creation and Evolution* in 1950. Mixter rejected a worldwide flood and thought it likely that organisms had evolved "within the order," thus not going as far as a total evolution. He held to a Progressive Creationism similar to that held in the decades before Darwin. The anti-evolutionist contingent was unhappy and the younger more progressive ASA members were losing patience with sops to anti-evolutionists. Kalthoof argues that the watershed on evolution was at the annual convention of 1957, when Mixter stated that he felt that gaps in the evolutionary record were being closed and that he could no longer emphasize those gaps. That could not have been anticipated when the ASA was founded fifteen years previously. Two years later the collection of essays edited by Russell Mixter *Evolution and Christian Thought Today* (Mixter, 1959) was published. The authors were competent scientists and their general tenor was accepting of evolution. However the essay by the theologian Carl Henry "Theology and Evolution" was skeptical of both the validity of evolution and whether it could be reconciled with evangelicalism. Then, as is often the case today, American evangelical scientists disagree with their theologians!

The most important book to be published from the ASA stable was written by a theologian. That was Bernard Ramm's (1916–1992) *The Christian View of Science and Scripture* (Ramm, 1955). Since the mid-forties Ramm had been involved with the ASA and in 1946 began to teach a course on science and the Bible at BIOLA, with one of Rimmer's works as the text. Ramm soon dropped it and devised his own course, which he turned into a book. This book has had a considerable twofold influence. First it has been almost programmatic for members of the ASA (and their British counterparts) in directing an approach more like that of evangelical scientists in the nineteenth century rather than the Rimmers and Prices. Secondly, it moved a young Old Testament scholar, John Whitcomb, to anger, who in collaboration with Henry Morris wrote *The Genesis Flood* (Morris and Whitcomb, 1961) triggering off the modern YEC movement.

The first twenty years of the ASA had resulted in a new openness to all science by evangelicals and unwittingly created the environment for a new and vigorous form of YEC to take root. This was totally different to the evangelical engagement with science in Britain in the same period to which we now turn, which had few concerns about flood geology and evolution.

SCIENCE AND THE POSTWAR EVANGELICAL RENAISSANCE IN BRITAIN

The first flickering of a greater interest in science by evangelicals occurred place during the Second World War, through the instigation of Oliver Barclay (1919–). In 1942 Barclay, who had just completed a Ph.D. in biology at Cambridge, was appointed to the Inter Varsity Fellowship (IVF), the evangelical group working among students. Over the next 54 years he had an immense influence in the IVF, the nascent RSCF and the later *Christians in Science*. As well as his work in the IVF he encouraged many to develop an interest in science and religion, including myself. In 1944 Barclay wrote an article on Evolution and Christianity, in which he made a distinction between evolution as a scientific theory, which he accepted, and evolution as a world-view, which he did not. Barclay opposed YEC when it was introduced to Britain in 1968, but he was not temperamentally suited to fighting fire with fire. What turned out to be the birth of the RSCF was a weekend conference with twelve participants in 1944. Most were Ph.D. students but included R. E. D. Clarke. The sixteen-page report of the conference cost 7d.

Initially the group had no name and was often referred to as "Oliver Barclay's Group"! Soon it was named—Research Scientists Christian Fellowship and through Barclay was affiliated to the Inter-Varsity Fellowship (IVF). Since 1944 there has been an annual conference for every year except 1946, which covered almost every possible religion and science theme from evolution (of course), moral responsibilities of the scientist, the environment, bioethics, indeterminacy, and so on (http://www.cis.org.uk/conference/past.htm). The caliber of papers presented varied as lecturers have been a mixture of experts, young scientists, and the nondescript.

The RSCF broadened its membership quite rapidly and drew in the doctor-turned-pastor Dr. Martin Lloyd-Jones. The main participants and leaders were Clark and Barclay, the latter using his position in the IVF to ensure its growth. In 1948, Prof. Reijer Hooykaas of the Free University, Amsterdam, was invited through Lloyd-Jones. Hooykaas was a noted historian of science, who directly or indirectly inspired many younger historians of science like Martin Rudwick.

Initially members of the RSCF published very little, but very little was published on science and religion in the 40s and 50s. Almost the only authors were Clark, Coulson, Mascall, Smethurst, Raven, and Yarnold in Britain, and Pollard in the United States. Only Clark was an evangelical. From the late 50s a growing number of books were produced by evangelical and non-evangelical scientists. Notable among evangelicals were books by Hooykaas (including the excellent *Natural Law and Divine*

Miracle (Hooykaas, 1958) on the Principle of Uniformity), Clarke and Donald Mackay. When the IVF published the *Christian Graduate* from 1947, it contained a section for the RSCF. Due to restraints of space, the RSCF produced an annual newsletter of sixty-four pages from 1983 to 1988, until after being renamed *Christians in Science* in 1988, they joined forces with the VI to publish the journal *Science and Christian Belief*, which has a high standard.

The RSCF attracted many evangelicals, some of whom left as they could not subscribe to the IVF basis of faith. Theologically it was conservative evangelical of the English variety. Even so, they had strong links with their American counterparts in the ASA, and almost used Ramm as their textbook. To the end of the 60s evolution was unquestioned, so much so that the young Clark Pinnock expressed his concern to Francis Schaeffer when he visited L'Abri in Switzerland after attending a joint RSCF–ASA conference in Britain 1965. During the 60s and 70s Oliver Barclay continued as the effective leader of the RSCF though he published little. It is to his credit that he brought together a number of scientists like Donald Mackay, Robert Boyd and Malcolm Jeeves. Barclay's gift was in organizing and inspiring.

The publication of *The Genesis Flood* in 1968 began to shatter the peace, despite a devastating revue in *The Christian Graduate* by A. N. Triton, the nom-de-plume of Barclay. The rapid effect can be illustrated by my own experience. I was at Oxford University studying geology from 1965 to 1968 and was in and out of the Oxford Inter Collegiate Christian Union (OICCU), part of the IVF, which was founded by my college chaplain, F. W. Dillistone, in his evangelical youth in the 1920s. By the 60s he was a respected liberal theologian! A combination of Dillistone and OICCU started me on my Christian pilgrimage and despite the strong evangelicalism of OICCU, not one ardent evangelical told me that I should not accept evolution or geology! From 1968 to 1971 I worked as a geologist in Africa and when I met old friends on my return many had become YEC. Those three years marked the crucial change in Britain.

POISED FOR SUCCESS OR . . .

If we could go back in time to 1959, the Centenary of *The Origin of Species*, and considered the prospects for an evangelical understanding of science, we would probably conclude that the interwar years of Fundamentalism and the Scopes trial were past history and that Evangelicals would largely return to the openness to science of their mid -nineteenth century forbears like Gray, Dana, Warfield, and even Charles Hodge. In the United States, we could look to the transformation of the ASA, and the influence of two books, one by Ramm and the other edited by Russell Mixter. In Britain we would note the evangelical renaissance headed by Stott and Lloyd-Jones and the work of the RSCF and their leaders like, Barclay, MacKay and Boyd and the bevy of young evangelical scientists including Rudwick and Polkinghorne, who, though they drifted away from a narrower evangelicalism were so significant in the history of geology (and its relationship to Christianity) and the revival of science and religion.

In 1960, U.S. evangelicals could have been expected to follow the direction of Ramm and much of ASA, with a slow acceptance of evolution. However that was not to be. The flood was about to break and like Noah's contemporaries they were unaware that they were about to be inundated, first slowly and then catastrophically, by a Deluge of biblical proportions.

A later quotation of Billy Graham sums up the general perception of evangelicals in 1960:

> I don't think that there's any conflict at all between science today and the Scriptures. I think that we have misinterpreted the Scriptures many times and we've tried to make the Scriptures say things they weren't meant to say, I think that we have made a mistake by thinking the Bible is a scientific book. The Bible is not a book of science. The Bible is a book of Redemption, and of course I accept the Creation story. I believe that God did create the universe. I believe that God created man, and whether it came by an evolutionary process and at a certain point He took this person or being and made him a living soul or not, does not change the fact that God did create man. . . . whichever way God did it makes no difference as to what man is and man's relationship to God. (Frost, 1997, pp. 72–74)

NOTES

1. This section is very largely based on Edward Larson's *Summer of the Gods*.
2. Personal communication with E. Davis, January 20, 2004.
3. http://www.answersingenesis.org/docs/3986.asp.
4. Not always so, at an Anglican Sunday School (St. Nathanael's Walton) in Liverpool in the 1930s children were taught the story of Snow White in preference to the Bible! My source, Edna Hull, who was sent to a Baptist Church, later became a churchwarden of that church.

Chapter 7

The Rise of Creationism: Young Earth Creationism and Intelligent Design, 1961–2007

In recent decades "Creationism" has caused immense controversy in America and throughout world. Many people, when they consider evangelicals and science, immediately think of creationism. My purpose in this chapter is to put it in context of the history of Evangelicalism and consider its attraction today. My historical account is very brief as Ron Numbers (Numbers, 1992, 2006) gave it in detail and the scientific aspects are expounded by Eugenie Scott (Scott, 2004). Most states in the United States have had attempts to mandate the teaching of *creationism* or *design*. The most significant case was in Arkansas in 1981 followed by Kansas in 1999 and 2005 and Dover, Pennsylvania, in 2005. None have had more than temporary success. Two presidents have indicated some support and polls indicate that nearly half of Americans believe in creationism. Often *creationism* is seen as purely American, but it is now worldwide. *Creationism versus evolution* attracts extremes of hostility or loyalty and since Henry Morris launched the movement with the book *The Genesis Flood* in 1961, it has produced more heat than light. "Scientists" cannot understand how *creationists* can be so stupid and many evangelicals see *creationism* as defending their faith in a hostile world. Good manners are at a premium and there is often little attempt to *empathize* with what *creationists* believe and, more importantly, why. This letter puts it in the perspective of an evangelical parent, who sent this e-mail in 2004 to "Ask a geologist."[1]

Dear Christian Geologist:

As a Bible-believer, I accept God's own narrative of how the world started, but my 5th grade boy is being subjected to ridicule for trying to hold to the worldwide

> Flood that killed them [dinosaurs] and every other living animal less than 5,000 years ago.
>
> They are being taught that these animals existed up to about 95,000,000 years ago, as "proved" by fossil evidence and radiocarbon dating! Clearly not a Biblical teaching. They say that any flood was local in nature and there is no "scientific support" for a worldwide flood as seen in fossils and geology.
>
> I'm sure true science and God's Word can't be at odds. Is there a single book you could recommend that gives the Bible Christian an answer to these and other questions of science or pseudo-science? I can't afford to home-school my kids but am terribly worried about the erosion of their faith in such a godless atmosphere where the most bizarre non-Biblical ideas are accepted as true!
>
> Thanks very much.
>
> In the Lord,

A hostile response guarantees its survival as no attempt is made to understand its appeal. The initial reaction of many when they first come across Young Earth Creationism (YEC) is to deride it as it runs counter to all accepted ideas of geology and evolution. As a result flame wars occur. So we need to understand its beguiling attraction.

I use the long-winded Young Earth Creationism deliberately, although YECs prefer to call themselves as *Creationist*. My reasons are twofold. First, prevents the use of *Creationism* to describe forms of creationism held before Darwin. Secondly, it implies that Christians who are not YEC are not creationists. That is a good debating tactic, as it excludes the middle ground and so non-YECs can be seen as a *compromisers* (Sarfati, 2004). The word "creationist" for YEC dates back to the 1960s and was rarely used in that context before that (Numbers, 1998, chap. 2). In 1859 a "creationist" was usually someone who believed in the special creation of the soul. Darwin used the undefined "ordinary view of creation" to discredit his opponents. As few held to a 6,000-year-old earth then, Darwin was referring to nineteenth century progressive creationism. The term was adopted by McCready Price and slowly took hold. Even so Mixter used creationism to describe his own mildly evolutionary views, in that he and others were creationists and "in this sense creationists can be called evolutionists." In 1954 Ramm used the term "Progressive Creationist" of himself. However the result was that YECs, and they alone, came to be called "creationists" with the resultant polarization.

WHAT IS YOUNG EARTH CREATIONISM?

It is best to start with a general summary of YEC beliefs, though YEC is not monolithic.

The earth and universe are no more than 10,000 years old and this is supported by the best modern science.
Most of the fossiliferous strata from the Cambrian (550 my) to the Pleistocene (10,000 years) were laid down in the Noachian Deluge.
Evolution from the primordial sludge (goo) to humans (you) did not happen and contradicts true science.
During the Creation Week, God originally created "kinds," for example, horse kind, which have micro-evolved into related species.
Standard "evolutionary–uniformitarian" geology, biological evolution and cosmology are flawed and based on false assumptions.
"Evolution" and "evolutionary geology" are based on atheistic assumptions stemming from the Enlightenment, including an insistence on randomness and chance, which excludes the possibility of a Creator God. Geology, with its long ages, is based on the assumption of evolution.

When the Bible is read correctly, without atheistic and Enlightenment presuppositions, the Book of Genesis only makes sense when read literally with a Creation in six solar days, a Fall resulting in the introduction of pain and death to the animal world, and that there was a world-wide deluge lasting a year during the lifetime of Noah. Only Noah, his family and two of every "kind" survived the flood. (This is alleged to be the traditional view of Christians.) There are many minor tenets. A few YECs are also geocentrists, for example Bouw and Bowden. For further details consult almost any YEC Web site.

THE APPEAL OF YOUNG EARTH CREATIONISM

The appeal of YEC cannot be understood without grasping the deeply felt reasons for believing what many scientists think nonsense. YEC provides the "scientific" capping to a "biblical world view," which provides an all-embracing outlook on life and integrates every aspect of life. It enables one to oppose non-Christian world views and to be confident in the "Culture Wars." Recently many evangelicals have stressed the uniqueness of the biblical world view against the secular world view, which may look to science for its justification. This is expounded, for example by John MacArthur of *The Master's College* in California in *Think Biblically* (MacArthur, 2003) and on the AIG Web site.

The reasons for accepting YEC are interrelated and are threefold being theological, moral and anti-reductionist. These predicate the scientific objections to "evolution" and are more than adherence to a literal Bible.

The most important reason for accepting YEC is a concern for salvation through Christ. The heart of evangelical faith is redemption through the death of Christ, expressed as Substitutionary Atonement in that Jesus' death on the cross forgives sin and takes away the penalty of death. This goes back to St. Anselm of Canterbury in the eleventh century. Since the late eighteenth century, the corollary of an ancient earth was that animals were living and dying long before humans, thus most evangelicals have limited the "death" brought about by the Fall to humans, and regard animal death as of no consequence to the atonement. However some evangelicals in the early nineteenth century, for example George Bugg, and YECs today argue otherwise and that physical death came in at the Fall (Genesis 3), and that the Fall resulted in a Curse over all creation, and before that no animal died or suffered. If T. Rex had actually attacked and killed herbivores 100 million years ago, then the whole Christian Faith will collapse like dominoes, hence the geological timescale *must* be false. This is at the heart of YEC arguments as expounded by Sarfati (2004, pp. 195–224), and Whitcomb in the appendix to *The Genesis Flood* (Morris and Whitcomb, 1961). Carefully presented (with evangelistic overtones) this is crucial.

The authority of the Bible is central to evangelicals, who often interpret it in its plain or literal sense. For early Genesis, that means creation in six days and a worldwide flood. A Young Earth model supports this "scientifically," so YEC is the *only* valid interpretation. A further "biblical" appeal is the Sabbath as the day of rest. The Fourth Commandment reads, "Remember the Sabbath day, to keep it holy. Six days you shall labor, and do all your work. . . . for in six days the LORD made heaven and earth, the sea, and all that is in them, and rested the seventh day" (Exodus 20 vs 8–11). Hence the Sabbath is dependent on a six-day creation and thus "billions" of years must be wrong. To reject this is to reject the Commandments.

Hence the biblical arguments for YEC are threefold, first, upholding the plain sense of scripture, which is inerrant in history and science, second, buttressing salvation through Christ's death, dependent on no death before the Fall, and third, defending the Commandments. As these are essential to evangelical belief then a Christian *must* be YEC. The appeal is irresistible. Well, almost.

We have already seen how moral concerns motivated the anti-evolutionists of the 1920s. YEC has amplified this position and stress that evolution leads to immorality of every kind. In his book *The Genesis Solution* (Ham and Taylor 1988, p. 97). Ham argues that evolution leads to a decrease in marriage, an increase of suicides, euthanasia, pornography, abortions, promiscuity, sexual abuse, homosexuality, theft, violence, racism, etc. Hence evolution is contrary to family values. The concern to counter teaching evolution partly stems from this.

A further appeal of YEC is the opposition to *Reductionism*, or *Nothing-buttery* as Donald MacKay called it. This is the view that everything is *nothing but* physics and chemistry and that there is nothing distinct about humans. Reductionism often stems from a scientific materialist philosophy. Opposition to reductionism is by no means restricted to YECs. Many oppose reductionism. Arthur Peacocke, the British biochemist and clergyman, opposed reductionism for decades from a liberal theology and founded the *Society of Ordained Scientists* in 1986 to facilitate this. John Polkinghorne, Donald MacKay, and many evangelical members of the CIS and ASA also oppose Reductionism. However YEC is extreme anti-reductionism.

When these arguments are put before an evangelical audience the appeal of YEC becomes compelling. Anyone who to challenges them, and "scientific arguments" for YEC are compromising the Gospel. That is why such beliefs are so tenacious. The argument is more over deeply held religious convictions than intellectual ones.

Some question that the motivation of YEC is religious. Charles Israel believes it to be the desire to control the education of children, and argues this especially for Austin Peay, Governor of Tennessee and Jennings Bryan in the 1920s (Israel, 2004). Israel gives several quotes from Peay, Bryan, and others to support this. I would argue that the primary motivation of all Bryan's beliefs stemmed from his faith, which inspired his politics. Recent anti-evolutionism is often bound up with the Religious Right and family values, but one must ask whether the motivation is the control of education or religious belief. All YECs I know of are so because of religious rather than political or educational convictions. It may be hard to understand their outlook if one is not "religious" and thus one may look for a non-religious explanation in line with the secular outlook of Western academic culture. But this often fails to understand their motivation.

One must grasp the religious and moral appeal of YEC in order to understand the movement and how it has developed. The scientific arguments are beyond the wit of most people, but the average evangelical will understand why the blood of Christ washes away his sin, even if he cannot evaluate the arguments for and against the decay of the speed of light. Now for the genesis of YEC.

THE GENESIS FLOOD: THE BOOK THAT STARTED IT ALL

Though there had been antecedents (see Chapters 4, 5, and 6) the most significant event was the publication of *The Genesis Flood* by Henry Morris and John Whitcomb in 1961. Morris was a hydraulic engineer, who was head of civil engineering at Virginia Polytechnic Institute from 1957, and wrote several engineering textbooks. In 1970 he joined Tim LaHaye at San Diego to become co-founder of Christian Heritage College. (LaHaye is a well-known evangelist and the author of the best-selling *Left Behind* series of novels based on his Dispensationalist interpretation of the Rapture.) From 1970 to 1995 he was President of the Institute for Creation Research.

Since the 1940s he had been advocating a recent creation and worldwide flood and his theology was strongly inerrantist and dispensationalist. John C. Whitcomb was Old Testament professor at Grace Theological Seminary in Indiana. *The Genesis Flood* is without question the most important Creationist work ever written and its influence on evangelical thinking in science cannot be over-estimated.

THE ORIGINS OF *THE GENESIS FLOOD*

Its genesis was due to Morris meeting Whitcomb at an ASA meeting in 1953 and sharing his concern over Ramm's book *The Christian View of Science and Scripture*. They began to collaborate in 1957, but soon Morris's chapters outnumbered Whitcomb's. Whitcomb provided the theology and Morris the "science," which was a largely unacknowledged adaptation of McCready Price's Flood Geology. Morris argued that all sedimentary strata were deposited during the Flood, the Geologic Column was based on a circular argument, radiometric dating was flawed, etc. Neither his criticisms of "evolutionary science" nor his alternative science models have gained acceptance by mainstream scientists. *The Genesis Flood* has formed the basis of all subsequent YEC thinking. The book had a checkered prepublication history but finally Presbyterian and Reformed published it in February 1961, due to the recommendation of Rousas Rushdoony, which was unusual for a book by a dispensationalist Baptist. Most criticisms have centered on scientific aspects and often Whitcomb and Morris have been accused of misquotation. Theological criticisms from an evangelical theology are less common because of Morris and Whitcomb's commitment to Evangelicalism.

RECEPTION

The Genesis Flood was ignored by the secular world for two decades. *Christianity Today* selected it as one its "Choice Evangelical Books of 1961," despite a negative review by Wheaton geologist, Donald Boardman. The ASA procrastinated for three years, possibly to avoid controversy. Finally in 1964 the ASA published two hostile reviews in their journal by Ault and Frank Roberts, only to be rebutted by Whitcomb and Morris. Arguments continued until a devastating review in 1969 by Prof. J. R. van de Fliert (van de Fliert, 1969), the evangelical geology professor at the Free University, Amsterdam. As well as pointing out the geological misunderstandings, he also criticized the authors' method of citation, He wrote, "To lift a certain sentence out of a publication, and to use it for something quite different than the original author meant, is scientifically dishonest." This is not the last time that YECs have been accused of misrepresentation in their citation, as examples are legion in YEC literature. Not all would agree with Van de Fliert, who continued, "I realize that the authors of *The Genesis Flood* did not intend to do this. . . ." None of these criticisms stemmed the YEC tide.

The Evangelical Press published *The Genesis Flood* in Britain in 1968. A. N. Triton (a.k.a. O. R. Barclay) courteously damned the book in the *Christian Graduate* in 1969 for the RSCF. As an aside, I read this review in a caravan in the Namib Desert while carrying out geological exploration for a mining company. I could not take the book seriously, but a year later had to change my tune when I was studying under Francis Schaeffer at L'Abri in Switzerland. When I began to read it, I had a horrible suspicion that it might be right.

EARLY DEVELOPMENTS OF YOUNG EARTH CREATIONISM, 1961–1981

The development of YEC up to 2005 has been chronicled by Numbers (1991, 2006) and by Morris (1984). YEC has grown rapidly in America and beyond, including in former Communist lands. It is the dominant opinion of evangelicals in South America, Africa, and India, if only by default. YEC grew linearly in the 1960s, and thereafter exponentially.

Whereas before the war YEC societies sprang up and withered away, many of the societies following *The Genesis Flood* have survived and new ones are still springing up. The first was the *Creation Research Society* in 1963. To ensure credibility to be a member one had to have an M.Sc. degree or equivalent. The society published the *Creation Research Society Quarterly*, the flagship YEC journal and worth consulting to understand YEC science. Creationist societies now exist in every continent. Most are in the United States and a Web search will indicate the plethora.

At the same time the YEC publicity machine was cranking up to full speed. YECs organized conferences and speaking engagements. Some, especially Whitcomb, sought to persuade evangelical colleges to rewrite their statements of faith to include YEC views. Before this time most colleges allowed staff to accept an old earth, but not evolution, which was part of the prewar fundamentalist inheritance. Several creation research institutes were founded in the seventies. Foremost was the *Institute of Creation Research* (ICR) in San Diego. In 1970 Morris helped Tim LaHaye (b. 1926) found a Bible college and creation research center at San Diego. Initially they partnered with Nell and Kelly Segraves, but in 1972 the Segraves took their Creation Science Research Center across town. After this Morris started the Institute of Creation Research and hired Duane Gish, a biochemist, Thomas Barnes, and Harold Slusher, both physicists with honorary doctorates of dubious worth. The intent was to focus on research, writing, and speaking, but not political action. In 1979 Steven Austin, a geologist, joined the staff and from there it grew.

Though Morris wished to major on education and research, he also wished to challenge the teaching of evolution. Their strategy was the *Two Model Approach,* in which they argued that students should be taught the two models of Creationism and Evolutionism, first discussed in the September 1971 issue of the *Creation Research Quarterly.* They argued that there are only two alternatives; Creation by God in six days, and Godless Evolution by chance and mutation. The strength of the two-model approach was that it excluded the middle ground, which most Christians occupied, which was a good debating tactic as it lured evangelicals into YEC. This challenge to science education was given political clout, first by Wendell Bird (b. 1954). Bird studied law at Yale and from there joined Morris at ICR for two years where he gave the Two Model approach for equal time a sounder legal basis. Within a few years it was adopted by the Arkansas and Louisiana statutes, and considered elsewhere. The two-model approach failed and another tactic had to be adopted.

In early 1981 the ICR announced graduate degrees in various creation sciences. Initially they offered M.S. courses in biology, geology, geophysics, and science education. Several geophysics graduates went into the oil industry and gradually unlearnt all they had been taught. The history of the ICR is one of changing personnel and conflict. In the late 80s the school was being investigated by the California State Department of Education and in 1990 had its license to teach removed only to have it returned in 1992. A further problem was to obtain accreditation as standard accreditation was denied to them. An alternative system was devised and ICR is now accredited by TRACS, which supports YEC and gave ICR another five years accreditation in January 2005.

CONSOLIDATION IN THE 80s

The 1980s were the most significant decade for YEC. In 1980, the profile of YEC was raised by attempts to introduce laws on the teaching of creationism in several states. The action in Arkansas in 1981 made YEC well-known throughout the western world and goaded the scientific community into action. In early 1981 Senator James Holsted introduced an act in the Arkansas Senate arguing for the teaching of both "evolution" and "creationism," following the two-model approach. Governor White, who had earlier ousted Bill Clinton, signed it into law on March 19, 1981. A re-run of the Scopes Trial was inevitable. Two months later the ACLU filed a lawsuit and the trial was held from 7 to 17 December.

At the trial both sides produced expert witnesses. Most significant for "evolution" was the philosopher of science Michael Ruse, who convinced

Judge Overton that "Creation Science" failed to show "the essential characteristics of science." Witnesses included the scientists Stephen Gould and Brent Dalrymple and clergy from several denominations including Southern Baptists. "Creationist" witnesses included the cosmologist Chandra Wickramasinghe, who stated that no scientist could accept that the age of the universe was less than a million years, and R. V. Gentry of polonium halo fame. The main question turned on whether "Creation Science" was science. Judge Overton in his thirty-eight-page ruling on January 5, 1982, declaimed that creation science was not science but religious doctrine and so the bill was overturned.

This seemed to be a resounding defeat for YEC, but even as Judge Overton gave his verdict, doubts were raised. The philosopher Larry Laudan (b. 1941) charged Ruse with failing to recognize the disagreements between scientists over scientific boundaries. Neither the euphoria nor the despondency lasted. The trial also drew in more evangelicals in the United States and abroad. Perhaps they took heart from Ronald Reagan, who during the 1980 election said that, "if evolution is taught in public schools, creation should also be taught." It may be significant that the atheistic evolutionism of Richard Dawkins and others gained momentum in the early 80s. In an invert way they also adopted a two-model approach. The Arkansas judgment proved a minor setback. During 1982 the commitment of YECs was unabated. State after state in the United States had statutes put forward for the teaching of equal time for "Creationism" and "Evolution" but these were never passed. Even so the result was frequently close.

ICR continued to grow in the 80s, though Slusher left abruptly at that time. Ken Ham, an Australian science teacher, began his thirteen-year association with ICR in 1981. In 1984 he returned on a speaking tour and then in 1986 he came to work for *Films for Christ* and in 1987 with Paul Taylor of *Films for Christ* produced *The Genesis Solution.*

GROWTH IN THE 90s

The most significant development of the nineties was the formation of *Answers in Genesis—USA* (AIG) in 1993. The founder was Ken Ham, who before he left his native Australia was involved first with the Creation Research, working with John MacKay until they fell out, and then with the original AIG in Australia. Thus at the end of 1993 Ham left ICR amicably to set up AIG-USA and in early 1994 headquarters were set up in Florence, KY. Their work rapidly expanded and they shared resources with ICR. AIG is a more popular and aggressive version of ICR, and has a higher profile. They carry out a varied ministry with lecture tours and publications. AIG takes no prisoners and many of their publications and Web articles make personal attacks on others; Ron Numbers is described as an apostate, Davis

Young as "supposedly evangelical" and all evangelicals who espouse the vast age of the earth or universe, whether an anti-evolutionist like Hugh Ross, or typically ASA/CIS evolutionists, are condemned as "compromisers." This is developed at length in Sarfati's book *Refuting Compromise* (Sarfati, 2004), though since then Safarti left AIG in acrimony.

AIG now has branches in many countries including New Zealand, South Africa, and Canada, with a British group based in Leicester. Elsewhere I have discussed the history of science promulgated by Terry Mortenson, who obtained a Ph.D. at the University of Coventry in 1996. The British director is Monty White, with a Ph.D. in chemistry, who after twenty years in university administration joined AIG-UK in 2000.

AIG has a vast, well-produced Web site. Much is well-written, well-argued, and superficially plausible. Many of the articles have been reproduced from their journals *Creation* and *Technical Journal* (TJ), which have now reverted to *Creation Ministries International* with much acrimony. The articles cover the whole range of subjects of interest to a YEC and are excellent in giving a contemporary picture of YEC arguments. Lecturers from AIG have a heavy itinerary and give lectures in the United States, Britain, and the Antipodes, as well as Creationist conferences. They seek to influence both the churches and evangelical seminaries, and to further this, list those colleges and seminaries, which either teach or fail to teach "creation."

This detail on AIG may seem excessive, but at present they are the most significant and effective YEC group, with a worldwide influence. They also work closely with ICR. As I write today the YEC movement is so large and has a presence in almost every country in the world, so that it is not possible to give a brief survey. In North America there are vast numbers of YEC organizations, not all of which have the professionalism of AIG and ICR. These are proliferating throughout the world and have a considerable impact. Rather than give a list of organizations it is better to understand a few. In the United States that means ICR and AIG, with AIG and Creation Ministries being the first port of call in much of the world. In Britain the two main groups are Biblical Creation Society and Creation Science Movement. Beyond that the best thing to do is to simply search the Web and see what turns up. The number of organizations and sites and reflect the fact that more and more evangelicals are coming to think that YEC is Christian orthodoxy.

PROSPECT FOR THE NOUGHTIES AND TENS

After Arkansas in 1981 many thought that the demise of YEC was imminent, yet YEC has continued to grow. It has made inroads into mainline denominations like the Anglicans and Methodists. There are several reasons for this and is partly due to the growing proportion of evangelicals within those churches. Within the Church of England, which is not as liberal as its American counterpart, between 5 and 10 percent of the clergy are YEC, compared to almost zero between 1860 and 1970. In March 2005 Sir John Polkinghorne debated John MacKay of *Creation Research* in Liverpool Cathedral. According to reports from *Creation Research,* MacKay won and was thanked by several Anglican clergy. According to Creation research's e-mail newsletter *Evidence News,* this was "A great victory for Christianity on the subject 'Is Evolution Compatible with the Christian Faith?'" (*Evidence News* 31 October 2007). From this report one would not have realized that Polkinghorne has conservative theological views, not far removed from many evangelicals in *Christians in Science.* The Dean of the cathedral personally gave me a different perspective. As the majority of growth among Christians is of evangelicals that means that more will be YEC. There are no signs of any concerted opposition or reaction to YEC from mainstream churches. Within the growing Third World churches the evangelicals are very susceptible to YEC and in 2004 Archbishop Peter Akinola of the Anglican Church of Nigeria was very outspoken against evolution. However at present this is not a dominant theme, possibly because they are more concerned about the gay issue, financial insolvency, or other issues. Trainee pastors in the Anglican churches of East Africa tend to be YEC. There has been no survey of how far YEC is held in third-world churches but the anecdotal information that I have acquired indicates that it is very widely held. Meanwhile in the formerly Anglo-Saxon countries YEC continues to grow, even though major splits occur between YECs as in the AIG, when the Australians hived off to form Christian Ministries International (CMI) in 2005. In May 2007 CMI has filed a lawsuit against AIG and Ken Ham. Jim Lippard has related all the problems that date back to 1987, when John Mackay was ousted from AIG and set up his own organization. It makes sorry reading (Lippard, 2007).

THE WORLDWIDE EXTENT OF YOUNG EARTH CREATIONISM

Very often YEC is portrayed as a dominantly American phenomenon. Of course, it started in the United States. with *The Genesis Flood* and spread first through the English-speaking world and is now highly significant in Britain, Canada, and Australia. It is also now the dominant understanding of "Origins" among Christians in Latin America, Africa, and Asia and has a wide appeal in the old Soviet-bloc nations and among the growing numbers of evangelicals in Western Europe. It appeals also to some Roman Catholics and Orthodox Christian. Conservative Muslims also hold it. Consequently it is a worldwide phenomenon and not restricted to small-town America.

YEC SCIENTISTS

Early "Creationists," like McCready Price and Rimmer, had minimal scientific education, but today's YECs are well-qualified. In North America, Britain, Australia, and New Zealand there are now many YEC Ph.D.s, supporting the claim that YEC is a new scientific "paradigm." However not all YEC doctorates are properly accredited degrees. Examples include Thomas Barnes, Harold Slusher, Edward Bliss, Carl Baugh, and John Blanchard, who seem to have acquired their degrees from "diploma mills" (www.talkorigins.org).

YEC scientists often have professional positions in industry and education. Many are engineers or computer scientists. There are far fewer biologists and geologists, who are mostly employed in evangelical or Adventist colleges. The book *In Six Days; Why 50 Scientists Choose to Believe in Creation* (Ashton, 1999) gives a useful snapshot of qualified YEC scientists. Each gives either a personal account or a justification for YEC. There is a good representation of most sciences. None of the geologists work for secular concerns, though Snelling did briefly.

In most English-speaking countries there are YEC scientists employed in universities or research institutions. In Britain there are at least ten full professors in physical science or medicine, but no biologists and geologists. It is, of course, impossible for a YEC to teach conventional geology. The two most well-known geologists are Kurt Wise and Steve Austin, both with Ph.D.s. Wise studied under Stephen Gould and used to teach at the YEC Bryan College at Dayton and Austin, who has a doctorate from Penn State on the deposition of coal in the Pennsylvanian, works for the Institute of Creation Research, as does John Morris, son of Henry, who has a Ph.D. in engineering geology, and the Australian Dr. Andrew Snelling. The reason is simple. Engineering, the physical sciences, medicine, and a high proportion of biology can be studied and practiced without any reference to the vast age of the earth. However geology is impossible without vast ages.

In 2003 when Paul Lauterbur and Peter Mansfield were awarded the Nobel Prize for Medicine for MRI (magnetic resonance imaging), many YECs thought that Dr. Raymond Damadian was denied a share of the prize, because of his YEC views. Damadian had first suggested that MRI could be used to differentiate between healthy and cancerous tissue in an article in *Science* in 1971. In 1972 he filed a patent on MRI scanning and built an MRI scanner making the first image of the skull of a live person in 1977. In their book James Mattson and Merrill Simon (Mattson and Simon, 1996, Chapter 8) stress that his work was fundamental to MRI. The history is not without controversy, especially over Damadian's patents. As three scientists can receive a Nobel Prize, both YEC and secular commentators, including Ruse, have suggested that his YEC views were the reason for being overlooked. The exact reasons are unknown and the deliberations of the Nobel Prize committee are not revealed for 50 years. It is difficult not to come to the conclusion that Damadian's YEC views were a factor in his missing the Nobel Prize. Whatever the rights and wrongs of the Damadian affair, it has been used to support the YEC claim that there is a culture war.

Though YEC scientists form less than one percent of all scientists, they show that some scientists have affiliated themselves to YEC. Many YEC scientists can boast competent research publications, but no YEC scientist has made any significant contribution in geology or biology. Even though the majority is from engineering, computer science, and physical science, the numbers should make all pause for thought. It is not possible to dismiss YECs as pseudoscientists as did Martin Gardner in his *Fads and Fallacies in the Name of Science* (Gardner, 1957).

ROOTS OF CREATIONISM

The common perception is that YEC is old orthodoxy revived, which opposed every advance of science. Critics of YEC, as does Philip Kitcher in *Abusing Science* (Kitcher, 1982), often portray YEC as a throw-back to Bishop Samuel Wilberforce and Archbishop Ussher. This misunderstands the historical situation (see Chapter 3) and stems from the Conflict Thesis of Science and Religion set forth by Andrew Dickson White and others in the 1890s.

The prevalent YEC understanding of their historical roots is that they go back to the founding fathers of science, who believed in God and rejected the atheistic ideas which were introduced into science during the Enlightenment (Mortenson, 2004). Most Christians accepted a science, which was young earth until non-Christians like James Hutton and other Enlightenment figures began to suggest Deep Time in about 1780 and undermine biblical teaching. Henry Morris presents this in a popular book *Men of Science, Men of God* (Morris, 1982) and lists all the sciences "founded" by "men of God." Superficially this is plausible as these scientists were theists, but few would be welcome in YEC circles today!

Previous chapters make it clear that (educated) Christians never dogmatically held to a six-day creation, though many, especially during the Reformation, favored creation in about 4000 BC. As Deep Time became apparent in the late eighteenth century it was accepted by most Christians, including evangelicals (Roberts, 2007), so that by the 1820s only a few, the anti-geologists described in Chapter 4, tried to insist on a "literal creation" and by mid-century had almost disappeared except for Seventh Day Adventists, Mid-western Lutherans, and a few others.

Present day YEC began with *The Genesis Flood* in 1961, which developed the ideas of McCready Price, a Seventh Day Adventist. This is expounded by Ronald Numbers in *The Creationists* (Numbers, 1991, 2006). Thus YEC *does not* have significant historical roots in orthodoxy whether Protestant or Catholic, nor Pre-Reformation or Early Church. Its roots are to be found in the Seventh Day Adventist teaching of Ellen White and the Fundamentalist reaction to all things Modernist and Evolutionary in the early twentieth century.

EVALUATION OF YOUNG EARTH CREATIONIST CLAIMS

At the beginning of the chapter I gave a brief summary of YEC beliefs. To assess each YEC argument would require expertise in every science from astronomy to zoology. Much ink has been spilt in refutation of YEC science. The most useful Web site is http://www.talkorigins.org but I question some of the religious and sociological judgments. Another useful anti-creationist site is that of the National Council for Science Education (NCSE) http://www.ncseweb.org. Of course, the AIG www.answersingenesis.org or ICR sites www.icr.org "refute" these (see Pennock and Ruse, 2007). The better YEC publications are well produced with many scientific references. They invariably find suitable quotations to support their arguments. Often YECs can cite a scientist who doubts an aspect of "evolution." That is not difficult as scientists robustly disagree.

Perhaps the most serious and persistent charge against YEC publications is that many quotations are taken out of context. (See the review of TGF by van de Fliert cited earlier.) It is often accompanied by accusations of lying. Thus on the talkorigins Web site, http://www.talkorigins.org, documents many examples, which in turn are denied. A commonly cited example is lavas from Hawaii, which were "dated" in the 60s. "In 1968 scientists applied radiometric dating to some rocks which *known* to be less than 170 years old. [1801 eruption on Hualalai.] The radioactive ages determined for those 170 year-old rocks ranged from 160 million to 3 billion years" (Ackermann, 1991, p. 81). Ackermann then commented, "Obviously, something is wrong with this method." However if one reads the paper cited (Funkhouser and Naughton, 1967, pp. 4601–4607) a very different picture emerges. The material dated were ultramafic *inclusions* in the lava of mantle material and thus not lava. The geochronologist Brent Dalrymple (a witness at Arkansas in 1981) made this clear in 1982, but the Hualalai example is still cited today; e.g. www.answersingenesis.org/creation/v27/i2/cause.asp. During these twenty-five years Dalrymple's criticisms were simply ignored. This is one of many examples. On a personal note, this type of misrepresentation in all the YEC books I read in 1971 convinced me of the falsity of YEC.

Because of this misrepresentation many critics of YEC have naively assumed that if it were exposed, then proponents of YEC would simply be forced to change. That has not been the case as the arguments are rarely corrected. One is faced with a paradox. Here are a group of Christians who are emphatic that they stand for family values and the Ten Commandments, who in their writings habitually misquote. Abortion, adultery, and homosexuality are out, yet critics assert that they break the Ninth Commandment (thou shalt not bear false witness). This is incongruous. I speak both of proponents and followers. No one seems to have made sense of this and questions are asked whether YECs activists are deliberately dishonest or simply deluded. Yet they often seem very sane, balanced, and upright people. I offer no answer.

ANTICREATIONISM

As "creationism" whether YEC or ID has mushroomed over the last forty years, so have the critics of Creationism. Critics vary from evangelical to scientific atheist. The content of criticism varies from gently critical to the militantly strident and, sadly, accuracy also varies. Until the 70s, YEC was ignored, except by a few scientific evangelicals who attempted to give a restrained critique of *The Genesis Flood*. Probably the first outside the evangelical fold to air concerns was William Mayer, who devoted the November 1972 issue of the *Biological Sciences Curriculum Study Newsletter* to YEC. Slowly during the seventies, many began to realize that YEC was an issue, but did not know how to deal with it. Many regarded YECs as hill-billies, who could be dismissed with ease. That was not the case. As the YECs consolidated their position, some academics began to address the issue, which was coming up in school boards, as did Dorothy Nelkin in *Science Textbook Controversies and the Politics of Equal Time* (Nelkin, 1977).

The first major works of refutation were Philip Kitcher's *Abusing Science* and Niles Eldredge's *The Monkey Business* both published in 1982 in the wake of the Arkansas trial. Kitcher was a professor of philosophy with a zoological background and Eldredge, cofounder with Gould of Punctuated Equilibrium. Along with scientific rebuttals Kitcher sought to explain the motives of YEC and brought out the "moral" appeal on how Darwinism encouraged the whole gamut of evils from communism to Nazism. Kitcher specifically avoids religion in the wider context, but assumes (wrongly) that today's YEC is a continuation of the Darwin controversy of the 1860s and Scopes. To draw a parallel between YEC and Samuel Wilberforce overlooks the fact that the good bishop was no young earther. Eldredge wrongly considers Wilberforce, Bryan and the early Fundamentalists as forbears of YEC. By failing to see the difference between late twentieth-century YEC and the creationist Christians of the nineteenth century, these authors reinforce the belief that orthodox Christianity before the rise of geology was dominantly literalist. Even two decades after these two books, this argument is still common, despite the careful scholarship of historians like Rudwick, Ruse, Numbers, Livingstone, and Larson. A recent example is McCalla's *The Creationist Debate: The Encounter between the Bible and the Modern Mind* (McCalla, 2006), which attempts to put the *Creationist Debate* into its historical context, but has bought into the conflict thesis of science and religion. The volume has many flaws of both of interpretation and factuality. He claims (p. 139) that Gosse wrote *Omphalos* in 1857 to counter *The Origin of Species* written in 1859—a clear case of prochronism! (His grasp of science is very poor as is shown by his confusion on radiometric age-dating claiming it began with Carbon 14 dating in 1950 whereas it began with Uranium-Lead in 1907 (McCalla, 2006, p. 137).) For interpretation he reckons that the essence of a liberal Christian is to accept geological time and its implications for Genesis (pp. 83, 118). He is unaware that Adam Sedgwick was an evangelical.

Some critics of YEC also present the choice of biblical understanding between a literal Genesis written by Moses and a mythological Genesis compiled from the sources J, E, and P in sixth-century Babylon, as in John Moore's *From Genesis to Genetics* (Moore, 2002, pp. 20–53). By focusing on extremes of interpretation, polarization is encouraged. Little mention is made of evangelical scholars like Wenham, Blocher, Lucas, and Walton, who eschew literalism.

Since Arkansas many evaluations of YEC have been published. Some deal purely with scientific issues and others attempt to put YEC into its social and religious context. There is little difference between secular and evangelical critiques of YEC science. Religious and sociological critiques are far more contentious. Some lump all "conservative" Christianity with YEC, and overlook that many evangelicals oppose YEC. Where secular critics fail to understand either Evangelicalism or Christianity, then, no matter how good their scientific critique is, their arguments often descend to a harangue against religion in general. An example of this is Richard Dawkins who dismisses any kind of religious belief as mindless and viral, and regard liberal forms of Christianity as "contrary to evidence" as YEC.

As YEC has grown so have opposition groups. These vary from individuals, like Dick Fischer (an evangelical), to well-organized groups. In the United States the most significant anticreationist organization is the National Center for Science Education (NCSE) based in Oakland, California headed by Eugenie Scott. The NCSE produces an immense literature of a high standard on very aspect of YEC and ID and has a useful and wide-ranging Web site. Though NCSE is often criticized for being atheistic, its policy is to be religiously neutral and it has the support of many Christians for example Keith and Ken Miller, an evangelical and Roman

Catholic scientist respectively. They also employ a staff member to deal with religious issues. It carefully avoids the strident atheism often associated with anticreationism.

In contrast to the careful and balanced approach of the NCSE, many anticreationists allow their opposition to creationism to give vent to stridency and frequently have a wrong understanding of the past and present relationship of evangelicals to science. This can be seen on the Web blog *The Panda's Thumb,* where some contributors show little discernment, in contrast to contributors from the NCSE, who combine pungency with precision.

THE SCIENCE OF YEC

The tendency to dismiss YEC as *pseudoscience* or *antiscience* overlooks the fact that YEC emerged from a scientific culture. McCready Price and Morris were products of a technological education, which they revered. YECs use an extreme version that science is empirical and experimental to support a literalistic faith. This goes beyond the witticism of Ernest Rutherford (1871–1937), and Nobel Prize winner in 1908, who said, "All science is either physics or stamp collecting." Like many aphorisms it is a half-truth, but it explains why experiment is thought to be the essence of science. This is reinforced as many studies on the philosophy of science focus on empirical science and ignore historical science. Historical science includes geology and archeology which deal with unique past events not open to empirical test. However that does not mean that historical science cannot be tested rigorously. This extreme experimentalism is reinforced as many, especially males, tend to have studied more physical science than geology or biology. Also, the study of rock strata and their relationships is observational rather than experimental, but that does not mean that it lacks scientific rigor. This can result in the hierarchy of sciences with physics at the top and the "soft" social sciences at the bottom and geology and biology somewhere in-between.

In *Scientific Creationism* Henry Morris develops this and questions the reliability of any historical science because past events are unrepeatable. He states, "At the same time, it must be emphasized that it is impossible to *prove* scientifically any particular concept of origins to be true. This is obvious from the fact that the essence of the scientific method is experimental observation and repeatability" (Morris, 1974, p. 4) He then develops his two-model approach. To him neither is provable as both are "faith positions" and dependent either on the Bible or materialism. As the Bible is true so must be "creation," that is creation in six days. Thus briefly Morris presents the case for rejecting all geological and cosmological arguments for deep time. Thirty years on Morris's argument is still held. On a popular level Ken Ham developed this with his question, "were you there?" about anything relating to the deep past, which is recommended for schoolchildren, claiming biblical support from Job 38 vs 4, "Where were you when I laid the foundation of the earth?" Similarly, John Morris of ICR says that not once has a rock "talked to him" and explained its history. The YEC extreme experimentalism also questions whether historical sciences can be science, as nothing in historical science can be tested experimentally. In a sense, that is true. However, rather than recognizing the difference between empirical and historical science YECs aim to show that historical science cannot demonstrate anything about the past. This would also nullify historical arguments for the existence of Jesus Christ!

Norman Geisler and Kerby Anderson tried to resolve the problem of "Creation and Evolution" in their book *Origin Science, a Proposal for the Creation-Evolution controversy* (Geisler and Anderson, 1987). Geisler was a witness at the Arkansas trial in 1981 and a well-respected conservative theologian at Dallas Theological Seminary. Following the suggestions of Thaxton et al. (1984, p. 204) that a science about past singularities should be termed *origin* science, the authors tried to resolve the controversy by distinguishing between *operational* and *origin* science, and "If both evolution and creation honor these principles, then proponents of each can at least engage in meaningful discussion" (Blurb to book). *Origin Science* deals with the unrepeatable events of the past and *operational* science deals with repeatable present events. This goes beyond the common distinction of empirical and historical science in that there is a possibility of divine action in *origin* Science. They regard geological science as dealing with historical regularities, but *singularities* can be explained by special creation or macroevolution and that scientific evidence can show "that there is a constant conjunction between a primary intelligent cause and a certain kind of event" (p. 17) which points to a supernatural cause. The authors are critical of the development of a "modern naturalistic approach" in astronomy (Descartes, Buffon, and Laplace) and geology (Lyell) and biology (Darwin). They overstate their case as they fail to realize how close the naturalistic geology of Lyell and much of Darwin's biology (and geology) is to that of their "theistic" counterparts like Sedgwick (Roberts, 2004, pp. 280–285) and others mentioned in Chapter 4. Their conclusion is that *origin* science, which allows for divine intervention is better than today's *naturalistic* historical science.

This and *The Origin of Life* (Thaxton, Bradley, and Olsen, 1984) has provided much of the basis for Intelligent Design (ID), with its attack on *Naturalism.* The distinction of *operational* and *origin* science is now well established with YECs and ID, with *origin science* now including *all* historical science. However despite the avowal of geological time by Geisler and Thaxton, *origin* science is often used to question geological time. This is combined with a rejection of conventional geology as *uniformitarian* and *naturalistic* and based on the antitheistic beliefs of Lyell (in fact a theist) and Hutton, who both had roots in the Enlightenment. The combination of *origin* science and anti-naturalism in its various guises, whether YEC or ID, provides the basis for a powerful rhetorical attack on geology and evolutionary biology. First, all "uniformitarian" geology can be charged with being naturalistic and secondly that as geological arguments for great age are *historical*, they cannot be verified *empirically*. Thus deep time or the short timescale of YEC are equally valid and both ultimately *faith-positions*, based on naturalism and theism respectively. It can also be used to dismiss all geology, by claiming that as an evolutionary *origin science* its conclusions are entirely dependent on its presuppositions, without defining what those are as does the AIG speaker Paul Taylor (Taylor, 2006). Opponents of evolution used this during the campaigns of 2002/2003 in Ohio, when SEAO published the following statement.

> Historical science. Most sciences, including chemistry and physics, are empirical (or experimental) in nature; theories can be tested by experiments in the laboratory and/or by observations of the world. Some disciplines, like origins science, are historical in nature; that is, they attempt to explain events and processes that have already taken place in the distant past. Theories in historical sciences cannot be verified experimentally, so the explanations are always tentative. Biological evolution (like creation and design) cannot be proven to be either true or false. The historical nature of evolution/design theory should be explained in the standards.

Had that passed into the statutes in Ohio it would have been impossible for conventional scientists to teach either geology or biology without breaking the law. At present, this distinction of *operational* and *origin* science is widely used in both ID and YEC circles, including on the teaching of science. Many find it plausible and so; if "creation" and "evolution" are equally valid "faith" positions, why not teach both?

Uniformitarianism has long been criticized by YEC writers. In *The Genesis Flood* Morris criticises Uniformitarianism at length, and this is now a basic tenet of YEC. When Morris wrote the book in the late 1950s, there had been little serious study of the history of geology, except that by the evangelical Hooykaas (1957) and the then *received* account of the birth of geology was that the founders, notably Hutton and Lyell, struggled heroically against the constraints of the church. Historians of geology have overthrown this "heroic" understanding of the history of geology during the last three decades (Rudwick, 2004). Hutton and Lyell were only two of many geologists, rather than the founders of geology with an anti-Christian bent. It is, of course, impossible to do any geology without a basic assumption of the uniformity of past physical processes. Otherwise one would simply invent changes to physical processes in the past to explain the unexplainable.

Steve Austin greatly misuses Uniformitarianism in his comparison of the Grand Canyon and gorges carved out Mt. St. Helens after the 1980 eruption. Mudflows carved out thirty meters gorges in soft volcanics in one day. YECs falsely claim that uniformitarians argue that this would have needed millions of years. They then claimed that the Grand Canyon could be carved out rapidly. This argument is found in the AIG tract *The Voice of the Volcano* and might convince the uninformed, but it fails to recognize that unconsolidated volcanic ashes can be eroded rapidly but not the hardened rock of the Grand Canyon. This can be shown by turning a garden hose on a pile of loose sand and then on the brickwork of the house.

On the philosophy of science YECs make use of Thomas Kuhn's *The Structure of Scientific Revolution.* Kuhn's thesis of paradigm shifts in scientific theories is well known, but is not the last word. By using Kuhn's Paradigm Shifts writers try to demonstrate that because of the new YEC scientific paradigm, the old evolutionary paradigm is crumbling and needs to be replaced by a YEC paradigm. Throughout his book *Creation and Change, Genesis 1.1–2.4 in the light of changing scientific paradigm* Kelly argues that the new evidence (for a young earth) is crying out for a paradigm shift. There is an incongruity in the YEC use of Kuhn as he reckoned paradigms were changed because of scientific consensus rather than a closer approximation to scientific truth. Kuhn rejected realism in science, whereas YECs (and Dawkins) are naïve realists and tend to absolutize what they consider true science.

THE APPEAL TO CRITICAL THINKING OR "TEACHING THE CONTROVERSY"

By now it should be clear that YECs are wellinformed and that, when it comes to evolution, students must be taught to think critically. This argument is used on both sides of the Atlantic and is very appealing educationally.

However that is not what Creationists understand by "teaching the controversy" about evolution. This came to the fore in 2002 during the campaign to introduce ID into Ohio schools. Here, as subsequently in Kansas and Dover, Pennsylvania, in 2005, there was an alliance of YEC and ID. The argument is beguiling and presented in such a way that only those with closed minds could oppose it, as it "appeals to popular values of fairness, openness and equality of opportunity. It thus plays well the public" (Scott and Branch, 2003). Hence students must be given both sides and think for themselves. As Stephen Meyer put it in the *Cinncinnati*

Enquirer in March 2003, "When two groups of experts disagree about a controversial subject . . . students should learn about both perspectives . . . " Educators call this "teaching the controversy." However it was a ploy to teach ID. Proponents of the bills took this up and their depositions emphasized the need for "teaching the controversy": This was forcefully made in the publications of Science Excellence for All Ohioans (SEAO), who were behind the bill.

> *Teaching the controversy.* Dr. Stephen Meyer, at the Panel Presentation in Columbus on March 11, proposed that the standards should "mandate mastery of the scientific evidence and arguments *for* and *against* Darwinism. Students should know the scientific case for modern Darwinism and contemporary scientific critiques of the theory as well." Dr. Meyer also proposed that the standards should "enact no definition of science that would prevent teachers from discussing alternative evidence-based theories" about biological origins. The standards should thus "teach the controversy" surrounding the evidence for evolution and the definition of science.
>
> We are not opposed to the teaching of evolution; we just want origins science to be presented objectively—that is, in a fair, reasonable, and unbiased manner. This is not an "either/or" situation; there is room for both biological evolution and alternative theories in the standards. We believe that teachers and students should be permitted to (a) criticize the weaker aspects of evolutionary theory, and (b) discuss alternative theories that have been proposed.

This was rejected in Ohio in 2003, but the tactic has continued to be used, including in the recent British attempt of *Truthinscience.* The argument is seductive, but presumes that the handful of YECs and Intelligent Designers should be considered as expert as the myriads who aren't. Superficially this is an extremely good educational method as by being given the arguments for and against, students will think critically. *But,* this depends on the validity of that evidence. Thus, for example take the age of the earth. For over half a century geologists state, from radiometric age dating that the earth is 4.6 billion years old. Hopefully they will have been taught how geologists came to this figure, starting with notions of great antiquity in the 1770s through to the development of radiometric age dating since 1905. However to "teach the controversy," they will be taught all the YEC arguments against an old earth and for a young earth, as if those were as valid as conventional science. These are repeated in many YEC publications and include charges of false assumptions in radiometric age-dating, the cavalier discarding of "inconvenient" dates, the rapid deposition of strata and the carving of a canyon on Mt. St. Helens in 1980, etc. The impression is that these arguments are as geologically sound as any other. It is impossible to "think critically" if one is not equipped to understand the arguments in the first place or if one is fed inaccurate information.

The appeal to "teach the controversy" has been strongly opposed, as mainstream scientists stress that there is no controversy over the fact of evolution. Having said this, there is not unanimity among scientists as there are many controversial issues, but there is a consensus over the vast age of the earth as 4.6 billion years, the development of life over the last 4 billion years and many other matters. Much "creationist" literature claims to present both sides, but their presentation of evolution is inaccurate, as is the book *Icons of Evolution* by Jonathan Wells. Though he is a member of the Unification Church (Moonie), his book is widely recommended by evangelicals.

Strategy and Rhetoric

When YEC came to public notice in about 1970, many opponents assumed that any competent "evolutionary" scientist could debate a YEC and win. Several became croppers and continue to do so. Henry Morris points to Dr. Dolittle's "devastating defeat in a formal debate with Dr. Gish at Liberty Baptist College . . . on October 12, 1981." YECs are keen to debate with prominent evolutionists for good reason. The essence of these debates is slick tactics rather than a reasoned presentation. Further the audience may have decided the result beforehand, as often it is largely YEC, so that the "evolutionist" labors under a hostile audience. The winner is the one with the best sound bite, who buses in most supporters or who records the debate, as there are several examples of inaccurate recording of debates. The strategy for lectures has changed over time. Here I write of the British situation having attended YEC lectures by major speakers for the last twenty-five years. In the 80s there were often opportunities for verbal questions. That changed to written questions, which could be edited preventing any no contrary viewpoint. In recent years, no questions are allowed and the meetings close with prayer and an evangelistic challenge.

Though YECs present themselves as scientific, their strategy means that they target potential sympathizers. Hence their presentations are geared to evangelicals. Thus, though their presentations contain much "science" a high proportion is on the moral and theological appeal of YEC. First, they defend the inerrant Word of God. Secondly they make much of the "fact" that evolution is based on 4 billion years of death and suffering and that this is neither moral, nor in accord with the Fall, and thus acceptance of evolution or even geological time has a skittles effect on Christian belief. Thirdly, they often argue that evolution has detrimental effects on ethics from Nazi racism to promiscuity, homosexuality and abortion. The net effect is for a Christian to accept YEC, without considering the scientific aspects.

This is coupled with a "hermeneutic of suspicion" so that the hearer is already halfway to rejecting evolution and thus halfway to becoming YEC. I have observed this to be highly effective among my colleagues in the Anglican ministry. The Church of England is not the most conservative church, yet I frequently find that non-fundamentalist colleagues doubt evolution and/or geological time because doubts have been persistently pressed on them and they do not have the scientific expertise to evaluate them. Consequently it is highly effective to throw out a series of questions against "standard evolution" as did Wells in *Icons of Evolution*, as even if these ignore many aspects, which are impregnable (viz. the great age of the earth). Then, evolution will be *perceived* to be dubious.

That, combined with a binary approach, whereby one has either to accept Evolution or Creation, reduces the argument to a choice between atheistic evolution and creation in Six Days. Though for over 150 years there has been a spectrum of opinion, YECs stress that there are only two choices as was made clear in an *ICR Impact* article in 1983, "creation scientists maintain that there are only two basic explanations—creation and evolution—all explanations can be included within one or the other of these two basic explanations." Hence once one has demonstrated that evolution is wrong (or cast doubts on it) then creation (in the YEC sense) must be right. In his classic article *Why Creation Science?* George Marsden (Marsden, 1991, pp. 153–181) demonstrates that it is simply not true to the understanding of evolution in relation to creation over the last 150 years.

CREATIONIST LITERATURE

There are many high-selling creationist books, which are usually well written and produced. A cursory look will show that the arguments are appealing to a Christian with little science, or one whose science is limited to physics and chemistry. Essentially they are based on two arguments; first an "evangelical" view of scripture will result in taking Genesis literally (forgetting that most evangelicals both past and present did not do so) and secondly geological and evolutionary science is fatally flawed. Most books are variations on the same theme. There are many well-produced children's books and short tracts of varying quality. The most popular is *Big Daddy*, which has had an immense distribution and influence (http://www.chick.com/ look for bigdaddy). The amount of YEC material on the Web is almost infinite and of variable quality.

EVANGELICALS AND YOUNG EARTH CREATIONISM

Owing to the highly successful public relations of YECs, there is the perception, especially in America, that YEC *is* the evangelical position on science. Thirty years of "witnessing for creation" has ensured this. As a result other views, be they Progressive Creationist or Theistic Evolutionist, are now regarded with suspicion by many evangelicals and open to the mantra, "You don't believe the Bible." There is much anecdotal evidence of Christians receiving hostility in evangelical churches for "believing evolution," resulting in heartache and members leaving churches. YEC insistence that an evangelical must be young earth also has its sinister side as less than Christian methods have been used to silence critics.

One of the earliest victims was Daniel Wonderly (d. 2004), who taught science at Grace College, Indiana, from 1966 to 1973. Wonderly was a conservative Baptist, who accepted geological time but not evolution. That was common among American evangelicals up to the Sixties and thus these beliefs, though seen as *ultra* by secular scientists, were acceptable with evangelical colleges. He regarded himself as a biblical creationist and adopted a Concordist view of Genesis. When applying for the post he did not hide his views from the college. This coincided with the time when YEC was gaining momentum after the publication of *The Genesis Flood*. While he was there Whitcomb, coauthor of *The Genesis Flood*, was persuading his colleagues that they must adopt "his extreme position if they were going to be true to the bible." Wonderly wrote to me in 1984, "By 1970 he [Whitcomb] persuaded the academic dean of the college that I would have to be either dismissed or entirely silenced . . ."[2] He was finally forced out in 1973 because of several unpublished papers giving geological evidence for age. In 1977 he published *God's Time-Records in Ancient Sediments*, which is a fine summary of standard geological arguments for great age, excluding radiometric age-dating. Wonderly also argued later that YEC was making thinking evangelicals inclined to adopt theistic evolution, resulting in the double marginalization of his intermediate position.

PERVERSIONS FROM YEC

In the nineteenth century, the Rev. W. J. Conybeare, son of the geologist W. D. Conybeare, wrote a novel entitled *Perversion*. It was not about kinky sex, but how Christians converted, or perverted, from their original faith. The hero had perverted from Protestantism to Roman Catholicism. There are many, who, after being brought up in a conservative church, move off, reject young earth ideas and move either into a broader Evangelicalism, liberal Christianity, agnosticism, or atheism. Some, like E. O. Wilson, were brought up in a fundamentalist church and on studying science reject the faith of their youth. This is not unique to those who go to college with YEC beliefs. More significant are those who were hard-core YEC, who later pervert. Over the last thirty years, I have come across many who were at one stage YEC. By that I mean a committed YEC, rather than one who adopted it as a temporary stage in their life. At present most are Americans, and include several who were YEC during their university days and then later rejected what they were taught in their conservative churches. The destination of those who left YEC varies. The common theme in all the stories is rejection by former soul-mates, coupled with a sense of loss.

One of the most high profile converts from YEC in the United States is the geophysicist Glenn Morton, whose story is on his Web site (http://home.entouch.net/dmd/gstory.htm). He was brought up a YEC, graduated in physics in 1972, and worked as a geophysicist in a seismic company. He wrote,

> This was where I first became exposed to the problems geology presented to the idea of a global flood.... I would see buried mountains, which had experienced thousands of feet of erosion, which required time. Yet the sediments in those mountains had to have been deposited by the flood, if it was true. I would see faults that were active early but not late and faults that were active late but not early. I would see karsts and sinkholes (limestone erosion), which occurred during the middle of the sedimentary column (supposedly during the middle of the flood) yet the floodwaters would have been saturated in limestone and incapable of dissolving lime. It became clear that more time was needed than the global flood would allow.

He contributed twenty papers to the Creation Research Society Quarterly, but after 1980 had doubts. In 1986 he presented a paper entitled *Geologic Challenges to a Young-Earth* at the First International Conference on Creationism. It was not well received. John Morris challenged him, and according to Robert Schadewald,

> went to the microphone and identified himself as a petroleum geologist. He questioned Morton's claim that pollen grains are found in salt formations, and accused Morton of sounding like an anticreationist, raising more problems than his critics could respond to in the time available. Morris said that the ICR staff is working on these problems all the time. He told Morton to quit raising problems and start solving them. Morton chopped him off at the ankles. Two questions, said Morton: "What oil company did you work for?" Well, uh, actually Morris never worked for an oil company, but he once taught petroleum engineering at the University of Oklahoma. Second, "How old is the Earth?" If the earth is more than 10,000 years old then Scripture has no meaning. Morton then said that he had hired several graduates of Christian Heritage College, and that all of them suffered crises of faith. They were utterly unprepared to face the geologic facts every petroleum geologist deals with on a daily basis. Morton neglected to add that ICR is much better known for ignoring or denying problems than dealing with them.

As Morton wrote,

> It appeared that the more I questions I raised, the more they questioned my theological purity. When telling one friend of my difficulties with young-earth creationism and geology, he told me that I had obviously been brainwashed by my geology professors. When I told him that I had never taken a geology course, he then said I must be saying this in order to hold my job. Never would he consider that I might really believe the data. Since then this type of treatment has become expected from young-earthers. I have been called nearly everything under the sun but they don't deal with the data I present to them. Here is a list of what young-earthers have called me in response to my data: "an apostate," (Humphreys) "a heretic" (Jim Bell although he later apologised like the gentleman he is) "a compromiser" (Henry Morris) "absurd," "naive," "compromising," "abysmally ignorant," "sloppy," "reckless disregard," "extremely inaccurate," "misleading," "tomfoolery," and "intentionally deceitful" (John Woodmorappe), "like your father, Satan" (Carl R. Froede—I am proud to have this one because Jesus was once said to have been of satan also), "your loyalty and commitment to Jesus Christ is shaky or just not truly genuine" (John Baumgardner).

By 1994 Morton was finished with YEC saying that,

> Nothing that young-earth creationists had taught me about geology turned out to be true. I took a poll of my ICR graduate friends who have worked in the oil industry. I asked them one question. "From your oil industry experience, did any fact that you were taught at ICR, which challenged current geological thinking, turn out in the long run to be true?" That is a very simple question. One man, Steve Robertson, who worked for Shell grew real silent on the phone, sighed and softly said "No!" A very close friend that I had hired at Arco, after hearing the question, exclaimed, "Wait a minute. There has to be one!" But he could not name one. I cannot name one. No one else could either. One man I could not reach, to ask that question, had a crisis of faith about two years after coming into the oil industry.

Since then Morton has continued in the oil industry and has written a book against YEC, and contributes robustly on geological objections to a young earth. Theologically, he has remained very conservative and strives for a concordist position on Genesis and geology, and thus claims to take Genesis literally. His concordism is strained, as he believes that Adam lived 5 million years ago. Morton is vigorously opposed to any allegorical and accommodationist interpretation. This means that he is an exposed position, as his theology is not far removed from the ultrafundamentalism of YEC. He seems to find it harder to reject YEC theology than YEC science.

Of a similar age is Edward Babinksi, who graduated in biology in 1974. Like Morton he began to have doubts about YEC teachings and in time moved to agnosticism. In his account, Babinksi traces out his route. He first moved to an old earth position and found Wonderley's writings very helpful, then rejected conservative theology and became agnostic. Like Morton he has received much harsh criticism from YECs.

Another with impeccable evangelical credentials is the geologist Davis Young (b1941), emeritus professor of geology at Calvin College. He has written several fine works on the history of igneous geology. Young is the son of E. J Young (1907–1968), who was Old Testament professor at Westminster Theological seminary from 1936 until 1968 and published conservative works on early Genesis. After studying geological engineering at Princeton he became convinced by *The Genesis Flood* but later questioned it, to Morris's disappointment. He remained committed to the biblical inerrancy of his father, but argued against YEC in *Creation and the Flood* (1977). A few years later he published *Christianity and the Age of the Earth* (1982) which, dealt with both the ways Christians have understood the age of the earth in the past, and summarized arguments for the age of the earth. It is now being revised. In 1995 he published a major historical study on *The Biblical Flood*, which considers how the flood has been understood since the Renaissance and its place in the rise of geology. Young has moved from the biblical literalism of his early years to a position now, which sees little history in early Genesis. Though his Christian convictions have not changed, he is perceived to have rejected orthodoxy for a liberal version of Christianity. As well as upsetting Morris by his defection Young has been disparaged by YECs with Mortenson describing him as "the professing evangelical old-Earth geologist at Calvin College."

I describe these at length to give a flavor of the way in which scientifically informed YECs may break out of the straitjacket. There have been many others, who have rejected YEC teaching. While researching this, I came across a good few who were prepared to give their testimonies, including several members of the ASA. Their perversion could be sudden or slow, but often it resulted in a rupture of old friendships or leaving the church of their youth.

INTELLIGENT DESIGN

Some readers will consider my discussing Intelligent Design (ID) along with YEC, with the implicit equating the two. as historically and theologically inaccurate, as ID and YEC should be considered as totally distinct. That was my opinion until a few years ago, as I thought that all the evidence pointed to a distinction, especially in that ID tended to be old earth. Writers like Pennock and others referred to ID as IDC (Intelligent Design Creationism) and regarded them as different varieties of the same species. That seemed wrong as Dembski was clearly old earth and Behe accepted common descent. Two things, which came to light since 2000 have demonstrated their close equivalence. The first is the alacrity in which ID associate with YEC on issues such as science education as in Kansas, Ohio and Dover, Pennsylvania and *Truthinscience* in the United Kingdom and the second is that the origins of ID have roots in YEC as was shown by Forest and Gross over the replacement of the word "creation" by "design" in the creationist text *Of Pandas and People* in 1987. Thus it is now not possible to distinguish the two as one could in the 90s.

Until the 90s evangelicals tended to adopt one of the three approaches of Old Earth Creationism (OEC), YEC or Theistic Evolution (TE). About 1990 a fourth player appeared—ID. Initially this seemed to be part of OEC or even TE, but since the late 90s it has become stridently anti-evolution and confrontational and more and more associated with YEC. From the 70s, the motivating force of all attempts to limit the teaching of evolution in schools came from YECs, with their insistence of the two-model approach—either Creation or Evolution. As the millennium approached the two-model approach was replaced by the demand to teach ID. This has resulted in a new tactic in educational campaigns as the emphasis has shifted to teaching "design" or to "teaching the controversy." Three of the recent attempts to change educational policy in Ohio (2002–2003), Kansas (2005), and Dover, Pennsylvania (2005) and *Truthinscience* in the UK (2006) have centered on teaching design rather than creation. This may appear to be a climb-down as gone are the YEC arguments. Those opposing such proposals are as vehement against the new approach as they were of the old. In their opposition several critics of ID have suggested that it is just old-fashioned creationism, which has evolved into a new form. As with YEC, ID produces more heat than light.

The chief argument of ID is that many biological features are so complex that they are "irreducibly complex," as is the mechanism of blood-clotting or the flagellum and thus could not have evolved and so are the result of an intelligent designer. Proponents claim that this is a purely scientific argument and that it has nothing to do with theology. Hence to teach "Design" causes no problem with law against teaching religion. ID does not claim to be an evangelical stance, but most of their proponents are, as are Steve Meyer, Bill Dembski, Paul Nelson, Phillip Johnson, and others. Michael Behe, author of the best-selling *Darwin's Black Box*, is a Roman Catholic, Jonathan Wells, author of *Icons of Evolution* belongs to the Unification Church or "Moonies." Many of these are members or fellows of the *Discovery Institute*, a right-wing think tank in Seattle.

Since the late 80s ID has been put forward as an alternative to YEC and "Evolutionism" and leapt to prominence with the publication in 1996 of *Darwin's Black Box* written by Michael Behe of Lehigh University. Though ID claims to be independent of any theistic approach, ID must be considered as an aspect of evangelical perspectives on science because, first, many

ID exponents are evangelical and secondly supporters of ID have worked alongside YEC evangelicals in various states in antievolution education bills. At the present time many evangelicals who wish neither to support YEC or TE have opted for ID and ID has the blessing of Inter-Varsity Press, a leading evangelical publishing house, who have published several works by Johnson and Dembski. As ID has come to the fore in recent years, I shall give an account of its history and development.

Origins of Intelligent Design

Though ID came to prominence in the late 90s, its roots go back to the early 80s, especially in two books; *Origin Science, a Proposal for the Creation-Evolution controversy* (1987) by Norman Geisler and Kerby Anderson and secondly, Bradley, Olsen, and Thaxton *The Mystery of Life's Origin* (1984). The two books rejected both a young earth and theistic evolution. The latter three authors, all scientists, argued that the self-organization of molecules was incapable of producing life, thus pointing to a non-naturalistic origin of life. The common YEC/ID division of science into *Operation* and *Origin* science stems from Geisler and Anderson. This is not accepted outside creationist circles, though it is part of the "controversy."

In 1981 Bradley and Olson were involved in the *Chicago Declaration on Inerrancy*, when they argued that inerrancy allows geological time but not evolution. It seems that their search for a non-naturalistic explanation of life was predicated by a belief in inerrancy. Much too has been made of Michael Denton's *Evolution; A Theory in Crisis* (1986), which challenged evolution from a secular standpoint, but he has recently back-tracked. Politically, the most significant event was the lawyer Phillip Johnson's sabbatical visit to England in 1987. He came to question "Darwinism" after visiting the British Museum of Natural History and reading Dawkins' *The Blind Watchmaker*, and also visited Stephen Meyer, who was then doing a Ph.D. on the philosophy of science at Cambridge. Four years later Intervarsity Press launched *Darwin on Trial*. The book is a sustained critique on neo-Darwinism, which he considered to be totally dependent on a naturalistic philosophy. He wished to challenge this by providing a "wedge" between scientific empiricism and naturalism. A common and justified criticism of Johnson is that he is inaccurate in his depiction of Darwinism. This movement grew and attracted the philosophers Plantinga, Moreland, and Craig, who introduced the concept of *theistic science*.

Soon after, in 1992 a conference was held at the Southern Methodist University, with Johnson, Behe, Meyer, and Dembski as speakers. After that conferences were held at Biola (1996), which gave rise to the book *Mere Creation*, (Dembski, 1998), Austin (1997), Baylor, Concordia in Wisconsin and Yale (2000) and Calvin (2001). At Concordia and Baylor some of the participants were strong critics of ID, including Conway Morris, Schermer, Ruse, and Ken Miller. The Concordia conference gave rise to the book *Debating Design*, but only a few of the chapters were given as papers at the conference. In 1996 *Darwin's Black Box* was published and also that year the Center for the Renewal of Science was formed as part of the Discovery Institute (DI) in Seattle. This gave both political and financial support for the movement. The DI began as a radical Republican think-tank and has remained right-wing. With the financial backing of the DI, fellows are able to research alternatives to Darwinism and have produced a great volume of literature, most published by Christian publishers. Despite their prodigious output, virtually no papers on ID have been published by peer-reviewed science journals.

Since 2000, attempts in the United States to limit the teaching of evolution in schools have argued for "Design" as an alternative rather than Creation. The influence of ID may be seen in their arguments. In 2000 Jonathan Wells published *Icons of Evolution*, which criticized several textbook examples of evolution; the Cambrian Explosion, the peppered moths and Haeckel's embryo diagrams. These are often presented as though they were deliberately fraudulent, for example the pinning of peppered moths to trees. Just imagine going out at night, waiting for a moth to land on a tree and then photograph it with a 1950s flash camera!! ID began to be involved in challenges to educational policy. All previous attempts to downgrade the teaching of evolution had failed, partly because YEC was presented as the alternative. A different tactic evolved with the emphasis on teaching the Design in living forms and ignoring or playing down the tenets of a Young Earth. This had a far wider appeal as those who accepted the vast age of the earth but not evolution could support it. Thus an alliance of YEC and ID was formed. Along with the teaching of Design as an alternative to evolution, it was also proposed that evolution should be taught critically, hence the cry, "teach the controversy." During 2002–2003 there was an almost successful attempt to introduce the teaching of ID in Ohio, followed by the disastrously unsuccessful attempt in Dover, PA in 2005. In all the discussions and hustings, various proponents were brought in from outside—Ken Miller and Stephen Meyer for example.

RELATIONSHIP OF ID WITH YEC[2]

The $64,000 question is whether ID is an evolved version of YEC. Intelligent Designers like Dembski emphatically deny this but many, whether Pennock in *The Tower of Babel*, or Barbara Forest and Eugenie Scott of NCSE claim that the two are genetically related. To confirm this ID is often called the *New Creationism*, which is most unhelpful. There are notable differences. ID does not explicitly base its ideas on the Bible and thus makes no use of either Creation or Flood. Neither does it make an appeal to the Judaeo-Christian God. As discussed above they eschew discussion on the age of the earth, though most IDers do accept an old earth. The two leading YEC organizations, ICR and AIG, are highly critical of ID for being neither Biblical nor Young Earth, but value their respect for design.

So there seems to be no connection. However in recent educational cases, as in Ohio, Kansas, Dover, Pennsylvania, and *Truthinscience*, the tactic is to enforce legally the teaching of "Design" rather than "Creationism," but the proponents are dominantly YEC. This highlights the change by YEC on the teaching of evolution. In the 80s they used the *Two Model Approach* of Creationism and Evolutionism, which was defeated in 1982 in Arkansas and in 1987 when the *Edwards v. Agouillard* case overturned Louisiana's creationist legislation. After several defeats a new tactic was needed. Here Design fitted the bill, especially after Philip Johnson's *Reason in the Balance* (Johnson, 1995) dealt with educational issues. Focusing on "design," "critical thinking," and, later, "teaching the controversy" seemed far more likely to succeed. Thus, in recent cases, young earth arguments are ignored. Hence some see ID as a Trojan Horse for YEC.

Superficially it may seem that ID—the "New Creationism" resulted from the *Edwards v. Agoullard* judgment and is clearly "descent with modification" from the old creationism, that is YEC. But it is not the whole story as there has been the transference of ideas as Barbara Forest and Paul Gross have demonstrated in the replacement of the term "creation" by "design" in the biology text *Pandas and People* in 1987. This was part of the plaintiffs' presentation at Dover and I cite from the Memorandum Opinion of December 20, 2005:

> As Plaintiffs meticulously and effectively presented to the Court, Pandas went through many drafts, several of which were completed prior to and some after the Supreme Court's decision in Edwards, which held that the Constitution forbids teaching creationism as science. By comparing the pre and post Edwards drafts of Pandas [in 1987], three astonishing points emerge: (1) the definition for creation science in early drafts is identical to the definition of ID; (2) cognates of the word creation (creationism and creationist), which appeared approximately 150 times were deliberately and systematically replaced with the phrase ID; and (3) the changes occurred shortly after the Supreme Court held that creation science is religious and cannot be taught in public school science classes in Edwards. This word substitution is telling, significant, and reveals that a purposeful change of words was effected without any corresponding change in content, which directly refutes FTE's argument that by merely disregarding the words "creation" and "creationism," FTE expressly rejected creationism in Pandas. In early pre-Edwards drafts of Pandas, the term "creation" was defined as "various forms of life that began

abruptly through an intelligent agency with their distinctive features intact—fish with fins and scales, birds with feathers, beaks, and wings, etc," the very same way in which ID is defined in the subsequent published versions. (P-560 at 210; P-1 at 2-13; P-562 at 2-14, P-652 at 2-15; P-6 at 99-100; P-11 at 99–100; P-856.2.) This definition was described by many witnesses for both parties, notably including defense experts Minnich and Fuller, as "special creation" of kinds of animals, an inherently religious and creationist concept. (28:85–86 (Fuller); Minnich Dep. at 34, May 26, 2005; Trial Tr. vol. 1, Miller Test., 141–42, Sept. 26, 2005; 9:10 (Haught); Trial Tr. vol. 33, Bonsell Test., 54–56, Oct. 31, 2005). Professor Behe's assertion that this passage was merely a description of appearances in the fossil record is illogical and defies the weight of the evidence that the passage is a conclusion about how life began based upon an interpretation of the fossil record, which is reinforced by the content of drafts of Pandas. The weight of the evidence clearly demonstrates, as noted, that the systemic change from "creation" to "intelligent design" occurred sometime in 1987, after the Supreme Court's important Edwards decision. This compelling evidence strongly supports Plaintiffs' assertion that ID is creationism re-labeled. Importantly, the objective observer, whether adult or child, would conclude from the fact that Pandas posits a master intellect that the intelligent designer is God. (http://www.yale.edu/lawweb/jbalkin/cases/kitzmiller.pdf [pp. 14-15])

There is much in favor of this claim but it is only a partial explanation, and it is not correct to state that "ID is solely creationism re-labeled." For a start, against that, Philip Johnson had no YEC roots and became convinced of ID sui generis in Britain in 1987. Several other leaders of ID have no roots in YEC like Behe, Dembski, Thaxton, Bradley, and Pattle Pun and most continue to distance themselves from YEC. But Nancy Pearcey and Paul Nelson are clearly YEC as well as ID.

However the replacement of "creation" by "design" in *Pandas and People*, the refusal to come clean over the age of the earth, and the association of YEC and ID makes it difficult for observers to distinguish between the two. I hope that by dealing with the historical order of events, rather than an assessment of ID arguments, has indicated both how ID came about in the last twenty-five years and its relationship with YEC. ID may not be an evolved version of YEC, but many of its genes have been spliced in from YEC.

WHAT IS INTELLIGENT DESIGN?

The term "Intelligent Design (ID)" is frequently used, but often there is no clear understanding of what it actually means. Design has a long heritage going back to the Greeks, but the most famous exponent was William Paley. Before defining ID it is best to consider the variety of Design. A rough typology gives four types of design, these are

1. The structure of the universe shows order and thus in the teleological argument its Design demonstrates a designer. More recently this has been expressed as the "fine-tuning" of the universe.
2. The succession of life and natural processes throughout the planet's history show so much order and design, that it suggests some being has guided and designed it. Historically this is how Asa Gray responded to Darwin's chancy ideas of Natural Selection in the 1860s and is explicitly or implicitly held by many Christians today who hold to evolution. This includes Popes John Paul II and Benedict XVI as well as many evangelicals, especially those in the ASA or CiS.
3. The ahistorical recognition of ordered biological structures as designed. This finds its epitome in William Paley and Robert Hooke and John Ray a century earlier.
4. The fourth example is to claim that some biological features could not have come about by natural causes and "abruptly appear." Behe and others make this argument over the flagellum and blood-clotting. It is defined as irreducible complexity and is considered to demonstrate Design. Here IC only applies to some biological features whereas Paley considered all features to show Design.

With Design being ascribed to such a variety of ideas, Design is a fluid concept. Intelligent Design focuses on the fourth—irreducible complexity and this is at the heart of ID. However fine tuning is often claimed as part of ID though it has a different provenance and the originators have nothing to do with ID. Hence ID must be seen as the claim of IC, which has been challenged by many especially Ken Miller and charged, justifiably in my view, as God of the Gaps. Further ID is given mathematical support by Bill Dembski in his many publications and in many ways he is the leader of ID, though Philip Johnson has been the leading protagonist.

There is now and immense literature by exponents of ID and it has attracted much critical comment and some have made their academic reputation by dissecting ID! The consensus of the latter is that ID has no scientific worth and this was borne out by Judge Jones' summing up at the Dover trial in 2005. Though ID is rejected by most scientists, a good proportion of evangelicals buy into it, especially if they eschew YEC. Though ID was initially an American phenomenon, it is gaining in significance elsewhere especially in Britain, where its main supporters seem to be YEC as are the leaders of *Truthinscience*, like Andy McIntosh.

THE APPEAL OF INTELLIGENT DESIGN

Most critics of ID argue that ID is "scientifically vacuous" and worthless. Undoubtedly several of their core arguments are lacking in substance, for example the irreducible complexity of the *flagellum* and blood-clotting, which Behe claimed as conclusive (Behe, 1996). Miller and Doolittle have challenged most of these arguments (Miller, 1999). Further Lamouroux and others have argued that the representation of "Darwinism" put forward by Johnson and others is seriously flawed and inaccurate. We are thus faced with the same question as over YEC. Why do so many Christians, especially evangelicals, accept ID arguments with alacrity, though they have been "demolished" by most scientists?

The slick answer is to say "Dawkins," but that does contain much truth. Dawkins has pushed his reductionist and antireligious agenda for decades beginning with the less-overtly antireligious *The Selfish Gene* in 1976 and recently overtly in *The God Delusion*. The evangelizing atheists who include Dawkins, Dennett, Pinker, Atkins, Crick, and others push an agenda which seeks to eliminate religion in any form, dismissing it as "contrary to reason" and since 9/11 as positively dangerous. It is these, who are often skilled authors and public speakers, who present the public front of science. However theirs is an atheism, which explicitly bases itself on science and above all evolution, which is considered to banish religion forever. Their followers are often even more strident as many blogs demonstrate (e.g. P. Z. Myers on *Pharangula*, and several others). Faced with that particular challenge, numbers of Christians are attracted by ID (or even YEC) as a way of opposing these atheistic claims. Hence as ID challenges Darwinism (whatever that is) at its heart, whether its naturalism or reductionism, the theism of ID has considerable attractions. Undoubtedly one should distinguish between the reductionist *Scientism* of these atheists and the *science* of evolution, but it is convenient both to the scientific atheists and ID to keeps this distinction blurred. Hence it can be said that ID and scientific atheism feed each other, and there is little room for moderation.

CONCLUSION

As Creationism in its many forms is the most visible and vocal expression of Evangelicalism and science, it often renders invisible those evangelicals, both past and present, who eschew any form of creationism. At the present time, and for the foreseeable future, YEC and ID are between them the high-profile responses of evangelicals to science in every country in the world, which has a significant Christian presence. With the continuing growth of Evangelicalism, it is inevitable that the type of controversy in churches and education will spread throughout the world, as has already begun to happen in Britain, Australia, and the nominally Catholic continental Europe and Orthodox Eastern Europe. Though beyond the scope of this study, it has also spread to the Muslim world.

Despite the prevalence of creationism today, there is a strong and vibrant strain of Evangelicalism in many countries which is accepting of mainstream science and to those we now turn.

NOTES

1. *Askageologist* is a service provided by the Association of Christian Geologists to provide answers either on geological or faith questions. "Answers" are provided by Dr. Ken van Dellen who used to teach geology at Calvin College.

2. This has been discussed at greater length by Matske (Pennock and Ruse, 2007) but was too recent to be considered.

Chapter 8

Evangelicals and Science Today

Because of the high visibility of Creationism, many, whether liberal Christians, atheists, or many scientists, often consider that all evangelicals are creationist. The media, left or right, often report such issues in a simplistic way. In the United States and the third world, many would assume that to be a real evangelical one must be YEC, as that is as the perceived logic of biblical inerrancy. Since the rise of YEC after 1961 there is some truth in this, but it does not do justice to the variety of Evangelicalism and science. From my historical survey, in previous centuries most evangelicals accepted mainstream science, whether the scientifically competent like Dana and Sedgwick or preachers like Cumming and Close. The anti-geologists, Seventh Day Adventists, and mid-Western Germanic Lutherans made little impact before 1900. In the first half of the twentieth century Price's Flood Geology began to make inroads into American Evangelicalism but only became a powerful force after the publication of *The Genesis Flood* in 1961. Since then there has been conflict between evangelicals who accept mainstream science and those who favor YEC or, more recently, Intelligent Design (ID).

As the center of the evangelical world is America, their situation tends to influence all world Evangelicalism, both in reality and perception. Many, especially evangelicals, would prefer to consider Evangelicalism and science in its purely religious aspects, but in America both Evangelicalism and science are highly politicized. Thus one cannot ignore the Religious Right and its understanding of science, especially evolution, the environment, and medical research. Closely related, overlapping, are right-wing Republican attitudes to science well chronicled in Chris Mooney's *The*

Republican War on Science (Mooney, 2005). Whereas many American evangelicals do not belong to the Religious Right, many do, like Pat Robertson and Jerry Falwell, and these have the most bucks. Robertson called down God's judgment on Dover, Pennsylvania, after the ruling against ID in December 2005. In the United States it is not possible to separate Republican hostility to science as in global warming, the WHO report on obesity, and stem cell research from YEC or ID. The Discovery Institute's and Bush's statements on evolution are good examples. Outside America this political aspect scarcely exists and so evangelical attitudes on science do not (yet) have political clout. Much of this is because there are relatively few as, for example, in Britain about 5 percent of the population are evangelical compared to 30 percent in the United States. However some groups, like *The Christian Institute* and *truthinscience*, are seeking to develop political clout.

Over the last forty years the evangelicals have come to prominence in many parts of the world, and, with Pentecostalists, are the fastest growing constituency of Christianity. There have been two evangelical presidents, Jimmy Carter and George Bush, who take diametrically opposed views on evolution. Their prominence and politicization in the United States is well known, as is the result of conflict. Much of this conflict is over science. Some evangelicals see this as part of the culture wars and an inevitable result of a biblical world view clashing with the secular.

However there is no one understanding of science among evangelicals, but rather a kaleidoscopic variety of opinions ranging from extreme YEC to being almost a theological liberal. The diverse opinions grade into each other. To illustrate the gradation of opinion, Eugenie Scott and Alan Gishlick (Scott, 2004, p. 57) attempted to catagorize reactions to evolution and religion as a *continuum* (with ID running through three to eight):

Flat Earthers
Geocentrism
Young Earth Creationists
Gap Creationism
Day-Age Creationism
Progressive Creationism
Evolutionary Creationism
Theistic evolutionism (spectrum from evangelical to liberal or catholic in theology)
Agnostic Evolutionism
Materialist Evolutionism

Scott stressed a gradation of views rather than distinct categories. If anything, it is a classification of a "splitter" rather than a "lumper" as it brings out the variety of evangelical views. Evangelicals are to be found in the first eight categories. Mainstream "liberal" Protestants and Anglicans are almost entirely theistic evolutionists, as are most Roman Catholics. However the few evangelical flat-earthers can be ignored. I would prefer to divide evangelical attitudes into three groups, which merge into each other. (I omit ID, because this forms a subset, which straddles the three, though ID tends to appeal more to Old Earth Creationists. Some TEs do support ID, as does the Roman Catholic, Michael Behe, and some YECs support ID, as do Paul Nelson and Nancy Pearcey. Some YECs, for example those from AIG, have scant respect for ID.) These three categories are:

1. Young Earth Creationism
2. Old Earth Creationism (accept geology and cosmology)
3. Theistic Evolutionism (accept evolution as well)

It is not possible to be precise on what proportion hold each of the three, as there have been no rigorous surveys. Surveys of the American population indicate that about half the nation rejects evolution and favors a young earth. From my knowledge of American evangelicals I suspect this reflects confusion in the face of a questionnaire rather than conviction. In the absence of a thorough survey I can give only my impression. About 40 percent of Americans are loosely evangelical and I would suggest that over half of those would respond by favoring a YEC position, with varying dogmatism. In Britain evangelicals number a few percent in a nation where less than 10 percent are churchgoers. My impression is that a minority of British evangelicals are YEC but this proportion is growing rapidly, including in the Church of England. In many ways these are irrelevant figures as most evangelicals are simply not interested in scientific matters. Many will reflect back what they had heard in the pulpit. Among evangelicals non-YECs are very much the silent majority in the UK and may be a large minority in the United States.

The attitude of evangelicals to science also varies according to geographical location and denominational history. In Chapter 5 I expounded how David Livingstone discussed the importance of place over the reception of Darwinism and how that can affect the response of those with almost identical theology. To my knowledge there has been no comparable study on more recent attitudes of evangelicals to science and particularly on evolution. The importance of geography may be illustrated by an evangelical shibboleth—drink. Among American evangelicals, especially those of a Holiness or Fundamentalist tradition, alcohol is taboo. That is less so for evangelicals of Lutheran or Dutch Calvinist roots or Anglicans. Among the British, Anglicans are least liable to be against drink, but those from independent churches more so. French evangelicals are typically French and happily drink wine and water it down for children. It does not take a genius to work out why America had prohibition in the 30s, the Temperance movement was strong in Britain in the Victorian era, and France is a major wine-growing country.

There have been no systematic studies of how evangelicals have responded to science either in particular traditions or denominations or places. A few general observations are in order. American evangelicals are far more likely to accept either YEC or OEC/ID than their coreligionists in Britain, Canada, Australia, or New Zealand. Thus Archbishop Jensen of Sydney, who is regarded as the most Fundamentalist of Anglicans, explicitly holds to an OEC, if not a TE, position, which is marked contrast to many church leaders in the United States. The reason for this can be summed up in the one word "Scopes" as this epitomizes the fundamentalist and antievolutionary course American evangelicals took in the 1920s. Britain and the Empire did not follow suit, but things are a-changing.

The majority of evangelicals throughout the world are hostile to evolution, coupled with a naïve appeal to Baconian empirical thought. Often this is not explicit, but even among evangelicals who would consider themselves more "liberal" there is often a suspicion that evolution has atheistic overtones and an unexpressed feeling that "We're no kin of the monkey." Among scientists who profess Evangelicalism there is a disproportionate number of engineers, physical, and computer scientists rather than biologists and geologists. Those who are bioscientists tend to be biochemists or biophysicists rather than evolutionary biologists. The ordered world of a biophysics or biochemistry laboratory tends not to deal with the messiness of biological forms in the field. One reason for opposition to evolution is the reductionism of some neo-Darwinists, especially Dawkins.

As well as considering how evangelicals respond to science one also needs to consider the number and variety of scientists, who are openly evangelical. There is no survey or register of them, but numbers are members of such groups as the American Scientific Association (ASA), Christians in Science (CIS), or equivalent groups in Australia and Canada, as well as those in YEC groups. Membership lists give some indication of the number and variety. Evangelical scientists are numerous with some outstanding scientists.

In Britain many evangelical scientists are members of *Christians in Science* (http://www.cis.org.uk/), which has some 400 members. Most have professional appointments, with a number of university professors (note, in Britain most university teachers are not professors but lecturers), as well as lecturers and scientists working for industry or research establishments. At least six are Fellows of the Royal Society (out of 1,300 fellows this is a low proportion), one Fellow of the British Academy, and at least three have been knighted for their services to science. A snapshot of some of the leading evangelical scientists of the early 90s may be found in *Real Science, Real Faith* (Berry, 1991). This showed how seventeen evangelical scientists considered their faith and their science. Almost all were TE, which is the dominant outlook of *Christians in Science* (and the RSCF before it). Most have made major contributions to their fields. Denominationally they included Anglicans, mainstream nonconformist and independent evangelicals.

Among the "leading lights" of *Christians in Science* are several significant environmentalists, who will be discussed in the next chapter, Sir John Houghton, R. J. Berry, and Sir Ghillean Prance. Other notables are Martin Bott and Bob White, both geophysicists, Malcolm Jeeves, a psychologist, David Livingstone, a historian of science, Derek Burke, geneticist, and Sir Brian Heap a biochemist and former Vice President of the Royal Society. To these may be added many more, who are almost as distinguished. From a previous generation could be added Donald MacKay and Sir Robert Boyd. All of these, and the many I failed to mention, have made good contributions to scientific research in their own fields and some have published on the relationship of science and religion.

So far I have described some who have achieved distinction and thus will be over fifty. Today, if anything there are more evangelical scientists of a younger generation, who are at present developing their careers. What has changed is the increase of YECs who hold university positions in science or engineering and who tend to be in the *Biblical Creation Society* rather than *Christians in Science.* Examples include Andy McIntosh, Phillip Burgess, and David Tyler, each of who have published YEC books, which sell well among evangelicals.

The situation in the United States tends to be similar. The American Scientific Association (www.asa3.org) has long attempted to be an umbrella group for evangelical scientists, and deliberately avoid making any policy statement on any aspect of science. We have already seen that over the shibboleth of evolution, as controversy has always been near, especially in the sixties when some left the ASA to form the Creation Research Society. However some YECs stayed, but the ASA has a more conservative complexion than the CIS, and their journal has some YEC articles. One competent observer, Ted Davies, reckons that 35–50 percent of ASA members are sympathetic to ID, which is a far higher proportion than their British counterparts. However, my observation is that ID is gaining ground among evangelicals in Britain.

Despite the far larger numbers in the ASA, there seems to be a lower proportion of top-ranking evangelical scientists than in Britain. These include Henry Schaefer, the Graham Perdue Professor of Chemistry at the University of Georgia, who was nominated for the Nobel Prize in 1994, and was cited for the period 1981–1997 as the sixth most-quoted chemist in the world. He is closely associated with ID and the Discovery Institute. Very different is Dr Francis Collins, the present Director of the Genome Project, a thoroughgoing evolutionist. Another is Owen Gingerich of the Harvard-Smithsonian Center for Astrophysics, a historian of science, whose life's work on Copernicus resulted in he popular science book *The Book Nobody Read* (Gingerich, 2006)

Charles Townes (b. 1915), who was awarded the Nobel Prize in physics in 1964 for his work on the maser and laser, is very sympathetic to evangelicals and on the advisory council of the ASA. In 1953, due partly to what Townes called a "revelation" experienced on a park bench, he invented the maser (Microwave Amplification by Stimulated Emission). By building on this work, he achieved similar amplification using visible light, resulting in the laser. He had chairs at Columbia, MIT, and Berkeley and was awarded the Templeton prize in religion in March 2005. After that he described himself, 'I'm a Protestant Christian, I would say a very progressive one. This has different meanings for different people. But I'm quite open minded and willing to consider all kinds of new ideas and to look at new things. At the same time it has a very deep meaning for me: I feel the presence of God. I feel it in my own life as a spirit that is somehow with me all the time.'

Several other ASA members have achieved distinction in science, for example, Richard Bube in materials science, especially photovoltaics, Ian Hutchinson in plasma physics, Ted Davis in the history of science, who is an authority on Robert Boyle, Fred Brooks in computer science, David Myers in psychology, and Cal de Wit and Richard Wright on the environment. Undoubtedly this list is incomplete, and some readers will be aware of some serious omissions on my part. It would be tedious to itemize every competent evangelical scientist working in a laboratory or college or university.

Since 1990 ID has made a considerable impact in America and most recent attempts to alter the teaching of evolution in schools have had considerable input form ID. Most of the leading lights are evangelical. A vast amount of ID literature has been published, but this is largely in the religious sphere or in philosophy. So far nothing to my knowledge has been published in peer-reviewed journals except for Meyer's notorious paper on the so-called Cambrian explosion in the *Proceedings of the Biological Society of Washington*, edited by the baraminologist Sternberg. There are questions whether the paper was properly peer-reviewed. (Numbers, 2006, p. 390 for references).

YEC has a tremendous following in America and though it is largely a grassroots movement has attracted a small but influential number of scientists. They tend to be engineers or teachers in Fundamentalist colleges, but count Raymond Damadian, one of the founders of MRI, among their number. YEC has received almost total rejection from the scientific community, but some support from Republican administrations and presidents.

However there are a large number of evangelical scientists in all echelons of scientific endeavor, whether in education from high school to university, or in industry. Many have Ph.D.s and are competent in their fields. Numbers teach science in evangelical liberal arts colleges in North America, where it is necessary to be both evangelical and scientifically qualified. At some, as at Liberty University, Cedarville, Bryan College, Bob Jones University, and others, it is mandatory to believe in a six-day creation, but others, such as Wheaton College, Illinois, Calvin College, Messiah College, are not so strict. What is taught in evangelical colleges varies tremendously. This will not be apparent in physics, chemistry, and computer science, but some colleges teach Biological and earth sciences from a young earth perspective. A recent survey of *Answers in Genesis* has shown that many colleges simply teach mainstream science accommodating those "billions of years."

If we look beyond Britain and the United States, we will find similar, but smaller, organizations in Canada (Canadian Scientific and Christian Affiliation [CSCA]) and Australia (Institute for the Study of Christianity in an Age of Science and Technology [ISCAST]), whose members exhibit a similar range of scientific expertise. These four associations (ASA, CIS, CSCA, and ISCAST) for evangelical scientists have a common purpose, operate in similar ways and work together. A prospective member has to sign a basic "Statement of Faith" acknowledging the authority of the Bible, the Trinity, and the person of Christ. Unlike YEC organizations little is done to enforce this, so that in the ASA a few "unbelievers" are members. Further there is a wide range of membership from YEC to moderately liberal theological views. However two groups of Christians tend not to be members. The first are liberal Christians, who in the United States would gravitate to the *Institute on Religion in an Age of Science* and in Britain would opt for the *Science and Religion Forum* or the *Society of Ordained Scientists,* though these include evangelicals. The second are YECs, who object the indifference or hostility to YEC from these organizations. Each of these does have YEC members, but many choose to belong to their own organizations. Several YEC members of the ASA, including Henry Morris, left to found the *Creation Research Society* in 1963. YEC members of *Christians in Science* have long been unhappy with the committee for failing to support them and some left after the annual CIS conference in 1992, because the tenor of the papers was anti-YEC.

Each organization produces a journal. The CIS, in conjunction with the Victoria Institute (see Chapter 5), have jointly published *Science and Christian Belief* (http://www.scienceandchristianbelief.org) since 1989. This journal has a rigorous peer-reviewing system. It is not sectarianly evangelical, as it has published articles by Arthur Peacocke, John Haught, Simon Conway Morris, and responses by Richard Dawkins. Its ethos is of open Evangelicalism and though the consensus of articles probably represents the sponsoring bodies, individual articles may not. The ASA journal, *Perspectives on Science and Faith* (formerly *Journal of the American Scientific Affiliation)* has been published since the 1940s and also has peer-reviewing.

The two North Atlantic journals publish articles, short contributions and book reviews on the whole range of science and faith issues. Some volumes focus on particular topics and over the years deal with the whole range of science and faith issues such as the environment, cosmology, genetic engineering, ID and of course, the ever present YEC. The last topic frequently gets short shrift. The ASA and CIS also hold annual national conferences along with more local regional ones. Every decade a joint ASA/CIS conference is held in Britain. Along with all the formal activities these organizations act as a support group, especially for younger scientists.

In 2005 two members of CIS (Denys Alexander and Bob White), who are science fellows at St. Edmunds College, Cambridge, founded the Faraday Institute with funding from the John Templeton Foundation to explore science and religion at depth. Since 2006 they put on many courses drawing on competent scientists throughout the world and attracting students from Muslim countries as well. The basis is a rigorous academic approach within a broad evangelical framework. Some evangelicals would say too broad.

This brief discussion should demonstrate that one cannot argue that cannot be an evangelical and a leading scientist.

THE DIVERSITY OF EVANGELICALS ON SCIENCE

By now, it should be clear that evangelicals take a wide variety of outlooks on science, which defies clear description and classification. We are liable to get different answers from evangelical scientists, clergy, and laity. Part of the problem is that it is rare for someone to be well versed in both science and theology. Many evangelical scientists are very competent in their field, but are only vaguely aware of other sciences and their theological understanding may not have progressed beyond teenage Sunday school.[1] Many clergy have no science beyond high school. Science scarcely figures in seminary courses, beyond a bit of apologetics and references to Darwin (frequently inaccurate!) in church history. It is dangerous to generalize for the average evangelical layperson, but it is tempting to say that many combine the weaknesses of the evangelical scientist and the evangelical clergy! They often reflect what they are taught from the pulpit and in bible class, with little reflection. As a result understandings of the relation of Christianity and science range across the spectrum from Young Earth Creationism to neo-orthodoxy, with a mythical Genesis and in theological thinking from the simplicities of saying "The Bible says" to highly sophisticated grappling with all issues with rigorous Biblical hermeneutics and a questioning, yet evangelical, approach to theology.

This is seen most clearly on interpretations of Genesis. There are many who interpret Genesis literally, however, I must re-iterate that, many who do so are not dogmatically committed to a Young Earth position, but uncritically reflect what they are taught. Then there are those who accept the vast time scale put forward by cosmologists and geologists and argue that Genesis does not preclude such a time scale or allow science to influence their interpretation. On this point there is often controversy as several YEC leaders, for example Ken Ham, Morris, and Jonathan Sarfati, accuse other evangelicals scientists of "compromise" and allowing science to dictate their biblical interpretation. Apart from the strict literalism of YECs, evangelicals hold a variety of interpretations summed up in decreasing conservatism as the Gap Theory, Day Age, the Framework hypothesis and the totally symbolic. Though the Gap Theory and the Day age interpretations have an illustrious heritage and were widely held before Darwin, today they are only held by the most conservative evangelicals who do not accept YEC. As a result there are often held by those who have recently "escaped" from YEC like Richard Fischer and Glenn Morton, who wish to see Genesis as historical in the strong sense, and strive to shoe-horn archaeological and paleoanthropological data into what they consider a literalist interpretation of Genesis. The absurdity is apparent when Morton attempts to prove that Noah's Flood occurred 5.5 million years ago.

A similar divide or rather gradation is to be seen over ethical issues whether the environment or medical and genetic issues.

TWO BOOKS

An old and time-honored way of considering science and Christianity is the doctrine of Two Books. This goes back to Francis Bacon in the early seventeenth century and up to the mid-nineteenth century provided the main method of reconciling the two. It is argued that God revealed himself in two books. The first Book is the Bible and the second is the Book of Nature. This had the great advantage of recognizing both spheres and allowing scientists a certain autonomy in their science, but it originated in a desire to keep science in its place, though clearly more than a handmaid of theology.

By the nineteenth century the doctrine of Two Books was only retained by Christian apologists for science and rejected by most scientists. The following century only some evangelicals, who wished to steer a middle course between "creationism" and theistic evolution, used it. Today it is rarely adopted as most theistic evolutionists have "moved on" and adopted some kind of *complementarity*, and YECs wish to have science under the dominion of biblical teaching. The clearest recent exposition of the Two Books is by Hugh Ross (Ross, 1994, pp. 56–57), where he prefers to call nature the sixty-seventh book after the sixty-six books of the Bible. To a YEC this denigrates the Bible and Sarfati argues that Ross's approach is to *compromise* (Sarfati, 2004, p. 41ff). Further strictures are given by the AIG historian Mortenson, who argued that the separation of the study of nature and of scripture by Galileo and Bacon has had dire results as it has resulted in the subordination of scripture to science (Mortenson, 2004, pp. 19–25, 228–230).

COMPLEMENTARITY

In the 1960s Donald MacKay developed his concept of *complementarity* to explain the difference between science and faith. He expounded this in several papers, which are found in his collected essays *The Open Mind* (MacKay, 1988, passim). This has been widely followed by evangelicals of the ASA or CIS mould, but is often criticized for totally separating science from theology, a charge MacKay thought unjust. In his Gifford Lectures of 1997–1998, R. J. Berry develops this theme, which was published as *God's Book of Works: The Nature and Theology of Nature* (Berry, 2003), which at first sight may seem to be in the Two Books tradition. He emphasized that *complementarity* does not lead to two independent models as both science and theology seek to understand the same reality, and provides the best way of relating science and theology.

OLD EARTH CREATIONISTS

This cumbersome term comprises those who accept Deep Time but reject evolution. Historically we saw these in the majority of immediate pre-Darwinian Christians like Sedgwick, and those after 1859 who were not able to accept evolution like Dawson and Duns. A goodly proportion of interwar Americans held it, as did the founding fathers of the ASA. Essentially the Progressive Creationism of Bernard Ramm is OEC. In Britain, OEC tended to be held by nonconformist evangelicals before 1970 as Anglicans tended to be TE.

After the rise of YEC, OEC was caught in the crossfire, and thus many of the rising generation became YEC, rather than move to the left. Thus Billy Graham was OEC with a nod to evolution but his son Franklin is YEC. Contemporary evangelical theologians in the United States are split between YEC and OEC, with a greater proportion of YEC than fifty years ago. Wayne Grudem, a leading systematic theologian is YEC, grudgingly tolerates OEC but has no truck with evolution (Grudem, 1994, pp. 273–309). Millard Erikson is OEC but of the more "liberal" evangelicals more and more tend to ID in preference to classic OEC, as does Bishop Fitzsimons Allison. The British scene is different as more accept evolution but even so, ID is becoming more popular and two leading evangelical bishops Tom Wright and Michael Nazir-Ali lean towards ID.

We have already considered the case of Daniel Wonderly, but probably the most high-profile OEC today is Hugh Ross (1945–), who is vilified by AIG but seen as creationist by many. (On *Panda's Thumb*, his site is listed as Creationist.)

HUGH ROSS, *REASONS TO BELIEVE*

Hugh Ross, a Canadian, gained a Ph.D. in astronomy from Toronto in 1973 and was a research Fellow in Radio Astronomy, California Institute of Technology, from 1973 to1978. Since 1976 he has been a pastor at Sierra Madre Congregational Church in California and founded *Reasons to Believe* in 1986. The *Reasons to Believe* Web site modestly describes itself as "The premiere science-faith think tank."

With his background in astronomy, Ross accepts conventional astronomy and geology with its vast ages, but does not accept evolution and inclines to ID. He also accepts the vast patriarchal longevity. To the strident atheist, Ross is no better than a YEC from AIG or ICR, but many YECs consider him as a total compromiser who has sold out to Godless evolution. Jonathan Sarfati, formerly of AIG and now of CMI, devoted the whole of *Refuting Compromise* (Sarfati, 2004) to refuting the "Progressive Creationism (Billions of Years) as popularised by Astronomer Hugh Ross" to refute the errors of Ross with considerable vitriol referring to "his ostensibly Christian apologetics ministry" (p. 13) This book is worth reading if only to see how YECs are as bitterly opposed to Progressive Creationists, who accept Inerrancy and all other *conservative* evangelical beliefs as they are to TEs or even atheists like Dawkins.

Reasons to Believe is probably the largest *progressive* creationist outfit and employs a large staff, is assisted by numbers of competent scientists and theologians, including Norman Geisler, Walter Kaiser, J. P. Moreland, and Allan Sandage. Its staff and supporters have extensive speaking engagements and encourage local chapters. Ross and others have published many books and articles and their Web site is very comprehensive and extensive, though its claim to be "The premiere science-faith think tank" may not convince all. However it does have a fulsome range of articles and audio clips. Mainstream scientists, whether Christian or not, would find little to object to on the scientific content except where it impinges on evolution and paleoanthropology.

THEISTIC EVOLUTION (TE)

A significant minority of evangelical scientists and theologians accept a more or less unqualified evolution, though this may be the majority among "whites" living outside the United States. The broadening effect of the ASA and the writings of Ramm, Kulp, and Mixter encouraged more of the rising generation of American evangelicals to accept evolution at the time when Henry Morris was reviving YEC. This meant that those who were OEC were caught between a rock and a hard place, and so OEC was squeezed by the pressure to be either YEC or TE. The term Theistic Evolution is unsatisfactory, but it is widely held and some try to change it by calling themselves Evolutionary Creationists, which has not caught on.

In Britain there had long been a tacit acceptance of evolution both by theologians and scientists, and thus many scientists and clergy, who were trained before the mid-1970s, simply accepted TE by default. There were a few exceptions, notably Dr. Martin Lloyd-Jones, a doctor turned pastor, but in the early postwar decades theologians like J. I. Packer, or John Stott, almost took evolution for granted.

Likewise the RSCF under the guidance of Oliver Barclay, who eirencially supported evolution when it was no cause for controversy. In the 60s evangelical scientists like Donald MacKay, Malcolm Jeeves, Frank Rhodes, a geologist who ended up as President of Cornell and Robert Boyd, took this up. This was expressed in their writings, which had much in common with the mainstream Christian writers on science like E. L. Mascall, A. Smethurst, G. D. Yarnold,[2] and C. A. Coulson. Apart from Coulson, a Methodist lay preacher who had chairs in mathematics, physics, and chemistry at Oxford, these were Anglican clergy. The consensus of the evangelicals in the RSCF took Genesis nonliterally but tended to insist on the first pair Adam and Eve and the historicity of Noah's Flood, even if it was a local inundation.

By the end of the twentieth century Theistic Evolution in one guise or another was put forward by numbers of evangelicals as by Berry, White, Alexander, and Wilkerson in Britain and Collins, Keith Miller, and others in the United States (Miller, 2003). Relatively few evangelical theologians explicitly support theistic evolution as opposed to more "creationist" positions, with the main ones being McGrath and Torrance in Britain and Murphy in America (Murphy and McGrath both have Ph.D.s in science). It is difficult not to conclude that the "Evolution Wars" have made evangelical theologians wary of supporting a position at loggerheads with many evangelicals.

Even such a scholar as the historian, Mark Noll seems to hover between ID and TE despite his book *The Scandal of the Evangelical Mind* (Noll, 1994), which is a critique of the intellectual flaws of evangelicals. He highlighted and expounded the major ones in politics, science and the preoccupation with "end-times." These are not separate issues but are inter-related. Noll brings together Evangelicalism's intellectual failings over the last 150 years, particularly the retreat into fundamentalism. His chapter *Thinking about Science* focuses on the change of evangelicals on science from Asa Gray and his contemporaries through the antievolutionists of the 1920s to modern Creation Scientists, which he sees as a disastrous decline. As Noll wrote in the early nineties, he does not deal with ID, which he now has some sympathy for.

However many evangelicals throughout the world are increasingly accepting of YEC almost by default. This needs to be clearly understood as often there is no dogmatic or doctrinal insistence on YEC, but now it is perceived as the most reasonable evangelical belief. In Britain this was not the case before about 1980. A study of statements of doctrinal beliefs, like that of the University and Colleges Christian Fellowship, the International Federation of Evangelical Students, the various Evangelical Alliances throughout the world, will show that most refer simply to God as Creator and make no mention of time. Many more recent (post-1980) statements of faith for churches or evangelical groups in all countries increasingly insist on a six-day creation, which is due to growth and influence of YEC.

So why do so many evangelicals seemingly accept YEC by default, without rigorous thought? It is more than a widespread naïve literalism, and a brief discussion is liable to oversimplify. Ultimately it is a result of the almost iconic status of the Bible among evangelicals. However much evangelicals insist that they follow sound interpretative principles and use a careful hermeneutic, what is most often heard is that, "The Bible says." That is the catch phrase of Billy Graham and many evangelical preachers and thus it is perceived that not only does the Bible have absolute authority in matters of faith, but also it becomes iconic. As the Bible is expounded to the "plain" man and woman, and, especially, child, the plain person will look for the plain meaning of scripture, and thus the Bible will be read as the account of football match, without any consideration of poetry or figurative speech. Simply accepting the plain meaning of scripture, instead of resulting in asking what kind of writing a part of the Bible is, results in a naïve literalist interpretation, especially on Genesis and prophetic parts of the Bible, which are most clearly poetic (try a Web search; *Bible, plain, meaning*). With a limited knowledge of geological and biological science and of their history in relation to Christianity, the simplest solution is accepted. This provides a ready and receptive audience for the YEC speaker who knows which buttons to press to capitalize on what "The Bible says."

The question behind all this is a hermeneutical one on what exactly the Bible *does* say.

EVANGELICAL COLLEGES AND SEMINARIES AND SCIENCE

North America

In the United States, and to a lesser extent in Canada, there are a great variety of evangelical colleges. These include Liberal Arts Colleges, Bible Colleges (often renamed as universities) and seminaries. These vary widely in both academic standards and faith perspective, but draw their students from teenagers from evangelical homes and churches. Most colleges teach science, often as a preparation for a career in nursing or similar medical profession, but relatively few teach the whole range of science. The perspective on science teaching depends on the faith stance of the college. Thus colleges which adhere to YEC in their basis of faith may teach most sciences but do so from a YEC perspective as do Bob Jones University, Cedarville College, Liberty University, Bryan College, and the various Concordia Colleges which are part of the Lutheran Church Missouri Synod.

Very different are the colleges like Wheaton, Calvin, Gordon, and Messiah Colleges, which teach the whole range of science. Their teaching reflects mainstream science, especially over geology and biology, even though many students come from a "creationist" background. The staff needs to be diplomatic as some parents are very concerned that their offspring are only taught "creation." Needless to say that this can cause crises of faith for students, with some opposing the staff and others rejecting their heritage. Some professors have published textbooks, but, because of the nature of Liberal Arts Colleges, relatively little research. A good number of students go on to do doctorates in mainstream universities, as the students receive a science education as least as good as most secular colleges.[3] Even so, some professors have a good publications record as do Davis Young on the history of geology, Richard Wright on ecology, and Edward Davies on the history of science.

The divisions and conflicts are reflected in some colleges like Taylor University and Dordt Colleges, which are betwixt and between the "liberal" colleges discussed above and overtly YEC colleges, not to mention universities, which favor ID like Biola, where the philosopher J. P. Moreland teaches. There college staff across the faculties vary in attitude to science at times causing tension. The YEC colleges like Liberty teach Creation Science and some object to any support of global warming as this is "junk science" peddled by evolutionists.

There is also a wide variation in evangelical seminaries. Fuller is probably the most "liberal" and has long been criticized by the more conservative (see Marsden, 1987). At the risk of overgeneralization most evangelical seminaries reject some aspect or another of mainstream science and incline to either ID or YEC and, at times, adopt antienvironmentalism. However, it would be wrong to be dismissive from a scientific angle, as, for example, some good hermeneutical work is be done at Westminster Theological Seminary, especially by Peter Enns, and Nigel Cameron, President of Trinity, Deerfield, has done valuable work on medical ethics. It is inevitable that seminaries will reflect the present ethos of American Evangelicalism.

Britain and Elsewhere

In Britain, Australia, and New Zealand there are no colleges to compare with American Liberal Arts colleges and all places of higher education are basically secular, though there are few church-based colleges and universities like Gloucester, Chester, and the University of Cumbria, where science is taught in a totally secular way. There is a wide range of seminaries (or theological colleges as the British call them). Half of the Anglican theological colleges are evangelical, but science does not figure much in their courses, except in apologetics. The present principal of St. John's College, Durham, David Wilkerson, has a Ph.D. in astrophysics and, unusually for an Anglican college, is a Methodist minister. The general ethos reflects mainstream science except for Oak Hill. Similar are the two Baptist colleges— Bristol and Spurgeons and the London School of Theology. Ernest Lucas, vice-principal of Bristol Baptist College and McGrath, former principal of Wycliffe, Oxford, both have science Ph.D.s, and have written widely on science and religion. There are a plethora of small independent colleges for evangelical denominations, which are frequently YEC, which may be affiliated to local universities so that they can give accredited degrees. The situation in Australia and New Zealand is similar.

KEY EVANGELICAL WRITERS ON SCIENCE

With the plethora of YEC and ID writings, which have the seal of approval of many evangelical theologians, like Wayne Grudem, the multitude of writings by scientific evangelicals who are neither YEC nor ID but follow, or rather agree with, mainstream science are often not heard in the noise of controversy.

The number of publications has been increasing exponentially over the last half century and are to be found in journals like *Perspectives in Science and Christian Faith* and *Science and Christian Belief* and the ever increasing number of books and Web articles by evangelical scientists or theologians.

Many are apologetic and may just be a scientist attempting to explain why a scientist can be a Christian. The quality varies immensely. Due to the controversial effect of YEC and ID, writings tend to fall into one of three camps YEC, ID, or TE and intermediate forms are rare.

As discussed previously the first and most significant evangelical book on science and religion was Bernard Ramm's *The Christian View of Science and Scripture* (Ramm, 1954), which laid the foundation how most evangelicals, apart from creationists, grappled with science and religion for several decades. Most of the writers up to about 1980 were following in his path whether from the ASA or the RSCF, and is thus THE seminal book for the immediate postwar era. Ramm has received heavy criticism for being semi-fundamentalist and James Barr in *The Fundamentals* (1977), criticized Ramm for dealing with such topics as Jonah, forgetting the fundamentalism Ramm was addressing.

As we saw in Chapter 6, immediate postwar evangelicals moved from fundamentalism to a more accepting view of science. In Britain the RSCF attracted scientists like Robert Boyd, Donald MacKay, and Malcolm Jeeves who produced a steady stream of publications from 1960. The next generation was more numerous and now most are in an active retirement. These include R. J. (Sam) Berry, Derek Burke, Sir Brian Heap, Sir John Houghton, Sir Ghilean Prance, and Colin Russell. However their writings have tended to be on Christian aspects of their own fields (ecology, global warming, genetic engineering, or history of science) rather than a theological overview integrating their theology and science as have Polkinghorne and Peacocke. In private, Peacocke has criticized CIS for tending to be too biblicist, but that reflects his own liberal theology.[4] However they probably had more in common with Peacocke than fellow evangelicals of a YEC persuasion. The next generation includes Ernest Lucas, Colin Humphreys, David Livingstone, Bob White and Denis Alexander.

The situation is similar in the United States, and over the last forty years there has been a tremendous interchange between the ASA and the RSCF/CiS, with a joint conference every ten years in Britain. Again we see a succession of contributors. As America was more anti-evolutionary than Britain this included more old earth creationists like Peter Stoner (1888-1980) whose OEC books were widely read and other founders of the ASA. In many ways, Hugh Ross is a successor to them as he takes a similar stance on both science and theology and it is easy to see why they and Ross are considered to be creationist and close to YEC by secular evolutionists.

However many members of the ASA moved on from the "gappist" OEC of their founders, who, as I argued earlier, were progressive in their own way. Hence by the 1960s there was an immense variety among members of the ASA and this was reflected in their publications and the articles in their journal. Most were scientific apologetics or short studies on particular themes. Initially they were mostly of the type "Can an evangelical be a Christian?" and dealt with the immediate theological questions, similar to Ramm in his earlier book. As in Britain after the 60s more writers considered ethical issues like the environment, genetics, and reproduction. Richard Bube, possibly the leading evangelical scientist of the 70s, wrote *The Human Quest* (1971) and *Putting It All Together: Seven Patterns Relating Science and Christian Faith* (1995) and edited *The Encounter between Christianity and Science* (1968) as well as writing many papers and articles often in the JASA/PSCF.

In recent years evangelical scientists have published widely with the PCSF as the main journal. Relatively few evangelical theologians have written on science, probably because few have the double training needed. One who does is George Murphy, a Lutheran pastor (ECLA) who has written several works on science, which start from the person of Christ rather than creation. Thus suffering is seen from the perspective of the cross rather than a possible event in Eden. The more conservative evangelicals regard Murphy as too liberal and his theology does resonate with that of Barth and Moltmann, on their nonliteral view of Genesis and the Crucified God respectively. Some start from creation and thus cannot make the linkage with Christ and redemption. This can be seen in the writings of the former evangelical Howard Van Til, whose theology emphasized a front-loaded creation, but said little about the cross. His volumes *The Fourth Day* and *Portraits of creation* are two of the most important evangelical works on science of the 80s and 90s, but are criticized by Murphy for being weak on soteriology. Toward the end of the 90s van Til moved from an evangelical faith to a progressive Unitarianism.

Most of the evangelical writing on science is by scientists rather than theologians. Many of these are either apologetic works of varying quality on the theme "Why I am a Christian and a scientist" or detailed Christian commentaries on their particular field. A fine example of the former is Francis Collins' *The Language of God,* which almost forms a riposte to Dawkins' *The God Delusion.* Collins is, of course, director of the Genome Project and converted from atheism to Evangelicalism when in his late twenties. He did not claim that his book was deep theology but rather an explanation why he, as a leading scientist, could be a person of orthodox faith. Reviews varied and the most critical were by either "creationists" or atheists. In 2006 *Time magazine* published a debate between Collins and Dawkins. Owen Gingerich emeritus professor of astronomy and of the history of science at Harvard, delivered the 2005 William Belden Noble Lectures, which have been published as *God's Universe* (Gingerich, 2006) almost covering the same issues from another scientific perspective. In this work Gingerich writes of God's purpose in the universe, which some regard as verging toward ID, but Gingerich rejects this preferring some kind of lower case *intelligent design* as he terms it.

Far more theological are the collection of essays *Perspectives on an evolving Creation* edited by Keith Miller (Miller, 2003), a geologist from Kansas, which from a broad evangelical perspective give theological reasons for evolution and approach the subject from a variety of disciplines theological, historical, and scientific. It does not give definitive conclusions, but shows how evangelicals are grappling with questions normally sidestepped.

Few writers have attempted a broad theological overview as did Peacocke from his liberal perspective in his many book from *Science and the Christian Experiment* in 1971 until his last, *Paths from Science Towards God: The End of All Our Exploring* in 2001. There are probably several reasons for this. Evangelicalism has never produced a probing wrestling theology, unlike the liberal variants typified by Peacocke and those attracted by the ethos of *Zygon*. However in recent years one writer has begun to do so, having spent two decades in preparation and that is Oxford professor, Alister McGrath.

ALISTER MCGRATH

Alister McGrath (1953–) is a relative newcomer to science and religion as despite having impeccable qualifications on science and theology from Oxford University and being a prodigious theological author since the early 80s, his first book on science and religion *The Foundations of Dialogue in Science and Religion* (McGrath, 1998) was only published in 1998. After schooling in Northern Ireland, McGrath graduated in chemistry at Oxford in 1975, and then obtained a D.Phil. in molecular biophysics under Professor Sir George K. Radda, FRS, in 1977 and a first in theology in 1978. He was ordained in 1980 and was appointed lecturer in Christian doctrine and ethics at Wycliffe Hall, Oxford in 1983 and has remained there ever since, relinquishing his position as Principal in 2006. He was made full professor in 1999 and awarded a D.D. in 2001. He is most definitely an evangelical in the tradition of Packer and Stott. To many Americans he might fall short on inerrancy, but his work is as widely read in America as in Britain and has recently joined forces with Ravi Zacharias to develop apologetics in Oxford.

Most of his writing has been on systematic and historical theology. His *Christian Theology, An Introduction* (McGrath, 2001) is rightly a standard work. His works on historical theology span the time scale from the Reformation to the present day. Unusually for one trained as a scientist his *Christian Theology* makes scant reference to science, which struck me as a major omission. That deficiency was soon to be remedied with a vengeance and it is best to let McGrath speak for himself. McGrath began to think of the relationship of science and religion in 1976, but allowed these thoughts to slowly mature for over two decades. As he says, he looked to the Scottish theologian Thomas Torrance for inspiration, who over several decades had related his orthodox, almost evangelical, faith to science. The first fruits of McGrath's quest were in 1998 followed in the next few years with his massive three volume study *A Scientific Theology* (McGrath, 2002–2003). The three volumes are entitled respectively, *Nature, Reality* and *Theory.* This trilogy is probably the most massive treatment of science and theology in the Anglo-Saxon world, and in perspective is conservative and orthodox and further removed from YEC approaches than it is from liberal theologies like that of Arthur Peacocke. Over the last half century many have tried to devise a "scientific theology," beginning with Coulson, Mascall, Yarnold, and Pollard in the 1950s. The three most significant writers in recent years have been the liberal protestant Ian Barbour, the liberal Anglican and process theologian Arthur Peacocke, and a more conservative Anglican, John Polkinghorne. All have been awarded the Templeton Prize for progress in religion. Of these Polkinghorne is closest to Evangelicalism, as evangelicals would consider both Barbour and Peacocke as being weak on atonement and salvation. To actually reflect theologically on science is a far harder task than to either carry out research science or to be a historian of science and religion.

In his Web article, McGrath perceptively and self-critically writes, "My weaknesses, however, must also be acknowledged. While I am completely at home in the worlds of mathematics, chemistry, physics and biology, I do not feel totally at ease in other natural sciences, and thus have not attempted to make correlations beyond my field of competence."

As well as his massive treatment McGrath has attempted to engage fellow Oxford professor Richard Dawkins in his small book *Dawkin's God* and *The God Delusion.* He has engaged him both in print and debates. He has also written a series of shorter books: *Science and Religion, An Introduction* (McGrath, 1998b), which is still a substantial volume, and two smaller volumes *The Twilight of Atheism* and *The Re-Enchantment of Nature.* It would not be unreasonable to expect McGrath to publish further significant works in the coming years, which may also act as a bridge between secular and liberal theological approaches to science and religion and the more conservative, which tend to dominate at present.

EVANGELICAL THEOLOGIANS AND SCIENCE

Over the last few decades evangelicals have published an immense amount of theological work, ranging from popular to technical articles, and all aspects of theology from biblical study, doctrine to apologetics.

There is an immense variety of perspectives from the most conservative, dare I say Fundamentalist, to the almost liberal. Even though the quality varies, many have excellent theological training and expertise, whether in Greek or Hebrew, history, philosophy or hermeneutics. Relatively few have scientific qualifications and thus McGrath is the exception. However this also applies to non-evangelical theologians.

SCIENTIFIC WARFARE BETWEEN EVANGELICALS

In the previous two chapters the "warfare" between some evangelicals and science was briefly documented, where battle lines were drawn between "creationists" and mainstream science. Since 1961 the conflict with mainstream science and first YEC and then ID has grown and is now a major feature of the relationship of Evangelicalism and the rest of society. It has attracted a vast amount of media attention, though often the reporting is of dubious accuracy. What is often unknown or ignored is the friction, which at times degenerates to warfare, between different groups of evangelicals over science.

Before 1961 this was fairly muted and as discussed previously began to appear following the ASA's rejection of YEC in the 50s. This was a non-issue in Britain and remained so until about 1970. As YEC gained ground in the 70s, more and more conflict occurred between the Young Earthers and Old Earthers, who were either OEC or TE. Soon the gradual spectrum from YEC to TE was replaced by hardened positions with YECs headed by Henry Morris, claiming orthodoxy and reviling others for succumbing to liberalism and by 1964 the split was irrevocable. Both the ASA in the USA and the RSCF/CIS in Britain attempted to be reconciliatory, but few YECs would countenance an old earth as an evangelical option. The placatory approach of most in the ASA and CIS had no effect and as ID and AIG got off the ground in the early 90s, the situation became more polarized. This resulted in a breakdown in discussions so that most TEs simply avoided the subject to prevent conflict. Relatively few evangelicals were as belligerent to YEC or ID, as the other way round. Consequently evangelical criticism of YEC and ID has been muted, and hard-hitting criticisms came from outside the evangelical fold, some of which was hostile to all things evangelical and often to all things Christians.

Even though proponents of ID or YEC are very strident in their criticism of "Evangelical Darwinists" both the ASA and the CIS are very muted in response, though the content of most of their publications whether in journals or that of individual members was strongly TE or OE. It seems that no one wished to offend their Christian brethren. If anything the criticisms from ID and YEC have become more strident since the beginning of this century, as exemplified by Ken Ham and Jonathan Safarti (AIG and

CMI) on one hand and Bill Dembski and colleagues on the other. As I write a recent charge by Dembski in his blog *Uncommondescent* is that the ASA has both attacked fellow believers of YEC conviction for their stance on the age of the earth[5] and has failed to take on the atheism of some forms of Darwinism, especially that of Dennett and Dawkins. This resulted in heated comments both on Uncommondescent[6] and the ASA listserve in April 2007. It makes sorry and inaccurate reading and it may seem remarkable that evangelicals can pour out so much vitriol on their brothers and sisters in Christ.

To the outsider, it must seem both odd and unseemly that evangelicals who are so close theologically should be at such loggerheads over certain issues. The friction is two-fold. The first is over the age of the earth—the essence of YEC and the second over the refusal of many evangelical scientists to condemn all forms of "naturalism"[7] combined with allegation that they are unwilling to engage with and seek to counter atheistic scientists like Dawkins.

Since the 1940s many evangelical scientists have sought to counter young earth arguments, but have often been rather muted. By and large conflict has been avoided and strident voices against YEC are rare, in marked contrast to the stridency YECs often adopt in tackling other evangelicals for "compromising" the authority of the bible. Even the most cursory survey of YEC Web sites or publications will produce many examples of how fellow evangelicals are accused of some kind of heresy. However in the Web sites of the ASA and CIS, belligerence against either YEC or ID is rare, and although most Web articles reflect one or both of an Old Earth and evolution, some are either YEC or ID. This reflects the purpose of both organizations to include all evangelical scientists, whatever stance they take on any issue.

A major plank of the ID project is to overthrow "naturalism" in all its forms. Clearly, naturalism as philosophical or materialistic naturalism precludes any belief in the supernatural, but most of science operates with a *methodological* naturalism. Philip Johnson (Johnson, passim) bases his critique of evolution by conflating methodological naturalism with metaphysical naturalism and maintaining that evolution is metaphysical rather than scientific. In *The Tower of Babel*, Robert Pennock discusses this and emphasizes the difference between metaphysical and methodological naturalism (Pennock, 1999, chap. 4). Since Pennock wrote this in 1999, this is still a major argument used against evangelicals who are in the scientific mainstream. A recent example is the book by Michael Rea *World without Design: The Ontological Consequences of Naturalism* (Rea, 2007), which argues that Naturalism leads to non–realism. He considers the alternative research programs to be intuitionism and supernaturalism, and that, often, intuitionism is self-defeating. That leaves only supernaturalism (or *theistic science)*, which is rejected by many evangelicals in favor of methodological naturalism.

If one visits Web sites like AIG, CMI, ICR or less well-known Creationist groups one will find a series of articles criticizing fellow evangelicals with as much hostility as atheists. The content of these articles is often repeated in books like Sarfati's *Refuting Compromise*. The success of this kind of attack by both YEC and ID becomes apparent as frequently evangelicals conclude that it is "evolutionary" evangelicals who are at fault and that if they respond, either gently or in kind (which is very tempting), it is they who are at fault for either "rejecting the Bible" or fanning the flames of controversy.

FROM SCIENTIFIC DIVERSITY TO ETHICAL DIVERSITY

The incessant controversy between evolution and creationism among evangelicals must seem wearing and pointless to many. Whether we like it or not it has been the dominant theme among evangelicals concerning science and often prevents adequate discussion of other issues.

The many issues raised by science impinge on a vast range of theological understandings and doctrines. These range across the nature and interpretation of the Bible, questions of theodicy, sin, atonement and redemption, miracles, and ultimately the nature of God. Few can be, or have been, studied at any depth in a brief introduction. Almost without exception evangelicals take their beliefs very seriously and objections to any branch of science are not from contrariness but from a heartfelt concern that their whole belief system is under threat. Conversely the acceptance of any science rarely derives from a "liberal" tendency, but from a theological perspective, which is only subtly different from those who come to different conclusions. The difference in theology between a Wayne Grudem and an Alister McGrath may not be very different and to a liberal theologian or an atheist or agnostic may be perceived to be almost identical, but the two come to radically different conclusions.

So far only passing mention has been made of ethical issues raised by modern science. These include the environment, bioethics, and various aspects of technology. Each of these merits a book length treatment and to them we now turn in the next chapter.

NOTES

1. This also applies to many scientists and other "educated" people, and contributes to the acceptance of the conflict thesis.

2. I have in my possession a letter from Coulson to Yarnold criticizing Yarnold's mistakes in chemistry in a book written for older high school students in 1951.

3. This is my personal observation having taught a course for Wheaton College in 2001, when I could compare what Wheaton taught with British universities.

4. Arthur died in November 2006, and this comment is based on interacting with Arthur for twenty-five years, rather than any public writing of his. He avoided any public criticism of evangelical scientists and there was mutual trust and respect. On a personal note in the 1950s Peacocke visited my uncle G. D. Yarnold, a priest physicist, as he began his work on science and religion.

5. This issue concerned was the project initiated by the ASA in 2002 to provide educational material setting forth the arguments for the vast age of the earth, both theological and scientific. For various reasons this never saw the light of day.

6. http://www.uncommondescent.com/philosophy/american-scientific-affiliation-whatever-happened-to-its-mission/

7. Most followers of ID regard any naturalism as wrong and do not agree with those who draw a distinction between *metaphysical* naturalism, which denies the possibility of God, and *methodological* naturalism, which is seen as a method, which leaves God to one side during scientific research rather than denying him.

Chapter 9

Evangelicals, the Environment, and Bioethics

In this chapter we will consider how evangelicals understand ethical issues raised by science. It is very easy to assume that the loudest voices from the evangelicals are typical and thus conclude that evangelicals are in the forefront of *The Republican War on Science* (Mooney, 2005), echo the opinions of Tom Bithell in *The Politically Incorrect Guide to science* and thus deny global warming, have no concern for energy usage (after all the Rapture is imminent), and regard embryo stem cell research as murder. That simple scenario is true for *some* evangelicals in America but it is most definitely not true for all American evangelicals nor for all evangelicals throughout the world. Evangelicals are divided over all these issues though several trends are discernable. Thus evangelicals have been both in the forefront of environmental concern and also in the denial of global warming. There is no one evangelical attitude to any of these issues, hence it is not easy to generalize.

First, many evangelicals, whether lay members or hard-worked pastors are simply disinterested in most of these issues and simply accept the lifestyle of those around them. The reasons for this indifference may be apathy, smugness or that they consider other things like evangelism are far more important.

Second, attitudes of evangelicals will depend on their religious and theological orientation. Certain evangelical theologies, for example forms of the Holiness movement and dispensationalism are more other worldly in their approach and put the emphasis on the Christian life as purely "spiritual." Hence concerns over "this" world, whether the environment or science are regarded as secondary unless they impinge directly on personal morality. Over the last 150 years both the Holiness and Dispensational movements have had immense influence on Anglo-Saxon Evangelicalism that these perspectives still influence many today. In Britain this has expressed itself in the Keswick Movement.[1] This has resulted in a greater concern for evangelism and personal holiness than for "creation care" or social action. Hence for most of the twentieth-century evangelicals have been weak on social action largely because of Holiness in its various forms and Dispensationalism. Part of this is a reaction against the Social Gospel in the early twentieth century. Many years ago David Moberg argued this in *The Great Reversal* (Moberg, 1973) which marked the time in the early 70s that evangelicals both in Britain and America looked again at the social dimensions of their faith as had Wilberforce and many evangelicals in the nineteenth century.

Third, Calvinists have long had a political and social side to their faith as may be seen in Abraham Kuyper and other Dutch Calvinists. Their influence in the United States has been out of proportion to their numbers and emanates, in particular, from Calvin College. The influence of Calvinists has been both to the "left" as in environmentalism, and to the right as Calvinism has been a source of both the right-wing theonomy associated with Rousas Rushdoony and the support for Apartheid by the Dutch Reformed Churches in South Africa from 1948 until the end of Apartheid in 1991.

Fourth, since the 1970s a good proportion of American evangelicals have been associated with the Religious Right in its various forms and look to such leaders as Jerry Falwell and Pat Robertson. As well as supporting every kind of anti-evolutionism, they are frequently militantly "pro-life" and have a tendency to oppose environmentalism, at times with charging it of being New Age or ultra-left. Much of this is documented in Mooney's *The Republican War on Science*, which makes a persuasive case that evangelicals, who are more right wing politically and conservative/creationist theologically, even if they are not Religious Right, are having a great influence on the scientific decisions of the Bush Administration, whether over climate, stem cells or the teaching of evolution. This can be verified by surfing the Web sites of ARN (Access Research Network), ICR, DI (Discovery Institute) and AIG, where even in 2007 these subjects are frequently mentioned.

Living as I do in Britain, I have found it strange to grapple with the peculiarly American aspects of Christian and political discussions on the whole gamut of scientific questions from the environment to stem cell research and GM crops. As I write, David Cameron, the new leader of the Conservative Party (Right Wing) is having his house redesigned to be eco-friendly, complete with a small windmill on the roof to generate electricity. In the late 1980s Margaret Thatcher, a right-wing Prime Minister, introduced a Green Agenda to the Conservatives. Though there is lively debate, almost the whole political spectrum is green tinged and there seems to be no political alignment to GM crops, stem cell research, etc. by Left or Right. Within the mainstream British churches there is little or no polarization on environmental issues between evangelical and liberal, both of whom largely sing from the same hymn-sheet, though a minority of British evangelicals question global warming.

ENVIRONMENT

Perhaps there is no issue in which evangelicals are more divided than over the environment. Though on both sides of the Atlantic and down-under there are numbers of evangelicals who are important players in the environmental movement, there is a body of evangelicals, especially in the United States who are not overly concerned for the environment and consider those who are to have rejected the Gospel for a potpourri of liberalism, leftism and New ageism. John McKeown delightfully calls these the Greens and the Browns respectively.[2] Because of this conflict of opinion, evangelical attitudes toward the environment provide an excellent case study to consider evangelical approaches to science as applied to contemporary issues. This is further complicated by the involvement of some evangelicals with American right wing politics.

It may be hard to accept this in the twenty-first century, but concern for the environment is very recent and started in the 1960s. Before then there was relatively little concern for the environment. This is partly due to the general lack of awareness of pollution or environmental damage on a large scale, and of species extinction. During the nineteenth century the effect of contaminated water causing cholera and air pollution caused by factory emissions became apparent and some campaigning for improvements were Christians for example William Buckland and Charles Kingsley, and not a few were evangelical. One of the great inspirers of the wilderness movement was John Muir, founder of the Sierra Club, and in Britain one of the founders of the National Trust, which was concerned in protecting both the countryside and old buildings were both Christians.

The great passion of nineteenth-century naturalists was collecting and this was done with abandon. Evangelical naturalists were as keen on collecting as any one else, thus H. B. Tristram was a "great sportsman and bad slaughterer" as Patrick Armstrong described him (Armstrong, 2000, p. 148) However in the latter part of the nineteenth century there was a sea-change among naturalists including Darwin and some evangelicals, and the gung ho attitude to collecting gradually went out of favor. One in particular was the Rev. F. O. Morris, an anti-Darwinian evangelical, who from the 1860s campaigned to protect seabirds as sporting gentlemen were massacring seabirds by the thousands. He was partly instrumental in forming the Society for the Protection of Birds, which was given its royal charter in 1904, and has long been one of the foremost British conservation societies.

These small beginnings and the rise of ecology as a science led to a greater environmental awareness, but it was only in the 1960s that environmentalism took the Western world by storm, beginning with Rachel Carson's *Silent Spring*, which catalogued the devastating effects of pesticides on wildlife. In 1967 Lynn White published his classic paper "The Historical Roots of Our Ecologic Crisis" (*Science*, 1967, vol. 155, pp. 1203–1207). There he firmly put the blame on all environmental ills on the Christian Faith and especially Genesis 1 vs 2 where mankind were given "dominion," which White argues has been taken in a rapacious sense by western civilization and the western church, and "dominion" has been taken to mean domination and exploitation as Western society went into technological overdrive since the Industrial Revolution. There was little immediate response from the churches but Christians were beginning to respond to environmental matters. Even so until the 1980s the environment was of minor concern to most Christians, evangelical or otherwise. Since then it has been of considerable importance to Christians of all traditions (see Oelschlaeger, 1994, for an account).

THE GREENS

However some Christians of all stripes responded critically to White. Francis Schaeffer in *Pollution and the Death of Man* (1970) is probably the earliest to defend Christianity against White. Twenty years before Schaeffer had founded L'Abri in Switzerland and from his fundamentalist background sought to tackle many intellectual issues. Conservative though he was Schaeffer was not quite YEC. However he had a great concern for the environment and this book expressed that. It was probably the first *popular* evangelical book on the environment and none the worse for that. Schaeffer was the intellectual guru of evangelicals in the 60s and 70s and encouraged many young evangelicals, including myself, to think intellectually about their faith. His promise was not fulfilled partly because of his superficial scholarship and also due to the rise of the Religious Right, which ignored his environmental concerns.

In the same year Richard Wright published a response to Lynn White's article in Bioscience (Wright, 1971 entitled "Responsibility for the Ecological Crisis." Wright is a marine microbiologist with a Ph.D. from Harvard who taught for three decades at the evangelical Gordon College in Massachusetts. He is the author of the textbook *Environmental Science* (7th ed., 2000) and the aptly named *Biology through the Eyes of Faith* (1989).

Since the early 1970s an increasing number of Evangelicals began to emphasize that the right care of the environment was an important Christian concern. This paralleled the growth of environmentalism in general and the concern of the Christian churches. However, as with society in general, there was limited interest and any concerned with such matters was considered an eco-freak, though now many of the teachings of the 1970s are now mainstream, except for the ultra-right and global warming deniers. A British example is that in 1976 Friends of the Earth were regarded as eco-freakish in their campaign to insulate the roofs of houses,[3] and now building regulations stipulate double the thickness for new houses. It was only in the late 80s that many Christians took the environment to heart and did so with gusto.

In Britain two of the earliest Christian environmentalists were Hugh Montefiore, the liberal Bishop of Birmingham, and Prof. R. J. (Sam) Berry (b. 1934) who was professor of Genetics at University College, London, from 1974 to 2000, and an evangelical. Berry has had a distinguished career spanning half a century on genetics, evolutionary biology, and, increasingly, environmentalism. Berry probably wrote the first publication on Christian environmentalism in Britain with his booklet *Ecology and Ethics* in 1972 at a time when concern for the environment was freakish and of no concern to the church. Looking back over thirty years later, it is almost unbelievable that this thirty-two-page booklet was then the *only* Christian offering on the environment, but mighty oaks from little acorns grow. Since that time Berry has been a significant environmentalist in Britain and beyond. Along with colleagues from *Christians in Science*, he has done much to galvanize evangelical opinion as well as in the wider world. In 1980 he contributed an article on ethics for the World Conservation Strategy of the United Nations Environmental Programme, in which he stressed that humanity is inseparable from nature, which the WCS had overlooked. In fact, both society in general and the church tend to separate humanity from nature for different reasons. At this point Berry agreed with the "deep greens" and New Agers for very different reasons, in that Nature is created by God and therefore good and should be "cared" for and not just seen as "resources" for humans. A quarter of a century on it is difficult to comprehend how novel Berry was on this. Evangelicals still tended to see people as souls to be saved from a wicked world and that this earth was the stage on which salvation was played out rather than humans belonged to the earth and that Christ affirmed creation in his incarnation. Even in the early 1980s mainstream Christians did not think that the environment was important in the concern for social justice. Hence, when in 1982 I attempted to introduce environmental issues onto the Social Responsibility agenda in Liverpool Diocese I was simply rebuffed. Yet now in Britain few Christians would dare deny the necessity of environmental awareness. The development of environmental concern in the United States has been similar, except that "Religious Right" evangelicals tend to dismiss environmentalism as "leftie" and damaging to the economy, though it would be interesting to see how this changes in the aftermath of Hurricane Katrina, or Bush's recent (May 2007) recognition of global warming. This attitude is amplified by the opposition to evolution as YEC tends to accentuate the difference between humans and the rest of the natural world and have a tendency to consider that humans are almost not really part of the natural world, as well as rejecting any science which argues for long ages, be it the age of the earth or 20,000 years for the height of the Ice Age.

Along with Berry, other significant environmentalists have been Sir John Houghton (b. 1931), Chairman of the Inter-Governmental Panel of Climate Change (IPCC) from 1988 to 2002, and Sir Ghillean Prance (1937), who was director of the Royal Botanic Gardens, Kew. As well as their professional involvement and international activities, they and others formed the *John Ray Initiative* (named after the seventeenth-century biologist; Web site: www.jri.org) in the mid-1990s to encourage a greater Christian awareness of the environment, based on an open evangelical theology of creation. They have possibly seen more tangible results than their American counterparts.

Evangelical environmentalism in the United States also began in the early 70s partly as a result of the influence of David Moberg, the author of *The Great Reversal* (Moberg, 1973) which called evangelicals back to the social concern of the their nineteenth-century forbears. This was also developed in the 1974 Lausanne conference which gave a view of mission as wider than just evangelism. Part of the same process was the Chicago Declaration of Evangelical Social Concern in 1973, which led to the formation of Evangelicals for Social Action (ESA). Environmentalism grew slowly in the 70s and in 1980 Loren Wilkinson published *Earth-Keeping* (Wilkinson 1980/1990). The same year saw the establishment of the Au Sable Institute for Environmental Studies to provide courses for the national Christian Colleges Coalition. Most of the staff were scientists and led by biologist Calvin DeWitt they helped develop theological engagement with environmental ethics. In 1992 the Theological Commission of the World Evangelical Fellowship (WEF) hosted a forum at Au Sable. As a result the ESA formed the Evangelical Environmental Network (EEN), which, in turn, collaborated with Au Sable and *Christianity Today* to produce the "Evangelical Declaration on the Care of Creation" which was signed by over 200 evangelical leaders in the United States. (An exposition of it was later published as *The Care of Creation*.) Significant Americans include Loren Wilkinson, Calvin DeWitt (*The Environment and the Christian*

[1991]), and Richard Wright all of whom have written profusely for many decades. Each have published important works.

A public joint declaration on the environment was made in 1994 as *An Evangelical Declaration on the Care of Creation* (see Primary Sources). It was one of the first fruits of the EEN. Both resulted in reaction to the World Council of Churches consultations on the "Justice, Peace and Integrity of Creation" at Seoul in March 1990 and from the 1992 WEF meeting at Au Sable. The EEN obtained support from several hundred evangelical leaders throughout the world and attracted support from John Stott (no mean ornithologist), Prof. O.T. O'Donovan from Oxford, and many others. By doing so they were able to present "care of creation" to the evangelical constituency in several parts of the world and have formed the basis of action by many environmentally concerned evangelicals over the following years. The book *The Care of Creation* (Berry, 2000) which contains significant essays by leading evangelicals summarizes many of their aims and was published by the leading evangelical publisher *Intervarsity Press*. Books by evangelicals on the environment are now very numerous and include Tony Campolo and James Jones, Bishop of Liverpool, who almost overstated the case for environmentalism in the gospels in his book *Jesus and the Earth* (Jones, 2003).

Whereas evangelicals have often been less concerned than most mainstream Christians over social issues, on the environment they have been in the vanguard both in Britain and America, despite the carping criticism of more conservative evangelicals that these evangelicals had sold the gospel for a pottage of leftism, new ageism, and junk science. Each of these accusations were made against Susan Drake Emmerich, who had worked on Tangier Island in Chesapeake Bay for the U.S. State Department trying to conserve fishing resources for the local community.

In recent years the greens have built support for a narrower and tougher venture, on climate change, calling for U.S. emissions targets. A turning point was the Oxford Climate Forum in 2002 organized by the Au Sable Institute and the John Ray Initiative, when the climatologist Sir John Houghton played a key role in gaining support among (largely Republican) conservative Christians. Richard Cizik, Vice President of Government Affairs for the NAE (National Association of Evanglicals) attended, "dragged there by Jim Ball of the Evangelical Environment Network and coordinator of the What would Jesus Drive campaign." As a result of what he heard, Cizik claims to have had a "conversion" so dramatic that he likened it to an "altar call." Since then, he has tried to convince American evangelicals that creation care is a core Christian responsibility.[4]

The result of this was the "Evangelical Climate Initiative" in February 2006 which was widely reported, for example in *Nature* and *Science* journals, and in U.S. news media, producing headlines such as "Evangelicals Go Green." This has made global warming literally the hotspot in American Evangelicalism.

THE BROWNS[5]

Whereas most Christians and many evangelicals have developed a great concern for the environment over the last forty years, significant numbers of evangelicals have not. Many of these are associated with the Religious Right in the United States and have a great suspicion of anything liberal or "leftie" Since 1990 evangelicals opposed to the approach of Calvin DeWitt and others have been gaining strength and have formed coalitions to express their understanding of environmental stewardship, culminating in the *Cornwall Declaration* (see Primary Sources) produced in 2000 which is not strictly environmental, though it questions much conventional environmental wisdom.

McKeown showed that an essential aspect of the opposition to mainline environmentalism, which he terms the Browns, came from free-market economics and this was linked to the upholding of conservative theological principles, both Catholic and evangelicals with the founding of the *Acton Institute for the Study of Religion and Liberty* in 1990 by Father Robert Sirico to, quote "promote a society that embraces civil liberties and free-market economics" It receives considerable funding from *ExxonMobil*. The *Acton Institute* and Beisner opposed *the Evangelical Declaration on the Care of Creation*. In 2000, the Acton Institute established the *Interfaith Council for Environmental Stewardship*, whose founders included leading evangelicals like James Dobson (Focus on the Family), James Kennedy (Coral Ridge Ministries), Bill Bright (Campus Crusade), and Charles Colson as well as conservative Roman Catholics and Jews. Their *Cornwall Declaration* was produced in 2000, which was posted to 35,000 churches. This ran counter to the *Evangelical Declaration*, but it was not entirely "anti-environmental" though it has been perceived as such. First, the declaration is anthropocentric and emphatic that humanity has dominion over the earth and criticize "Some unfounded or undue concerns include fears of destructive manmade global warming, overpopulation, and rampant species loss" (Concerns 3). Then in the section on beliefs, the fifth statement reads "By disobeying God's Law, humankind brought on itself moral and physical corruption as well as divine condemnation in the form of a curse on the earth. Since the fall into sin people have often ignored their Creator, harmed their neighbors, and defiled the good creation." This as we see claims that the Fall had an effect on the whole of creation, in that it was a "curse" and not just a "fall." Larsen wrote that, "the Cornwall Declaration represented the first acknowledgment of the need for environmental care" by politically conservative leaders. That in itself may be very significant for the future. In 2005 ICES was re-launched as the Interfaith Stewardship Alliance (ISA).

To understand the Browns and the ISA, it is best to focus on their leading theoretician, Calvin Beisner, who is associate professor of social ethics at Knox Theological Seminary (which is attached of James Kennedy's Coral Ridge Church) and has written three books on environmental stewardship: *Prospects for Growth: A Biblical View of Population, Resources, and the Future* (1990); *Man, Economy, and Environment in Biblical Perspective* (1994), and *Where Garden Meets Wilderness: Evangelical Entry Into theEnvironmental Debate* (1997). Beisner is not a scientist and studied under the economist Julian Simon, whose book *Resourceful Earth* 1984 advocated the "Cornucopia hypothesis" of unlimited economic growth, and who did not recognize the limited nature of natural resources. Hence Beisner has far more of a free market approach to the environment rather than a scientific one, whether on climate, pollution or material resources. This is in marked contrast to those like Berry or Houghton who base their arguments on science not economics.

McKeown argued that Beisner supports his understanding of environmental stewardship from his interpretation of early Genesis where he argues that there are two different mandates in chapter 1 and 2 and that the "curse" of Genesis drastically changed the natural world and that it was a real curse and not just a human Fall. Richard Wright argued in 1995 that "the emerging Christian anti-environmentalism . . . The presumed biblical support for this position is currently found primarily in Beisner's work" (Wright, 1995). Beisner rejected the common idea that the meaning of subdue and rule in Genesis 1 vs 28 and to till and keep in Genesis 2 are essentially the same to argue from the Hebrew, as he did in *Where Garden Meets Wilderness* that there are two contrasting cultural mandates. Genesis 2 vs 15 being gentle, and 1 vs 28 harsh. One is appropriate to the garden, the other to the earth outside the garden, that is the wilderness. Thus the wilderness must be "subdued" to become a "garden" and that includes taming wild animals. As Beisner expressed it, "the incremental transformation of wilderness into garden, bringing the whole earth under human dominion, taming the wild beasts, and building order out of chaos . . . while tender cultivation is suited to a garden, forceful subduing is suited to all of the earth that has not yet been transformed into the garden. In short, subduing and ruling the earth should metamorphose gradually into tilling and keeping the garden as the earth is progressively transformed into the garden." It would be interesting to see how this would apply to animals like bears, lions, and even eagles and condors.

Many would not accept this biblical interpretation, but it has serious implications in that Genesis 1 vs 28 becomes a command to tame the wilderness and as Mckeown put it "so the logical outcome of his reading of Genesis (though he mostly avoids it) is that it is a dereliction of duty to leave any wild area untransformed or any wild creature untamed." That goes completely contrary to any understanding of protecting the wildernesses of our planet and the ideal of National Parks. It does run totally counter to the Christian vision of John Muir. I will leave the reader to make up her own mind!

McKeown continues by explaining that Beisner also claims that "there is a difference between the Fall and the Curse. The fall is man's sin, and the Curse is God's response to man's sin. The Curse is on the earth." He points out that "most evangelical books on the environment never mention the Curse" but only the Fall and that "The only degradation that the *Declaration* mentions occurring to the earth is all through human action" neglecting God's direct action against the earth by curse and flood. Beisner judged that this silence was motivated by the greens' desire to identify environmental problems as human-caused, but the report of the 1992 WEF meeting (above) indicates that the reason was their uncertainty about whether the earth's physical aspects were actually changed by the curse. In other words, were earthquakes, storms, predation, death, and disease actually introduced after the Fall to be the Curse. This as we have seen in Chapter 7 is a basis premise of YEC. In his contribution to the Pontifical Council for Justice and Peace in April 2007 he wrote, "According to both the Bible and sound science, the great pools of oil and veins of coal formed from sudden, simultaneous deposits of vast numbers of plants and animals in a great geological cataclysm—what Christians recognize as the Flood of Noah's time" (http://www.cornwallalliance.org/). Many would not agree that this is sound science.

It is not possible to give a detailed discussion of Beisner's and the ISA's basis for environmental issues, but it is difficult not to conclude that they are based three contentious conclusions, first that there are unlimited resources in the earth, second on a rather idiosyncratic interpretation of early Genesis, and that it is predicated by a YEC view of the fall as a Curse, along with a rejection of the consensus of scientists on so any issues, whether environmental or basic science.

From the presentation to the Senate on religious views of Global Warming (June 2007) discussed below, it is clear that Beisner has convinced a large proportion of religious conservatives, including the Southern Baptists. His whole approach has been savaged by two environmentally informed evangelical scientists, Richard Wright and Jeff Greenberg of Wheaton. The latter wrote with complete exasperation! Despite, or in spite of that, many evangelicals and conservative Catholics have supported the Cornwall Declaration and under the guise of good stewardship of the environment reject many of the aims of most environmentalists, particularly those which are seen as junk science. This has caused a rift among American evangelicals, which does not follow the usual demarcation of creationist and evolutionist.

GLOBAL WARMING

As I write this there is immense controversy over global warming. The consensus of most scientists is that warming has been increasing since the 60s and the future is dire. The causes of warming are mixture of natural effects, for example solar fluctuations, and a general warming of the climate, and also the effect of human technology especially the use of carbon fuels. However global warming has its critics, who are often either right-wing Republicans or from the religious right (lower case) many of whom are evangelicals. This is often accompanied by charges of junk science. Yet one of the ironies is that evangelicals are well-represented on both sides.

Global warming was presented as a threat to the planet in the 1980s, hard on the heals of warnings of a nuclear winter following an atomic war. The basic thesis is that not only has the earth been warming since about 1820, but is now warming at a greater rate due to human activity and above all the emission of greenhouse gases. Suffice it to say that initially the thesis was questioned, but since the early 90s few scientists have doubted it. It has been a concern of many throughout the world of all faiths and none. Even so, the Republican Senator James Inhofe has called global warming "the greatest hoax" and tried to ensure that the Bush Administration will regard it as such. According to Mooney (Mooney, 2005, p. 78f) he has been fairly successful, and as we will see has the support of numbers of evangelicals.

The interest in global warming is that it has caused polarization in the evangelical community. Evangelical environmentalists have been proactive in their taking up of the cause and have been forceful in their attempted persuasion of the large American evangelical constituency. One of the leading scientists on global warming over the last decades has been Sir John Houghton, an evangelical. After being professor of Atmospheric Physics at Oxford was director-general of the Meteorological Office from 1983 to 1991. He was Chairman of the Royal Commission on Environmental Pollution (1992–1998) and Chairman of the Inter-Governmental Panel of Climate Change (IPCC) from 1988. He has published widely on the subject including a popular book for Christians *Global Warming* (Houghton, 1994). As a result he has been involved at the highest level on an international scale, including the Kyoto agreement.

Alongside his international and scientific activities Houghton has strived both to make Christians aware of environmental problems and the reality of global warming and has spent much of his "retirement" traveling the world either for international environmental groups or for church groups.[6] Without undervaluing the work of evangelical environmentalists in the United States, Houghton was probably instrumental in persuading many in the NAE (National Association of Evangelicals) of the dangers of global warming.

The "Evangelical Climate Initiative" of February 2006 has already been mentioned. It has divided the 30 million strong NAE as it stated that "as evangelical leaders, we recognize both our opportunity and our responsibility to offer a biblically based moral witness that can help shape public policy in the most powerful nation on earth and therefore contribute to the well-being of the entire world . . . Many of us have required considerable convincing before becoming persuaded that climate change is a real problem and that it ought to matter to us as Christians. But now we have seen and heard enough." This aptly describes how many, including Cizik, rejected Global Warming for many years. The statement made four claims:

1. human-induced climate change is real;
2. the consequences of climate change will be significant and will hit the poor the hardest;
3. Christian moral convictions demand our response;
4. and, governments, businesses, churches and individuals all have a role to play in addressing climate change—starting now.

Given the whole nature of American Evangelicalism this was a remarkable turnabout and document and now some two-thirds of evangelicals now accept global warming and that they as individuals have a responsibility because of their faith. A reaction was inevitable and soon the report was challenged by the Interfaith Stewardship Alliance as they consider that global warming will lead to increased agricultural productivity and that the poor would suffer most from attempts to slow climate change.

Since then evangelicals and others in the ISA have been very combative in opposing the "junk science" of global warming. One of their chief spokesmen is Calvin Beisner who gave Oral Testimony to the Environment and Public Works Committee of the United States Senate on October 20, 2006, despite having no relevant scientific training. In early 2007, James Dobson and others made representation to the NAE to stop the support for global warming. As well as those, the Acton Institute and the Interfaith Stewardship Alliance (interfaithstewardship.org/pages/listarticles.php) warn readers of "Scientific orthodoxies and politicized science, like those encountered in the global warming debate, make for dangerous waters, and evangelicals who want to swim in them should look carefully for rocks and riptides and be well prepared before they dive in." Beisner and the ISA have also had support from several leading anti-evolutionary groups, notably Answers in Genesis, Institute of Creation Research (ICR), the Discovery Institute and the Access Research Network (ARN), headed by Paul Nelson. Dembski's blog *Uncommon Descent* often contains strident articles criticizing global warming. On July 25, 2006, the ISA responded with *An Open Letter to the Signers of "Climate Change: An Evangelical Call to Action . . . and Others Concerned About Global Warming*, signed by more than 130 scholars, theologians, scientists, economists, and other leaders, including James A. Borland, D. A. Carson, Guillermo Gonzalez, Wayne Grudem, James Kennedy, Michael Oard, Joseph A. Pipa, Robert L. Reymond, and Jay W. Richards.

The simplest interpretation is to conclude that there is a very strong correlation between Creationism in any form and rejection of global warming and creation care. That was the case until a few years ago. Many of the signatories of *An Open Letter* are creationist and I cannot identify any who are not, but when one considers the signatories of Evangelical Climate Initiative, "non-creationists" are highly represented but a good number are creationist and even YEC. What has happened is that some "creationists" have now made common cause over the environment with mainstream scientists, who maybe Christian or not.

This division of opinion among evangelicals was clearly seen in the Senate hearings on "An Examination of the Views of Religious Organizations Regarding Global Warming" on Thursday, June 7, 2007.[7] Although this was clearly a purely American event, it actually reflects the worldwide issue over global warming from a religious perspective. Most contributions were Christian, though there was also Jewish submission but none from a Muslim. All shades of Christian opinion were presented. There was a contribution from the Roman Catholic Bishops appealing for action, which was shared by that of the Presiding Bishop of the Episcopal Church, which included statements from several mainstream protestant churches. The consensus of all these was that Global Warming was a reality and must be urgently addressed.

Against one Catholic, one Protestant and one Jewish depositions there were four by evangelicals. Three were essentially global warming deniers and only one was convinced of its reality. This was the Rev. Dr. Jim Ball president of the Evangelical Environmental; Network, which essentially reiterated the concerns of the Evangelical Climate Initiative of 2006, and in basic agreement with the mainstream churches. The three other evangelicals, Russell Moore, Dean of Theology at Southern Baptist Theological Seminary, David Barton and Dr. James Tonkowich, all present the case against global warming and largely based their presentations on Beisner's work. Moore included a new article by Beisner in his submission, and Senator Inhofe referred to him favorably.

The result of this is that submissions to the Senate hearing were even balanced between the two sides, but reflecting the partial shift by evangelicals. Also in June 2007 President George Bush also warmed to global warming in what must be seen as a step towards Kyoto.

As the significant events on evangelicals and global warming have occurred only in the last eighteen months as I write this, it is not possible to predict the outcome of the present dispute. As there are many powerful evangelical lobbyists on both sides, the controversy is set to continue. Part of the resolution may come after the 2008 Presidential Election. Global warming is clearly a storm centre for evangelicals at present and contains some very far-reaching implications for evangelicals and their relationship to science, not only in the United States but throughout the world.[8]

FROM THE ENVIRONMENT TO BIOETHICS

There is little doubt that environmental questions will remain high on the agenda of all peoples in the world throughout this century and that they raise questions which are ethical rather than scientific. Thus as many draw ethical understandings from their faith, be it Islam, Hinduism, Buddhism, or any of the varieties of Christianity, how evangelicals respond to environmental questions becomes important as within twenty years 10 percent of the world's population will probably be evangelical. The other significant ethical issue surrounding science are all the questions concerning biological and genetic questions such as Genetic Modification (GM),[9] cloning, stem cell research, etc. In the media these are often hyped up as Frankenstein foods, the specter of cloning humans and meddling in reproductive processes. The scientific advances here have been spectacular in the last half-century, with the sequencing of the human genome and the scientific possibilities which raise immense ethical questions, which were inconceivable a generation ago. On the former it is to be noted that an evangelical Francis Collins succeeded Craig Venter, an atheist, as the Director of the Genome Project.

When it comes to ethical issues over bioethics, evangelicals attempt to apply biblical principles to their ethical reasoning, so almost always tend to be conservative and cautious in their conclusions. Thus on abortion, evangelicals of whatever stripe begin from the Sanctity of Life and respect for the fetus and hence none will support abortion on demand or see it as the mother's right. Yet many evangelicals, especially in America, support Capital Punishment, which is almost in contradiction to the Sanctity of Life. However some evangelicals oppose abortion on every ground, rather like the traditional Catholic and others allow abortion in certain cases as the lesser of two wrongs. The amount written on this is vast but a good summary of a middle of the road evangelical opinion is by Stott (1999, chap. 15). Stott also deals with IVF. More accessible material of differing standpoints and quality will be found on many evangelical Web sites.[10]

There is often a distinction between what must be termed knee-jerk reactions by some evangelicals who have little expertise in ethics or the science involved, and those, who have considerable expertise and training whether in science, medicine or ethics, or even all three. It is the former who make all the noise! As these ethical issues are of such importance there are considerable numbers of serious evangelical ethicists dealing with them. These include Nigel Cameron, President of Trinity Evangelical Seminary, Deerfield, Illinois, Oliver O'Donovan, until recently Professor of Moral Philosophy at Oxford and Gareth Jones of Otago, all of whom publish in both religious and secular works and are respected as authorities way beyond the evangelical community. A useful survey is James Preston's *Genetic Turning Points* (Preston, 2001), by one who has expertise in molecular and clinical genetics and the field of ethics. In considering how evangelicals and others may respond to these issues, Gareth Jones (Jones, 2005, p. 214) gives four basic alternatives, which apply not only to discussions over stem cell research but all bioethical issues;

1. The bible alone gives a complete guide; science is irrelevant.
2. The bible is one of several sources, but is the major determinant where there is conflict.
3. The bible is one of several sources, but may not be the major one.
4. The bible is irrelevant.

Evangelicals may adopt one of the first three, but Jones argues that none in practice adopt the first as one has to use science as a source. This is because one cannot transfer even ideas of fertilization to the Bible as it was unknown to the biblical writers. Even more so are genes, blastocysts, and other things! Bryant and Searle (2003) warn against this "reverse transposition." The "popular" evangelical may claim to use the first and seek "what the bible says" but is obliged to take some cognizance of science. The more "traditional" ethicist will follow the second alternative, and thus the issues are far greyer than sometimes presented.

GENETIC ENGINEERING

The realization of the structure of DNA in 1953 revolutionized biological science and now the Human Genome Project has obtained our complete

DNA nucleotide sequence. This has opened up the possibility of the manipulation of an individual's DNA sequence, which potentially could alter that person's character or liability to certain diseases. Genetic engineering is commonplace in laboratories on plants and animals, but is clearly controversial when applied to humans, even when sensationalism is removed. The general concerns are widely known and discussed at all levels and a good place to start is Preston's *Genetic Turning Points* and the Web sites in note 6.

STEM CELL RESEARCH: THE STORM CENTER FOR EVANGELICALS

Evangelicals along with Roman Catholics have been at the forefront of the opposition to abortion in many parts of the world. As a result anti-abortion has become a fundamental for the Religious Right. In recent years attention has partially switched to the ethics of stem cell research and thus this is a good topic to consider how evangelicals "do" medical ethics and how this translates into possible scientific research.

Stem cells are unspecialized cells which under the right conditions can develop into very different body cells within the human body. Until recently most stem cells were thought to be "switched off" once development was complete, but this is not the case. The question of stem cell research has become contentious because the normal source for them has been from human embryos, but they can be obtained form adult tissue especially marrow or blood. The use of adult stem cells is more acceptable even though it is more difficult as a scientific procedure. Hence the question of stem cell research has reopened the question of the status of the human embryo and whether it is ethical to perform experiments of cells derived from an embryo, which must be destroyed in the process.

To those who argue that the human life, which begins at conception or fertilization, must be protected and to kill it is the same as murder, and on this view stem cell research is, in fact, murder. Hence an anti-abortionist opposes anything that harms the embryo including the morning after pill. As a result most evangelicals are opposed to any kind of embryonic stem cell research and in 2001 pressure from evangelicals persuaded President Bush to decide against embryonic stem cell research. He also claimed that there were "more than sixty" embryonic stem cell lines, which did not give confidence in his decision (Mooney, 2005, p. 185ff). Undoubtedly Mooney is sympathetic to neither the Bush Administration nor the Religious Right, but his book does indicate the tie-up between the two over science, of which stem cell research and global warming are only two areas of contention.

Conservative articles on stem cell research are legion and for a taster one can refer to articles on the AIG or "Reasons to Believe" Web sites. Writing for the *Answers* magazine on January 1, 2007, Dr. Georgia Purdom[11] begins by quoting "Thou shalt not kill" as the main argument against embryonic research and then suggests the use of adult stem cells. Dr. Fuz Rana makes a similar argument because "I regard a human embryo as a human life" and also argues "In fact, I think that the federal government should fund adult stem cell research at "Manhattan Project" levels"[12] The ethical argument remains on a simple application of the Ten Commandments.

Somewhat against these two, who represent the conservative understanding, is Prof. Gareth Jones, Deputy Vice-Chancellor and full professor of the University of Otago in New Zealand. Jones has contributed much on life and death issues in biology from an evangelical perspective over the last forty years, though some are concerned that he is too "liberal." In a recent paper, which reflects his expertise on all aspects, Jones attempts to lay out the Christian alternatives (he deals with Catholic and other Protestant views as well) and extends the discussion to the whole nature of pre-natal existence and at what stage an embryo has "personhood." He stresses that the view discussed above simply uses the bible as a complete guide to ways in which a blastocyst should be treated, whereas some use of biblical teaching is needed in conjunction with scientific evidence. To a conservative this may seem a slippery slope, but all do in fact make some use of science. Jones concludes by giving cautious approval for embryonic stem cell research. I feel I cannot give an adequate exposition of Jones but raise the issue, so that it can be considered by reading Jones (Jones, 2005) and also two other evangelicals, J. Bryant and J. Searle (Bryant and Searle, 2003) in their book *Life in our hands.*

PSYCHOLOGY AND EVANGELICALS

The study of psychology for evangelicals raises many questions on the nature of the human and also the existence of the soul. More recently additional questions have come from evolutionary psychology, which appear to be unacceptable to any theist. At present for evangelicals a major issue is whether there is an actual soul or whether the more psychologically monist view of Jeeves and Murphey is a better way. Malcolm Jeeves, who is a professor of psychology at St. Andrew's University, Scotland, gives an overview of psychology and evangelicalism in *Science, Life and Christian Belief* (Jeeves and Berry, 1998) and with several others presented a major study of human nature at the combined meeting of the American Scientific Affiliation and Christians in Science, Churchill College, Cambridge University, August 1998. Papers were presented by scientists and theologians like Joel Green. This was reported in *Science and Christian Belief* in 1999 (Brown and Jeeves, 1999) and written up as *Whatever Happened to the Soul? Scientific and Theological Portraits of Human Nature* (Brown, Murphey, and Malony, 1998), with Jeeves editing a further volume *From Cells to Souls* (Jeeves, 2004). The consensus was that humans are better considered as *being* souls, rather than dichotomous beings of *body* and *soul.*

In contrast to Jeeves and colleagues, more conservative evangelicals like the philosopher J. P. Moreland continue to maintain that the dichotomous nature of humans is truer to scripture and presented this on *Body & Soul: Human Nature & the Crisis in Ethics (Paperback)* (Moreland and Rae, 2000, vol. IV). In one sense this is another manifestation of the fault lines which run through evangelicalism between the more and less conservative, as Moreland and colleagues incline to ID and are opposed to evolution, though Moreland is, himself, old earth, whereas Jeeves and Murphy are TE.

MEDICAL MISSIONS

Over the last two centuries overseas missions from the west have not only focused on evangelism and church planting. This has resulted in the vast number of Christians throughout Latin America, Africa and Asia, many of whom are Roman Catholic as well as protestant. Today the majority of missions in those countries are evangelical. Mission work has also included education and medical missions. And so in many countries there are now thriving hospitals, which have a missionary foundation. The majority is probably evangelical and found in every continent. Until the end of the twentieth century these were controlled by the missionary society for example Church Missionary Society, Africa Inland Mission, etc., but two things have happened. First control has been handed over to local churches and secondly the emphasis of the societies has changed from sending out missionaries to lead the work to sending out mission *partners* to work alongside and under indigenous leaders.

Many of the hospitals date from early in the twentieth century and vary in size from a collection of huts to large complexes like Ludhiana in India. Initially hospitals were very basic and even in the 60s some were little more than a few huts with only a few hours of electricity per day and with only one doctor, like Kagando Hospital in Uganda. Others became large complexes with numbers of doctors and facilities for training. Many of these provided pioneer health care in countries with no facilities and thus provided the basis of health care ahead of provision of any government or colonial hospitals.

Though medical work may not be seen as strictly scientific, one cannot escape the fact that much early medicine was mission-based and often evangelical. They gave the initial foundation for tropical medicine, frequently under appalling conditions, and were hardly noticed. Two accounts may be found in *Heralds of Health* (Stanley G. Browne Christian Medical Fellowship 1990) and *Sent to Heal: Emergence and Development of Medical Missions* (Grundmann, 2005).

CONCLUSION

On almost every ethical issue related to science, there is a great divergence of evangelical opinion, except that there is a rejection of euthanasia, universal abortion, and unfettered genetic research on humans. There is no one attitude and if one avoids knee-jerk opinions of some popular Evangelicalism, there is a well-informed gradation of opinion, usually based on careful scientific and theological research, which is often predicated by one's theology and at times by one's political ideology, as these are not always kept separate.

At present the most vigorous debates occur in the United States but evangelicals are making their presence felt on both environmental and bioethical issues in other parts of the world.

NOTES

1. I need to confess that this is part of my religious heritage.

2. John McKeown, The Use of Genesis in Debates about Environmental Stewardship among Conservative Christians in the USA. Unpublished paper. Paper given at *Critical Perspectives on Religion and the Environment*, Birmingham, September 18–19, 2006. Most of this section, and the division into Browns and Greens, is based on his paper, which has been used with much gratitude.

3. I spent a couple of Saturday mornings in 1976 helping the Friends of the Earth to put roof insulation into terraced houses in Wigan.

4. This is largely based on McKeown.

5. This is largely based on McKeown.

6. In the late 1990s I heard Houghton speak at an evangelical church in Oswestry, where the minister was an YEC. It is interesting to see him put up overheads of temperature variations over the last 100,000 years.

7. Web site: http://epw.senate.gov/public/index.cfm?FuseAction=Hearings.Hearing&Hearing_ID=e39940af-802a-23ad-4371-252edd78194f.

8. The way this unfolds will most likely be reported on the web and one should be able to follow it by putting words, like Beisner, Cizik, and the various evangelical environment groups into a search engine.

9. Space forbids discussion of this subject, which though it gets popular knee-jerk responses is not as contentious to many evangelicals.

10. The evangelical literature on this is immense and a surf of websites will produce much, try the ASA (section on ethics), CIS and their journals, alongside

AIG, ICR, ARN, Center For Bioethics and Human Dignity: http://www.bioethix.org/ and Leadership University (www.leaderu.com), and one will find many articles on bioethics and demonstrate the diversity of opinion.

11. Web site: http://www.answersingenesis.org/articles/am/v2/n1/stem-cells.

12. Web site: http://www.reasons.org/resources/apologetics/fazale_rana_embryonic_stem_cell_research.shtml.

Conclusion

For the last 270 years Evangelicalism has been one of the most significant religious movements in the world, and in political and social influence is probably on a par with Islam. Its influence is felt in almost every country, whether through the growth through its evangelism, connections with Zionism in the Middle East, or its influence in politics, the environment, biological research, or education. Like Islam it is growing rapidly worldwide, in contrast to the Roman Catholic Church and the Orthodox traditions which are almost static in numbers. More liberal protestants are in decline, and thus in a mixed denomination like the Anglicans and Presbyterians membership is shrinking in the liberal West. However in much of the rest of the world these denominations are growing and becoming increasingly evangelical. Projections indicate that evangelicalism will continue to grow in the next few decades, becoming dominant in Africa and Latin America and of increasing significance in continental Europe and even possibly China, as religious freedom increases there. As a result as their numbers increase so will their influence, for good or ill. This has implications for the church and for society.

WHERE ARE EVANGELICALS NOW?

As evangelicals are such a diverse group, continually changing and evolving over time, with new subgroups and new beliefs, which can also vary considerably from place to place and within a denomination, it is very difficult or impossible to give a clear and succinct answer to that question. My first chapter brought out the incredible variety of beliefs and practices of evangelicals in general. My historical survey showed the diversity of evangelicals in relation to science during the last three centuries. Unless one draws artificially narrow boundaries for Evangelicalism there has always been this diversity, but the centre of gravity may change and at times swings like a pendulum. Thus in the late nineteenth century evangelicals were largely in the mainstream of science, but by the mid-1920s were outside, especially in America. During the 50s and 1960s, British evangelicals were totally within mainstream science and Americans were moving that way, only for the pendulum to swing again in the direction of YEC and ID during the 80s. That is where the balance is today. At best evangelicals are wary of research in controversial areas like stem cell research, and numbers oppose it entirely. Thier opinions on this have considerable political clout in the states and by collaborating with Roman Catholics could do the same elsewhere. Evangelicals have very disparate views on the environment, and it is difficult to say whether the Greens or the Browns, described in chapter 9 have the upper hand. After Cizik's espousal of global warming in 2005 there are signs of greening and that the grip of the Religious Right is being broken. This may seem to be a local American affair but the attitudes are easily exported. Already they are apparent in Britain.

Once we start considering the future state and influence of evangelicals it is dangerous to prophesy, as who would have thought in 1907 that the influence and power of evangelicals would have drained away in twenty years and then in 1945 evangelicals would rise like a phoenix from the ashes of fundamentalism. My own opinion, and that what it is, is that evangelicals will continue to grow at a steady rate in Europe, and with leaps and bounds in Latin America, Africa, and parts of Asia. I also think that the conservative, almost Fundamentalist, brand of evangelicalism will dominate. In the United States evangelicals will probably hold their own but they may become the losers in their self-made culture wars. One cannot say definitely for the future, but I would suggest that the pendulum will continue to swing in the conservative direction for several decades.

IMPLICATIONS FOR THE FUTURE

A few decades ago, Evangelicalism would have been regarded as marginal on either the national or international stage, and what they said about science was of no more concern to the Flat Earth Society. The concern of some evangelicals over the Genesis Flood, Jonah and the Whale was something which may have caused amusement but of no concern. That is not the case today as evangelicals force their views on science and its teaching and will be so in the foreseeable future. As evangelicalism grows throughout the world, evangelicals will make their presence felt in science In many countries, with considerable implications. One can expect that issues over the teaching of "creationism" will occur in most parts of the world and especially in Africa and Latin America, as the groundswell of evangelical opinion is already there. This will occur if evangelicals gain political clout based on their numerical advantage. In fact, as I write this some Kenyan evangelicals are very concerned about the inclusion of "Turkana Boy," a 1.5 my fossil *Homo erectus*, in Nairobi National Museum, on the grounds that "we are not descended from apes." A few years ago Archbishop Akinola of Nigeria was outspoken against evolution. This is understandable considering some of the actual connections of evolution with racism and false claims by some creationists on this. Further most third-world evangelicals tend to be conservative and rather literalist on the bible. As in America today, Christians will not want evolution taught. This will have implications on many aspects of both scientific research and teaching.

There are two other areas where evangelicals may exert their influence. The first is over the environment how some evangelicals in the United States have tried to rubbish global warming as junk science. The tie up of the Religious Right and the Bush administration has prevented global warming and the Kyoto agreement from being adopted in the United States. That weakens the argument for other countries to follow suit and thus the rapidly expanding economies of China and Japan are and will continue to be unwilling to control carbon emissions.

The second is research on aspects of human developmental biology, including stem cell research. There are some serious ethical issues here, but not all want to take these seriously. If research in these areas is totally banned in the United States, then it will probably be carried out elsewhere and research institutions will move to countries where their research will be unfettered. As both China and India have rapidly growing economies and scientific research facilities, scientists may choose to work there and take their research away from America. This is a distinct possibility if the more conservative evangelicals are successful in their opposition to this kind of research.

Evangelicals have had a long relationship with science, beginning with Jonathan Edwards' student experiments on insects and his death while living and dying on the cutting edge of scientific medicine. The relationship has had its ups and downs, with high points with many of the early geologists and, recently, the leadership of John Houghton over global warming. It has also had its downs, spectacularly over the Scopes trial, and, if mainstream science is right, over Young Earth Creationism and Intelligent Design. However many evangelicals see the latter two as evangelical science as its best. (Let the reader decide.)

Opinions will differ on whether an evangelical faith is compatible with science. Numbers of atheists regard them as incompatible, and make this very clear as do Richard Dawkins and Sam Harris to name two contemporary writers. Some Christians of a more liberal perspective also do, and reckon that the Christian faith must take its lead form science, whereas the evangelical will never to do so because of the authority of the bible. Most evangelicals believe science and evangelicalism can walk hand in hand, but are divided on how.

However one perceives the compatibility or otherwise of Evangelicalism and science, it should be clear from this volume that their paths have never been separate over the last 300 years and will not be in the future.

It will be interesting to see what the future holds for the continuing relationship of Evangelicalism with science.

Primary Sources

—1— Article from Volume IV of *The Fundamentals* (1910)

This article by Orr represents the happy coexistence of science and Evangelicalism at the end of the nineteenth century, which was to be shattered during the fifteen years leading up to the Scopes trial. For brevity this includes only the introduction and conclusion. The whole article is online at: http://www.xmission.com/~fidelis/volume1/volume1.php and in *The Fundamentals* (1910, Vol. IV, pp. 91–104).

Science and Christian Faith
Rev. Prof James Orr

In many quarters the belief is industriously circulated that the advance of "science," meaning by this chiefly the physical sciences—astronomy, geology, biology, and the like—has proved damaging, if not destructive, to the claims of the Bible, and the truth of Christianity. Science and Christianity are pitted against each other. Their interests are held to be antagonistic. Books are written, like Draper's "Conflict Between Religion and Science," White's "Warfare of Science with Theology in Christendom," and Foster's "Finality of the Christian Religion," to show that this warfare between science and religion has ever been going on, and can never in the nature of things cease till theology is destroyed, and science holds sole sway in men's minds.

This was not the attitude of the older investigators of science. Most of these were devout Christian men. Naville, in his book, "Modern Physics," has shown that the great discoverers in science in past times were nearly always devout men. This was true of Galileo, Kepler, Bacon, and Newton; it was true of men like Faraday, Brewster, Kelvin, and a host of others in more recent times. The late Professor Tait, of Edinburgh, writing in "The International Review," said: "The assumed incompatibility of religion and science has been so often and confidently asserted in recent times that it has come . . . to be taken for granted by the writers of leading articles, etc., and it is, of course, perpetually thrust before their too trusting readers. But the whole thing is a mistake, and a mistake so grave that no truly scientific man . . . runs, in Britain, at least, the smallest risk of making it. . . . With a few, and these very singular exceptions, the truly scientific men and true theologians of the present day have not found themselves under the necessity of quarrelling." The late Professor G. J. Romanes has, in his "Thoughts on Religion," left the testimony that one thing which largely influenced him in his return to faith was the fact that in his own university of Cambridge nearly all the men of most eminent scientific attainments were avowed Christians. "The curious thing," he says, "is that all the most illustrious names were ranged on the side of orthodoxy. Sir W. Manson, Sir George Stokes, Professors Tait, Adams, Clerk Maxwell, and Bayley—not to mention a number of lesser lights, such as Routte, Todhunter, Ferrers, etc.,—were all avowed Christians" (page 137). It may be held that things are now changed. To some extent this is perhaps true, but anyone who knows the opinions of our leading scientific men is aware that to accuse the majority of being men of unchristian or unbelieving sentiment is to utter a gross libel.

If by a conflict of science and religion is meant that grievous mistakes have often been made, and unhappy misunderstandings have arisen, on one side and the other, in the course of the progress of science,—that new theories and discoveries, as in astronomy and geology, have been looked on with distrust by those who thought that the truth of the Bible was being affected by them,—that in some cases the dominant church sought to stifle the advance of truth by persecution,—this is not to be denied. It is an unhappy illustration of how the best of men can at times err in matters which they imperfectly understand, or where their prejudices and traditional ideas are affected. But it proves nothing against the value of the discoveries themselves, or the deeper insight into the ways of God of the men who made them, or of real contradiction between the new truth and the essential teaching of the Scriptures. On the contrary, as a minority generally perceived from the first, the supposed disharmony with the truths of the Bible was an unreal one, early giving way to better understanding on both sides, and finally opening up new vistas in the contemplation of the Creator's power, wisdom, and majesty. It is never to be forgotten, also, that the error was seldom all on one side; that science, too, has in numberless cases put forth its hasty and unwarrantable theories and has often had to retract even its truer speculations within limits which brought them into more perfect harmony with revealed truth. If theology has resisted novelties of science, it has often had good reason for so doing.

It is well in any case that this alleged conflict of Christianity with science should be carefully probed, and that it should be seen where exactly the truth lies in regard to it.

I. Science and Law—Miracle
II. Scripture and the Special Sciences
III. Evolution and Man

The conclusion from the whole is, that, up to the present hour, science and the Biblical views of God, man, and the world, do not stand in any real relation of conflict. Each book of God's writing reflects light upon the pages of the other, but neither contradicts the other's essential testimony. Science itself seems now disposed to take a less materialistic view of the origin and nature of things than it did a decade or two ago, and to interpret the creation more in the light of the spiritual. The experience of the Christian believer, with the work of missions in heathen lands, furnishes a testimony that cannot be disregarded to the reality of this spiritual world, and of the regenerating, transforming forces proceeding from it. To God be all the glory!

EXTRACTS FROM THE *JOURNAL OF THE AMERICAN SCIENTIFIC AFFILIATION*

The next four sources are from the *Journal of the American Scientific Affiliation* (see chapters 6–9) and illustrate the controversies among (American) evangelicals over science in the immediate Post-War years.

—2—

The American Scientific Affiliation—The First Decade

F. Alton Everest

From: *JASA* 3 (September 1951): 33–38.

Part of an address given at the Sixth Annual Convention at New York City August 30, 1951

(This address is self-explanatory as it is an account by one of the founders of the ASA.)

The ten-year period, 1941–1951 was a time of great significance to the world. During this time a second world war was fought which quite definitely silenced voices raised in optimistic hope for a peaceful world'. International relations during this period were in a chaotic state. Sneak attacks, subversive activities, the lack of honor in international relations, the dishonest practices in high places within our own government, the deplorable flourishing of organized crime, juvenile delinquency–all these things have brought a feeling of insecurity and despair to our people and the people of the world.

In 1939 scarcely two years before the opening of this significant decade, there was announced the hypothesis of nuclear fission and its experimental verification. By the end of 1941 an extensive review of the entire uranium situation was completed end full impetus was given military exploitation of the fission effect. In 1942 the famous Manhattan District was organized culminating in that fateful birth of the atomic age in the gray dawn of the New Mexico desert on July 16, 1945.

These striking events in the military and political world had, perhaps at least some effect on the changing attitudes in the scientific world. The pendulum seemed to be swinging away from the rigid materialism of the past. Heisenberg's Uncertainty Principle pointed out that strict determinism at least did not apply in the realm of the quanta. While the leap from so-called "free will" of the electron to the freedom of the human spirit is too great a jump for man to make, these new concepts were having their over-all impact on modern scientific thought.

The equivalence of mass and energy–a concept dramatically demonstrated on that morning in the New Mexico desert–also seemed to pull some of the underpinnings from beneath materialistic philosophies.

Let us not be confused—a knowledge that all our material world is but a manifestation of energy does not necessarily make a spiritual-minded religionist of the physicist. Nor should we feel that this revolution has taken place when we see cosmologists put forth their hypotheses of a practically instantaneous condensation of all the elements from a sort of hot cosmic stew end equate the event to the creation of Genesis 1:1. All these things may have their cumulative effect, but we still have to deal with scientists who, in the main, do not know our Redeemer and yet who are somewhat more Inclined toward a consideration of spiritual things. Today we find less inclination among scientists to brush off the claims of the Bible with a wave of the hand. This is illustrated by the Professor of Zoology of a large Western university who each year shows all of the Moody Science Films to his large class—gospel message and all—as he puts it, "to present the other side of the question in all fairness." Such instances may be few and far between and the general trend away from rigid materialism may appear to be all but swallowed up in the familiar glorification of human wisdom but it seem incontrovertible that the trend is there.

This is a brief sketch or conditions existing just prior to and Immediately following the organization of the American Scientific Affiliation. There probably never was a more significant period in history—the growth of science and the opening of the atomic era—and a period that presents its own unique problems to an effective witness and defense of the faith.

It would be unfair and distinctly untrue to leave the impression that, in the midst of these great problems, the ASA was born In solitude. The ASA owes much to other similar organizations some of which rose briefly only to fade, others which are contemporarily functioning today.

The Victoria Institute of Great Britain was founded in 1865 to treat the problems arising between science and the Holy Scripture with the view of reconciling any apparent discrepancies. This organization has had a long and useful life and is actively continuing its work today. It as a total membership of around 600 and its Journal of the Transactions-contains many valuable papers.

The Religion and Science Association was organized about 1928. Dr. L. Allen Higley, then of Wheaton College, was one of the founders. Other name associated In this early endeavor are Ben F. Allen, and Mr. Clarence Benson, and others. Its activities were abandoned at least ten years ago.

The Kelvin Institute had its origin in Toronto, Canada) about 1935. This group was founded by Mr. Arthur C. Custance, formerly a member of ASA. Our own Dr. John F Howitt was one of the early members. At its peak it had some 50 members in England, Scotland, Australia, USA, and Canada and produced a number or very interesting papers. However, its activities ceased during the early 1940's.

The Evolution Protest Movement, active in England for more than two decades, did not find ready acceptance in the USA. Sensing this, Dr. Arthur Pierson Kelley thought that an indigenous group patterned along the same lines would be better received and proceeded to found The Creationist Society. This proved to be entirely abortive. For some time. however, Dr. Kelley issued a series of papers under the name The Landenberg Review which carried on the anti-evolution cause. As far as it is known, this effort has been abandoned.

In addition to these there have been a number of local organizations devoted to the study of the relation of science to Christianity. Typical of these was the Nature and Scripture Study Club of Grand Rapids founded in 1935 by Dr. John P. Van Hartsma. of Calvin College, one of the founder-members of the ASA. This group disbanded prior to 1942, at least partially because of Dr. Van Hartsma's failing health.

About 1938, the Society For Study of Deluge Geology and Related Sciences was formed. This group published -The Bulletin of Deluge Geology,

1941 through 1945. To an outsider, what then happened is somewhat obscure, but apparently reorganization resulted in the use of the names Natural History Research Group and also The Society For the Study of Natural Science. A publication called The Forum came out with one or two issues under the latter-named society. Many associated with these groups include Prof. George McCready Price well-known advocate of deluge geology, Dr. Cyril B. Courville, Dr. Moulerus Couperus, Mr. Ben F Allen, and others. Correspondence with members of this group and a perusal of its publications leads one to conclude that it was founded principally to defend the so-called deluge interpretation of the earth's stratigraphic history and to contend for a recent formation of the earth. As many of its leaders are also leaders in the Seventh Day Adventist group which strongly promulgates these particular interpretations, it is felt by many that the society, directly or indirectly, was influenced by the teachings of this denomination.

There are no doubt many other groups not named here that were formed during the 1930's and 1940's to correlate scriptural interpretation and now scientific facts. All these groups had one thing in common. They were anxious to demonstrate the harmony of God's Word end God's Work In nature. Some may have struck too negative a note, some may have been unwise in their emphasis, some may have been downright wrong in either their exegesis or their science–but their motives were commendable, and we shall leave to the great Judge of all things the ultimate value of their labors.

I believe it was in the wise providence of our God that the men that founded the ASA and who were responsible for its nurture during the early years were relatively ignorant of these other groups. For this reason their Influence on the ASA was negligible and it is only in retrospect that we can see how fortunate this was in the formation of the policy of the ASA which the testimony of ten years' growth indicates as sound.

What is this policy? For what purpose does the ASA exist? The constitution states that the object of the Affiliation is to "correlate the facts of science and the Holy Scriptures." We are vitally interested in determining the proper interpretation of the creation account, the flood of Noah, Joshua's long day, etc., but we consider it distinctly Improper for the ASA to become so enamoured by particular interpretations of these accounts that we shift our efforts from study to propaganda. Dr. Allan A. MacRae, prominent archaeologist and past Vice-President and member of the Executive Council of the ASA has wisely put it this way:

> To my mind it would be unfortunate for the Affiliation to go on record strongly in favor of any one of the various views. It seems to me that its purpose should be rather to show that the Bible as correctly and carefully interpreted, and without

any twisting whatever, leaves room for every scientific fact at present known, and does not contradict any scientific fact as yet discovered, however much it may be at variance with some particular theory built upon these facts.

Now this does not mean that ASA members do not have some pretty strong personal convictions concerning these Interpretations—they do. Furthermore, there Is a wide range of interpretations held among the membership. This is only natural and It Is a very healthful situation—one that causes one to search his case thoroughly before submitting himself to one of our justly famous discussion periods. Thus in the *Journal of the ASA* you will find a paper supporting a particular interpretation and a little later another one apparently demolishing it. We consider our job well done if we can present a Bible-teacher, a pastor, or a university student an adequate survey of the various views held on a given problem and the historical and scientific data pertaining to it. The problem of the church is principally, one of plain ignorance of the many and complicated factors entering into a wise interpretation of the Scriptural accounts.

Another member of the ASA, Dr. Barley Barnes, a geologist, stated the following views six or seven years ago which I believe to be a fair expression of the prevailing attitude of the majority of the present Associates and Follows of the ASA:

"The essence of my attitude towards evolution and the Bible as a Christian geologist is that Christians should be non-evolutionary because the Bible does not give unequivocal grounds for being anti-evolutionary. Few, if any, Christians who, are scientists now would say that the Flood of Noah produced all fossils or that the earth is only 6000 years old; yet there was a time when persons holding contrary opinions were generally labelled modernists. The Bible said the same thing then as now but our interpretation has changed because discoveries have narrowed the field of interpretation possible to rational men of faith, We are faced then, with the realization that the Bible allows numerous interpretations of the creation account, but our choice of 'the' interpretation has been limited to those which do not conflict with accumulated scientific observations. The viewpoint of scientists changes continually as additional data fills In the picture, shifting the perspective. Let us not repeat the mistake of earlier Christians by forcing the Bible to speak in the language of current or recently current theory.

"The devout 'Fundamentalist' persecutors of Galileo and Copernicus helped discredit the Bible to many thinking persons because they were morally certain it specifically taught the Ptolemaic astronomy.' After accumulated data overwhelmed these pious but bigoted spirits, a new crop of second guessers come forth with the 'obvious fact' that the Bible had agreed with the former heretics (Galileo and Co.) right along, so that 15 years ago I was impressed by the quotation of verses 'proving' that the earth was round. And if science should find sufficient information that the earth after all really has the shape of a dodecahedron, someone would find a verse in the Bible which had said so all the time!

"The Bible does not pretend to give a technical description of these matters. What it does say is accurate but generalized; some passages are figurative, but we are not always able to determine which ones-Isaiah 40:22 or Revelation 7:1. Let us not tie the Bible to any mode (or Victorian) scientific theory. If it can be interpreted more then one way, let us admit it and when all the pertinent facts are known rejoice in the Wisdom which bad been revealed but until that time not understood. Perhaps the Bible suffers less from too little literal interpretation than from too much literal interpolation."

In November, 1940, while the writer was on the faculty of Oregon State College, a message was received from Irvin A. Moon who was holding his famed Sermons From Science demonstrations in nearby Salem, Oregon. A meeting was arranged In Salem during which Mr. Moon told of the great need he had encountered among young people for an organization of reputable man of science who were also Christians. The young people flocked to him with their questions, eager for reassurance that modern scientific knowledge does not rule out faith. That afternoon Mr. Moon named several qualified men of science whom he believed would be interested in forming such a society.

Several months passed during which Mr. Moon was-in constant touch with Dr. Will H. Houghton, the late President of-the Moody Bible Institute of Chicago, concerning ways and means of bringing this group into existence. Early in 1941 at the request of Mr. Moon, the writer examined the constitutions Of Many existing scientific societies and drafted one which was considered suitable for such a society of Christian men of science as was proposed. Considerable attention was also given to a fitting name for the society.

In June 1941, Dr. Houghton addressed a comprehensive letter to a number of men who had shown their interest In the formation of a group for the correlation Of science and the Bible. In the recent biography of Dr. Houghton, *A Watchman on the Wall**, Dr. Wilbur M. Smith entitles a chapter "Through Science to the Souls of Men" in which Dr. Houghton's letter Is quoted in Its entirety. Dr. Smith states that this letter might "be called the birth certificate of the American Scientific Affiliation." Dr. Smith further say "The letter itself, I think, will prove in days to come a significant document-" In this letter the man to whom It was addressed were invited to meet In Chicago, September 2 to 5, 1941, to "canvass the possibilities of a larger conference" and eventually, perhaps, to organize a new society. Dr. Wilbur Smith writes, "Out of this letter has come a strong. vigorous organization which today is undertaking a number of projects which the Lord willing, in the days to come will be the means of great help to those who are troubled about the complicated relationships of contemporary science with the Word of God."

Five men journeyed to Chicago as a result of Dr. Houghton's invitation. Their expenses were borne by a donor unnamed at the time but who has since been identified and the saintly "Breakfast Table Autocrat" and intimate friend or Dr. Houghton, Henry Parsons Crowell. Those who attended were: Dr. Irving A. Cowperthwaite, Prof. Russell D. Sturgis, Prof. Peter W. Stoner, Prof. John P. Van Hartsma, Prof. F. Alton Everest, Dr. Houghton, Mr. Moon, and Mr. H. Coleman Crowell, the son of Henry P. Crowell and Vice-President of Moody Bible Institute conferred with the group from time to time but Dr. Houghton's statement In his letter of Invitation was faithfully carried out when he said "some or us are initiating this, but we haven't any desire to control it, and certainly ilt would be limited if it were known to be sponsored by any one educational Institution—. We want you to be very sure the group will be entirely free to make its own plans."

The American Scientific Affiliation was organized at this time although its constitution and new were not officially adopted until a few months later. The precious time together was spent in considering future programs, membership qualifications and prospects, constitution, aim and purposes, spiritual standards, and many other vital topics. Dr. Cowperthwaite was elected Secretary-Treasurer and Prof. Everest Chairman and the group adjourned.

The next few months were filled with intensive correspondence and the laying of plans for a large national meeting in 1942. Three months after the Chicago meeting, on December 7, the Japanese attacked Pearl Harbor and such grandiose plans as a convention were eliminated by travel restrictions.

Looking back on it, the Lord's overruling hand in interfering with our plans by the war was probably the best thing that could have happened to the embryonic group. We were forced to work and study independently during the 1941–1945 period and were given practically no opportunity to arouse each others ire in heated debate at close range. Rather, this time served to clarify the issues, streamline the organization, bring in new members and to begin work on the book later entitled, *Modern Science and Christian Faith* which is now in its second edition.

While war-time travel restrictions practically eliminated civilian travel, the chairman was not only able but was forced to do considerable travelling over the United States in the interests of his wartime work. This provided opportunities for dozens of conferences and contacts which increased the effectiveness of the work greatly over complete reliance on correspondence.

Articles of incorporation for the ASA as a non-profit Organization were filed with the Secretary of State of California In August. 1943. Also, during 1943 we began voting in new members after cautious consideration of their qualifications. By the close of the war, the membership was about 50, and by the end of 1950 about 130. The adoption of the new constitution with the new membership classifications has naturally resulted in a sudden spurt over the average rate of approximately 20 now members per year. At the time of this 1951 convention in New York the membership is about 220. The growth has been steady—we have never had ambitions for a large organization or we would have had lower standards for membership, After all, there are just so many scientists, and a very small percentage are interested in Christian things, but the conservative policies adopted are rapidly gaining the confidence of these qualified persons.

It seems entirely proper that the greatest strides of the ASA be taken during the tenth anniversary year. Without doubt, the adoption of the new constitution with no dissenting voices (one man expressed disagreement with one portion) was a remarkable expression of the members' confidence in the leadership of the Council. This in especially true when it is realized that, for those members not qualified for Fellow grade, It meant virtual disenfranchisement. This new constitution's distinguishing feature is Its placing in the hands of the Fellows the voting privileges and the policy responsibilities, yet opening the Associate grade for those whose interest in and enthusiasm for the work of the ASA is in no way measured by their lack of advanced scientific degrees. Thus, maximum protection is coupled with maximum protection for the high spiritual and scholastic standards of the group.

Another great stride was Dr. Mixter's selection for the presidency after nine years of service In this capacity by the writer. Sudden upsurge in all activities or the ASA is evidence cited to uphold this statement. One of the greatest experiences in the life of the writer has been the thrill of working side by side with man of the ASA—men whose faith has been tried in the crucible of spiritually sterile scientific criticism, men who have devoted their lives to the study of God's handiwork in nature and who see there the infinite resources of the One, in 'Whom we live and move and have our being. It is our priceless privilege to join together in proclaiming the salvation which He has offered to a lost world.

—3—
Fundamentalism and the Fundamentals of Geology

J. R. Van De Fliert*
Free University of Amsterdam, Amsterdam, The Netherlands

From: *JASA* 21 (September 1969): 69–81.

(A review of *The Genesis Flood.* This review caused a strong response. In the extract most of the purely geological argument is omitted.)

Introduction

With increasing astonishment, I read through the book *The Genesis Flood-The Biblical Record and Its Scientific Implications,* by Henry M. Morris and John C. Whitcomb, Jr.[1] If I had been told a few years ago that an apparently serious attempt would be made to reintroduce the diluvialistic theory on Biblical grounds as the only acceptable working hypothesis for the major part of the geological sciences I would not have believed it. I would have considered it just incredible that a professor of Old Testament and a professor of Civil Engineering would write it, and that the foreword would be written by a professional geologist.

It is almost incredible that such an effort, which must have cost an enormous amount of work and money, has been made for such a bad procedure as this. I have felt very reluctant to write against it, but finally agreed to do so, yielding to stress from different sides.

There are two main reasons for this article. The first is that the authors of *The Genesis Flood* have written on the basis of their belief in the Holy Scriptures as the reliable Word of God. This belief I share. Second, it is my sincere conviction that it is a fundamental and extremely dangerous mistake to think that our belief in the reliable Word of God could ever be based on or strengthened by so called scientific reasoning. Any attempt to harmonize the historical geology of today with the account of the first chapters of Genesis represents a colossal overestimation of science as well as a misunderstanding of the Genesis record an overestimation which is as great as that of those scientists who completely reject God as the Creator. If we thus overestimate science, we lose the battle before it is started. The Bible does not give outlines of historical geology nor accounts of scientifically controllable creative acts of God! If we think the Bible does provide these, we have brought God's creative work down to scientific control, down to the visible things, contrary to the teaching of the Bible that "through faith we understand that the worlds were framed by the word of God" (Hebrews 11:3a). We deal a death-blow to the Christian religion when we bring the Holy Scriptures down to scientific level by teaching that the Bible should give us a kind of scientific world-picture or axiomata of historical geology, or of Western science of history, or physics, biology, jurisprudence or whatever science it be. Thus, we lose the Bible as a reliable Word of God completely, because we then make its teachings dependent on the poor state of our scientific knowledge today . . . which will change tomorrow!

The overestimation of science fails to see its possibilities and its limits. It means the corruption of true scientific working, both in the evolutionistic thinking of those who do not believe in God, and also in the thinking of Christians who do believe in God. These latter corrupt scientific work thoroughly when they start from a pretended biblical (in fact, imposed by them on the biblical teaching) elementary historical geology, into which then the geological data will have to fit! This is no less pseudo-scientific than that kind of evolutionistic reasoning that ignores God, and therefore presents truly a very had case for orthodox Christianity today!

Scientific Pretension and Scientific Foundation

Before I start a more technical treatment of a few important geological questions, I want to make a few critical remarks of a general character concerning the pretended scientific value of *The Genesis Flood*.

First, writing a book with such significant claims or conclusions requires a thorough knowledge of the geological sciences and their principles. Neither author—one a theologian, the other a civil engineer—is a geologist. Everybody knows that in the present state of scientific development it is practically impossible for one person to master more than one branch of science. Now, the list of modem publications cited in the book is impressive but at the same time misleading. The way in which part of this literature is used proves that the real problems have often not been understood. A theologian should know how dangerous it is to lift a text out of the context and to treat it separately. This is true not only for interpreting the Bible but also for explaining scientific publications. To lift a certain sentence out of a publication, and to use it for something quite different than the original author meant, is scientifically dishonest. I realize that the authors of The Genesis Flood did not intend to do this at all, and in a few cases they even admit that the author they cite used his words in a slightly different way, but in others they give evidence of not having understood the exact bearing to which they refer. Thorough scientific work makes extremely high demands on professional knowledge!

The Essential Importance of the History of Science and Theology

(omitted)

Uncritical Criticism of Geological Principles

Third, the last general remark I want to make concerns the uncritical attitude of the authors regarding their own reasoning. The whole hook intends to levy a fundamental attack on the so-called uniformitarian principle in the geological sciences. They do not realize that, in part, their reasoning is based on the same starting point. In part, also, they fight against windmills, because most present-day geologists do not accept this principle exactly in the sense as it was understood by Lyell (who was no evolutionist when he wrote the first edition of his *Principles*[3]), but use it in the sense of a constancy of physical and biological laws. . . .

P15-21 omitted

The Trustworthiness of the Geological Time-Scale Disputed

Let us now turn to a few fundamental facts and principles of present-day geology. First of all, consider those that concern the stratigraphic column and the geologic (relative) time scale.

As an introduction, note a few quotations from the summary of the chapter, "Modern Geology and the Deluge" in *The Genesis Flood.*

We read on page 206:

> The geological time series is built up by a hypothetical superposition of beds upon each other from all over the world.

That this superposition should be "hypothetical" (which here clearly means "not factual") is argued with a quotation from a geological text book:[8]

> If a pile were to be made by using the greatest thickness of sedimentary beds of each geological age, it would be at least 100 miles high... It is, of course, impossible to have even a considerable fraction of this at one place. The Grand Canyon of Colorado, for example, is only one mile deep

Then the authors of *The Genesis Flood* continue:

> This frank statement makes the method by which the geologic time scale was built up quite plain. Since we have already noted that lithologic identification is unimportant in establishing the age of a rock, it is clear the "fossil successions" constitute the only real basis for the arrangement. And this means, in effect, that

organic evolution has been implicitly assumed in assigning chronological pigeonholes to particular rock systems and their fossils.

There follows a second quotation from Von Engeln and Caster, which apparently should confirm this conclusion:

The geologist utilizes knowledge of organic evolution as preserved in the fossil record, to identify and correlate the lithic records of ancient time.[9]

This is commented on as follows:

And yet this succession of fossil organisms as preserved in the rocks is considered as the one convincing proof that evolution has occurred! And thus have we come round the circle again.

The trend of this reasoning is clear: Historical geology is basically unsound because it has been trapped in circular reasoning. First, geologists determine the order of succession of fossils in the earth's crust on the basis of the superposition of the strata, but at the same time they declare the position of the strata reversed-by some tectonic process-when at another place the succession of fossils is found reversed! What is more, and even worse: Behind this is the 'hypothesis' of evolution, of "a gradual progression of life from the simple to the complex, from lower to higher" (pp. 132, 134).

Moreover:

quotations from outstanding evolutionary authorities both in geology and biology, demonstrate the great importance of the paleontological record to the theory of evolution. In turn, the principles of evolution and uniformity are seen to be of paramount importance in the correlation of the geologic strata. These principles are absolutely basic, both from the point of view of the history of the development of modern geology and from that of present interpretation of geologic field data. The circular reasoning here should he evident and indeed is evident to many historical geologists (p. 134).

How corrupted and preconceived present-day historical geology really should be is then formulated in the following words:

The basis for the apparent great strength of the present system of historical geology is here clearly seen. Provision is made ahead of time for any contrary evidence that might be discovered in the field. The geologic time scale has been built up primarily on the tacit assumption of organic evolution, which theory in turn derives its chief support from the geologic sequence thus presented as actual historical evidence of the process. Fragments of the sequences thus built up often appear legitimately superposed in a given exposure, but there are never more than a very few formations exposed at any one locality, occupying only a small portion of the geologic column. Formations from different localities are integrated into a continuous sequence almost entirely by means of the principle of organic evolution (p. 136).

I give these rather long quotations in order to show in what light such a sentence as "The geological time series is built up by a hypothetical superposition of beds upon each other from all over the world" should be read, and furthermore to give an example of the mixing up of truth and untruth in the way of arguing of the authors of *The Genesis Flood* when it concerns one of the fundamentals of geological science.

(pp. 24–7 omitted)

The Primary Superposition in Highly Disturbed Areas

However, much more is to be said. When discussing what they called "Methods of resolving contradictions", the authors of *The Genesis Flood* write:

Furthermore, even where superposed strata are exposed, it rather often happens that the fossils appear to be in reverse order from that demanded by the evolutionary history, which paradox is commonly explained by the assumption that the strata have been folded or faulted out of their original sequence (p. 135).

It is an old story which is told here. It was already elaborated in Professor Aalders' book[10]. And it seems that this favorite argument of professors of Old Testament is supported even by some geologists; the authors of The Genesis Flood give the citation of C. H. Rastall, lecturer of Economic Geology at Cambridge University, saying:

It cannot be denied that from a strictly philosophical standpoint geologists are here arguing in a circle. The succession of organisms has been determined by a study of their remains embedded in the rocks, and the relative ages of the rocks are determined by the remains of organisms that they contain (p. 135).

Now, Mr. Rastall may be a good economic geologist; he is definitely not a good philosopher because his statement is simply not true!

What are the facts? A reversed position of strata is the result of strong disturbing movements after deposition. Complicated tectonic deformation occurs when the sediments are deposited in an area which is or becomes highly mobile, in contrast with relatively stable regions.

(pp. 28–39 omitted)

Science and the Bible: Not the Fundamentalistic Way

It may seem as if I have written very little about fundamentalism so far. However, I was fighting against it all the time, but silently and indirectly until now.

The book of Whitcomb and Morris was written on the basis of what we usually call a fundamentalistic or biblicistic viewpoint. This standpoint implies the belief that the Bible teaches us principles, fundamentals or elements of human science in general and of historical geological science in particular.

For the fundamentalist, therefore, the reliability of the Bible as the Word of God is related to scientific reliability. For him this is particularly true with respect to the first eleven chapters of Genesis. This conception, however, implies inevitably that science and God's Revelation in the first chapters of the Bible are placed on the same (scientific) level, on the basis of which scientifically obtained data about the history of the earth and man will have to fir, into the 'Biblical scheme or framework.'

The 'question' of the reliability of the Holy Scriptures can thus be fought out on the scientific field, and, as a consequence, we then see theologians enter this field, as Professor Whitcomb now does, as Professor Aalders did in Holland a few decades ago, and as so many before them have done since the end of the Middle Ages. But these 'scientific' battles for an infallible Word of God have been lost right from the start. In constant retreat, the theologians have had to surrender every position they had once taken in this struggle. That's what the history of the warfare between science and theology should have made conclusively clear. The tragedy of men who wanted to defend the reliability of the Word of God 'scientifically' should have taught us that this entire approach was wrong. It should have convinced us that this science is a very bad ally, because its word had only temporal and no eternal value.

The most tragic aspect of the fundamentalist conception seems to me that his standpoint requires scientific proof, so that he must somehow live in fear of the results of developing scientific work, because indeed this development could then also disprove the reliability of the Holy Scriptures. And this leads to the cardinal question whether in this way the fundamentalist's conception does not reveal an implicit faith in science, which is far more dangerous for Christian religion than is the scientific development itself.

A few years ago, I was speaking to a conference of Reformed ministers in the Netherlands about some fundamental facts of geology. In the discussion, one of them arose and declared that, if he were convinced that what I had told them was true, he would immediately abandon his ministry. But I ask myself 'what kind of a religion is Christianity when scientific geological facts can prove or disprove the reliability of God's Revelation to man? What then do we really believe in? In our own 'image,' conceptions or ideas about an infallible Bible? In an interpretation of the first chapters of Genesis with the help of current natural scientific knowledge just as earlier theologians did with the help of a world picture, incidentally, usually already out of date in their own time?! Does the message of the Bible then really necessarily change with the changing world picture? It surely does as long as we continue trying to accommodate Genesis and geology.

Instead of giving human scientific work its proper place in the light of Scripture, fundamentalism indeed implies, as I indicated already in the beginning of this article, a colossal overestimation of natural science. Neither geology nor any other natural science can ever be a direct exegetical tool, as they have been used, and still are used in fundamentalistic conceptions.

However, the history of the natural sciences and the results of modern geology, for example, could play a far more modest role, the role of an indirect exegetical tool. Such would be not a tool to test, to prove or to disprove the reliability of Scriptures, but to test the reliability of our ideas and conceptions about the Bible, the inspiration, and the historicity of the first chapters of Genesis.

The reliability of the Word of God spoken in this world through His prophets and apostles is beyond the reach of scientific control, because the Bible is not a scientific book. As such, it is not vulnerable to the results of science. Therefore, Christian astronomers, geologists, and biologists can work without fear as long as they respect the limits of their own scientific field.

Our ideas and conceptions concerning the Bible may indeed appear to be vulnerable to the results of scientific development. This state of affairs seems to he difficult to accept, particularly for many evangelical Christians. It cannot be denied, however, that there is 'revelation' (be it of a different kind than that of the Bible) in the development of this created world, also in the results of human scientific and technical advances during the last centuries. It cannot be denied and should not be denied that, as a result of this development, our (scientific) world picture (Weltbild) has obtained huge dimensions, both in time and space and has become entirely different from that of the authors of the Bible. But, this is the world God has wanted us to live in, we and our children.

The fundamentalistic view, conservative in an erroneous sense, requires us to accept a so-called "biblical world picture" which should be normative for scientific work. This is a poor predicament indeed for contemporary Christianity, because it tends to transform twentieth century Christians into aliens, standing, as it were, in Old Testament times. Since this is, of course, not possible, the fundamentalistic view tends to deprive them of their belief in a reliable Bible. It alienates us from the Words of Eternal Life, which we understand through faith and not through science, and which stand firm in this rapidly changing world.

References

[1]Published by the Presbyterian and Reformed Publishing Company, Philadelphia, Penna., 1961.

[3]Charles Lyell, Principles of Geology, *being on attempt to explain the former changes of the earth's surface by causes now in operation*. 1st Ed. Volumes I-Ill, London 1830–1833.

[8]A. H. von Engeln and K. E. Caster, *Geology*, 1952, pp. 417, 418.

[9]A. D. van Engeln and K. E. Caster, *Geology*, 1952, p. 423.

[10]Dr. G. Ch. Aaldcrs, *De goddelijke openbaring in de eerste drie hoofstukken van Genesis*, Kampen, 1932.

—4—

Letters to The Editor of The JASA in Response to Van De Fliert's Review Pro and Con Van De Fliert From: *JASA* 22 (March 1970): 35–38.

(These letters illustrate the divide between YEC and evangelicals who accept "orthodox science. Some were supportive, but of great interest is the response by Henry Morris.)

I especially appreciated van de Fliert's article. I find about as little virtue in presenting arguments to establish the infallibility of the Bible as I do trying to prove the existence of God scientifically. I take these matters on faith. So I trust God and use the Bible. Since people have generally given up trying to prove God's existence, I think we may have reasonable hope that they will one day give up trying to prove the validity, infallibility, authority, spiritual power or whatever of the Bible. But until we reach that point, men like van de Fliert serve us well.

Henry Booms
Pastor, Ten-Cities Christian Reformed Church Pasca, Washington 99301

I greatly enjoyed the refutation of Morris and Whitcomb's debacle by van de Fliert.

James It. Moore,908 Country Lane Mount Prospect, Illinois 60056

For several years now we non-geologists have had to defend truth with our inadequate geological knowledge while the geologists in the A.S.A. either chose to remain silent or else were in such a small minority that their voices were not heard.

Dr. van de Fliert's article exposing the fatal flaws of Whitcomb's and Mortis' book on flood geology was a most welcome and long overdue breath of fresh air.

I would like to add these facts to his scholarly article:

1. Professor Kane, Chairman of Geology Dept., Ball State University, indicated to me that none of the geologists approving Whitcomb and Morris's hook have been involved in geological research in at least 2 and probably more likely 3 or 4 decades. This, he said, he can vouch for by personal knowledge of the geologists themselves as well as of the research literature. He said that they are essentially "arm-chair geologists."
2. On Whitcomb's third visit to Ball State University (Fall, 1967), several geologists, anthropologists, biologists, and archaeologists pointed out to him some of the more obvious errors in the book. Not once in the two hour meeting did he admit to the need for any correction in spite of overwhelming evidence from those research scientists to the contrary. However, it is encouraging to note that so far as many students were concerned, the meeting had beneficial results as they testified that no one had taken the trouble before to point out the dangerous grounds upon which flood geology rests.
3. The book's testimonies by non-geologists such as plant breeders, civil engineers, chemists, and the like are irrelevant to the primarily geological problem, are misleading to laymen, and appear somewhat (probably unintentionally) dishonest to be included.

   It is pitiful to see so many of our fellow conservatives resort to intellectual dishonesty, browbeating, pseudo-pietism, and even hatred-all in the name of Christ and fundamentalism. It would seem that they feel that the end justifies the means. It was good to see the testimonies of geologists Cuffy, Tanner, and Boardman regarding the article. May God give courage to other geologists in the A.S.A. to stand up and he counted for the faith!

Thomas Key Department of Biology Oglethorpe College Atlanta, Georgia

I appreciate very much the attention devoted to our book[1] by Professor van de Fliert[2] since most professional geologists have ignored it. However, I regret that he allowed himself to resort to emotional language in his discussion ("incredible," "flagrant nonsense," "extremely dangerous," "pretended scientific value," etc.). One evidence that evolutionary uniformitarianism is a religion rather than a science is the fact that its advocates almost invariably react emotionally whenever a fellow scientist questions it.

We agree completely with most of Professor van de Fliert's paper and are puzzled as to why so much that is in agreement with *THE GENESIS FLOOD* is included in a polemic against it. In many instances it seems to me that he is battling a straw man of his own preconception-like those evolutionists who forever are attacking the supposed creationist doctrine of fixity of species.

Thus we have always stressed the uniformity of natural law as a basic principle in science. Similarly we recognize abundant evidence of extensive earth movements in the past, including overthrusting, folding and other remarkable tectonic features which we do not see occurring at present. As a hydrologist and hyraulics engineer, I certainly believe that the same basic principles of hydraulics operating at present were in effect when the ancient lands and rivers were eroded and the ancient sediments were deposited. Furthermore we recognize the value of the standard geologic column as a taxonomic device and the fact that strata usually occur in the accepted order and that paleontologic criteria of identification are generally valid.

But the point of the discussion in *THE GENESIS FLOOD* (and not discussed by Professor van de Fliert) is that there are a great number of exceptions to the usual order in which the supposed physical criteria of overthrusting, reworking, etc., are not present, and that there are a great many geologic features which (on consistent uniformitarian principles) could not possibly be correlated with geologic phenomena actually observed by modern geologists, either quantitatively or qualitatively (e.g., regional volcanic terrains, continental glaciation, mountain-building, peneplain formation, fossil graveyards, incised meanders, regional alluviation, submarine canyon formation, and many others.) It seems to many of us that such things as these absolutely demand catastrophism of some sort, though within the framework of uniform natural law.

In the decade since *THE GENESIS FLOOD* was written (though I do not mean to suggest any connection) a significant reaction of orthodox geologists has emerged against the older uniformitarianism, with an increasingly frank recognition that local or regional catastrophism is fundamental in geologic interpretation. I have discussed this trend to some extent in two other papers.[3,4] Of course there is still as much antipathy as ever to the idea of a worldwide cataclysm such as the Biblical Flood.

In the book we attempted, in an exploratory way, to see how the actual observed data of geology and other sciences could be harmonized with the Biblical record of the Flood. We repeatedly stressed in the book that our proposed geologic interpretations are tentative and subject to revision with further study and evaluation. However, the one point we insisted on was that the basic Christian presupposition of the inerrancy and perspicuity of the Genesis record must he maintained. If this is not done, then the remaining system may possibly be theistic, but it can be neither Biblical nor truly Christian.

Now it is this fundamental requirement which not only van de Fliert but all other critics of *THE GENESIS FLOOD* have studiously ignored. Critics invariably dwell on certain supposed flaws in our geological perspective (e.g., our alleged failure to recognize the real nature of the geologists' concept of uniformitarianism, the supposed impropriety of documenting our ease with quotations from men who don't agree with it, our alleged ignorance of the fact that there really are some examples of overthrusting, re-working of sediments and faunal mixing and other phenomena whose universal applicability we questioned, etc.), but they always pass by the much more important and fundamental fact that the written Word of God unequivocally teaches that there was a world-destroying cataclysm in the days of Noah!

This reaction of course is to be expected from non-Christian geologists, to whom the Biblical record is utterly irrelevant anyhow. But it is disheartening and puzzling when evangelical scientists, who insist that they still believe in the divine authority of the Bible, also completely ignore this powerful Biblical evidence for the worldwide cataclysm, as presented in *THE GENESIS FLOOD* and many other places. That this is a fair statement of the situation has been thoroughly confirmed in a recent study,[5] by a man trained in both science and theology, who has analyzed all the reviews and criticisms of *THE GENESIS FLOOD* since its initial publication.

Professor van de Fliert admits, in fact, that "our scientific world picture has become different from that of the authors of the Bible."[5] To him, therefore, the fact that the writer of Genesis (as well as job, David, Isaiah, Paul, Peter, and even Christ Himself) believed in a global Flood is of no importance. He feels this issue can he settled simply by saying that "the Bible is not a scientific book." He even thinks (and one is almost startled to encounter this kind of circumlocution in a serious scientist and Christian) that to apply the Biblical doctrine of inerrancy to matters of historic fact is "a colossal overestimation of science."

Atheistic scientists and philosophers, on the other hand, reason much more directly. To them, if the Bible is unreliable when it deals with matters of human observation and experience (i.e., science and history) as it does with great emphasis and frequency-then it is surely not worth trusting when it attempts to treat intangibles such as sin and salvation, heaven and hell-and God!

Available space for this communication does not allow for a rebuttal to Professor van de Fliert's criticism of our discussion of hydrodynamic sorting as a partial explanation for the lithologic and paleontologic divisions in the strata of a sedimentary exposure, or of the highly uncertain growth rates and subsequent histories of ancient coral reefs, or of other geological problems. I can only say that he has not at all settled these questions.

But this is not the important thing. I again acknowledge that there are many, many problems in geology for which we do not yet have adequate answers in terms of the Biblical framework, even though we can at least see in a general way how many of the data can be reinterpreted to correlate with it. There are even more serious problems, on the other hand, for the dogmatic evolutionist and uniformitarian.

The real crux of the matter, however, is "What saith Scripture?" In THE GENESIS FLOOD, as well as in our other writings, Dr. Whitcomb and I have maintained, with a considerable number of straightforward Biblical arguments, that the Bible teaches a recent special Creation of all things and a worldwide Flood, and that there is no permissible interpretation of the Bible which can accommodate evolution and the geological ages. No one has answered these arguments to date.

How, for one example, can we harmonize the concept of a billion years of random variation, struggle for existence, natural selection, evolutionary dead-ends and extinctions without number, disease, confusion, disorder, decay, slaughter and death; with the fact of a God of perfect wisdom, order, power and grace who could easily have created all things complete and perfect from the beginning (as He has revealed in His Word), but who according to the consistent evolutionist and uniformitarian, chose the tortuous route of evolution instead? This is a serious theological problem, one that cannot really be settled by a quip or a platitude. It seems to me that each evangelical scientist and theologian owes it to the Christian community to do one of two things: (1) develop a sound Biblical exegesis of the fundamental chapters of the Bible (Genesis 1–11), consistent with the rest of Scripture, which will clearly warrant his acceptance of the geological ages and the general evolutionary world-view; or else (2) develop a re-interpretation of the observed facts of geology and other sciences to correlate with the facts of Biblical revelation concerning primeval earth history, centered in special Creation, the Fall, and the Flood.

There can be only confusion and danger in continuing to embellish the superstructure when the foundation has been destroyed.

[1]John C. Whitcomb, Jr., and Henry M. Morris: *THE GENESIS FLOOD* (Nutley, N. J., Presbyterian and Reformed Pub. Co.) 1961, 518 pp. 14th printing (paper back), 1969.

[2]J. R. van de Fliert: "Fundamentalism and the Fundamentals of Geology," *Journal of the American Scientific Affiliation*, Vol. 21, September 1969.

[3]"Science Versus Scientism in Historical Geology," *Quarterly of the Creation Research Society*, Vol. 2, October 1965.

[4]"Sedimentation and the Fossil Record: A Study in Hydraulic Engineering," *Quarterly of the Creation Research Society*. Vol. 2, December 1967.

[5]Charles C. Clough: A Calm Appraisal of *THE GENESIS FLOOD*. Th. M. Thesis, Dallas Theological Seminary, May 1968, 196 pp. This study is summarized in the *Creation Research Quarterly* for September 1969.

Henry M. Morris Professor of Hydraulic Engineering, Virginia Polytechnic Institute, Blacksburg, Virginia 24060

I have been a pastor for more than forty years. During that time nothing has grieved my heart more than to see a young man start out with great promise and later end up a shipwreck.

I have enjoyed the privilege of being an Associate Member of A.S.A. during most of its history. At the beginning it promised to do what I and many others had prayed for a long time, namely, to bridge the widening gap between science and the Bible. For some years you did well, But the past few years there has been a gradual drifting. To me you reached the ridiculous in the recent article, "The Three Storied Universe." Now you come forward with the article, "Fundamentalism and Fundamentals of Geology," by van de Fliert. All of his arguments are what you find in the average book on the subject. Mostly he vented his spleen on Morris and Whitcomb along with all the rest of us who still believe in the Bible as offering reliable historical facts as well as promises.

It is this kind of dividing reason (or futile effort) from faith that is causing most of the tragedy of our day. Of course, van de Fliert shows an abiding faith in his assumptions, his limited experiments, and conclusions. This is placing a Mighty faith in human reason and judgment. No wonder he shunts his faith in God and his word out into some mystical realm where reason can ask no questions nor expect any answer.

In my judgment this is a perfect example of the *Escape from Reason* (*Journal ASA*, 21, 54 (1969)) that Francis Schaeffer recently wrote so well about. This is a true picture of modem man-and a professed Evangelical Christian not only lends aid but defends it.

J. Vernon Wheeless
Pastor, Rice Temple Baptist Church Houston, Texas 77025

J. R. van de Fliert faults Morris and Whitcomb (*The Genesis Flood*) for dogmatism. It is true that they might have presented their case in a less judgmental way. However, van de Fliert then proceeds to be just as dogmatic for the historical geology approach, allowing no possibility that Morris and Whitcomb might be correct. This is strange, when in the same article he is able to speak of the "poor state of our scientific knowledge today . . . which will change tomorrow!" (p. 69) One then reads with wonder such statements as:

. . . this idea . . . was soon to be shown false by evidence accumulated as the science of geology began to grow" (p. 70).

"There is no doubt about the answer in the present state of our knowledge . . . (p. 73).

"This basis makes it possible indeed to say that the broad lines of present-day historical geology are correct, and are to be accepted as a well established fact" (p. 80).

Van de Fliert may be correct, but the very fact that there are scientists (including geologists) today who hold such a view, should call for a little more caution. Morris is faulted because he is not a geologist; it is true, this is not his main field. However, he is trained in geology, and he is so listed in the American Scientific Affiliation Directory.

Criticism, yes. Solid, first-hand research, yes. But dogmatism and stone-throwing, from either side, no. That will get none of us anywhere. Nor does it seem to me to he in the best Christian spirit.

Vernon A. Raaflanb Box 188 Nipawin, Saskatchewan

—5—
Towards a Christian View of Science
Richard H. Bube, editor
From: *JASA* 23 (March 1971): 1–4.

(In response to an article in a British magazine, Richard Bube, a leading figure in the ASA, presents the case for a "noncreationist" approach to science and religion. He brings out where the differences lie.)

The journal *Torch and Trumpet*, Vol. 20, No. 10, October 1970, page 18, carries the final installment of a reprint by the above title from The Banner of Truth. The author is Gordon J. Keddie. The conclusion of this article sums up the author's criticism of what he terms "neo-evangelicals."

It is the contention of this paper that the neo-evangelical "trend" is a down-grade movement from truly evangelical Christianity. We shall briefly summarize the characteristic views of this group and comment thereupon by way of conclusion.

Included in the company of "neo-evangelicals" are F.H.T. Rhodes, R.H. Bube, and M.A. Jeeves, as well as Bernard Ramm, N.H. Ridderbos, Jan Lever, Carl F.H. Henry, and J. van de Fliert. What the author calls "the characteristic views of this group" form such an important area, that this response is given in the hope that we may come to grips with the basic issues and cease to be shunted into blind alleys.

In what follows, the quoted items in the summary of Gordon Keddie are printed in italics. Comments on each point follow; I can speak only for myself.

1. *Scripture, while held to be divinely inspired and infallible, is nevertheless handled in such a way as to subordinate it to modern science where the Bible speaks on topics common to theology and natural science (e.g., creation, miracles).*

The Scriptures are not subordinated to modern science when the Bible speaks on topics common to theology and natural science, e.g., such topics as creation, miracles, the resurrection etc. Only the Scriptures have the authority to speak about the reality of the creation, the miracles and the resurrection. These questions cannot be touched by modern science, much less settled.

We may know that God created, for example, only through the revelation of the Scriptures. We may know that we are not the chance results of a "collocation of atoms," to quote Bertrand Russell, only by the authority of the Word of God. We may know that the entire universe depends for its very existence moment-by-moment upon the sovereign power and will of Cod only because He has spoken in the Scriptures. Divine Creation is always revealed, never discovered. Of the fact of creation there can be no debate.

What the Scriptures do not necessarily tell us, however, is the mode of creation. Scientific mechanisms were generally not within the purpose of Scriptural writers who aimed to present a revelation in terms understandable by all men in all times. It may well be that they did present the mode or mechanisms by which God worked. It may also well be that they had no intention of presenting such a mechanism, and in fact did not present such a mechanism. How are we to decide? It is at this point that we can admit the investigation of God's natural revelation in His creation (Romans 1:20) to act as a guide in our interpretation of the content of the Biblical revelation as to mechanism.

(a) *The Bible is assumed, a priori, not to reveal "scientific" information, that not being the purpose of special revelation.*

This is not an a priori claim. It is arrived at, when it is, only by the guidance of the interpretation of God's natural revelation as this bears on our interpretation of God's special revelation. That we may not expect the Biblical writers to be concerned with the revelation of specific mechanisms of God's activity (electromagnetic fields, DNA code, nuclear forces etc.) is a consistent reaction to the Bible's own commentary on its purpose. The Bible tells us that it was written to reveal the redemption of God in Jesus Christ (John 20:31), to comfort and strengthen the faith of God's people through the witness of men who had known and experienced God's life (Luke 1:3,4), and to present a guide for Christian living (II Timothy 3:16).

An analogy may be helpful at this point. In a way the situation to which some portions of the Biblical revelation address themselves, particularly the early chapters of Genesis, is like that of a mother attempting to explain the "facts of life" to her 5-year old daughter. To her she speaks of growth, love between man and woman, and the beginning of a new life that flows from that love. She does not give the technical details of sexual activity. When the child is older she will learn these things too. Her total understanding of the role of sex will depend on her appropriation of both these kinds of information, that dealing with interpersonal relationships of love as well as that dealing with technical details of reproduction and sex fulfillment. The mother has spoken truly to her 5-year-old daughter. She has told her what she needs to know, things that will always be the most important part of her knowledge in this area. So in his infinite wisdom, God has revealed to us the most important things-things that we could never really find out for ourselves. The details of mechanisms we will learn in His providence as we grow and understand more of such things.

(b) *General disregard for strict exegesis of the Scripture is evident. This also holds for the practice of comparing Scripture with Scripture before arriving at any interpretation.*

Exegesis is guided by hermeneutics. Presumably it is the purpose of the Biblical exegete to derive the content of the Biblical revelation, to set forth the revelation which God gives to us. He cannot do this without an understanding of the proper hermeneutics. These cannot be chosen arbitrarily, but must be consistent with the nature of the Biblical revelation. Before we talk of "strict exegesis," therefore, we should talk of "Scriptural hermeneutics." We cannot understand what it is that the Word of God says to us if we impose some arbitrary system of hermeneutics in our exegesis.

There remain Biblical areas where the appropriate hermeneutics need further work. Since God's truth is one, any guidance that can come to the formulation of these hermeneutics by an understanding of God's natural revelation in creation, must be welcomed and not rejected.

(c) *There is increasing use of the inductive method of studying Scripture-a methodology inconsistent with the Biblical doctrine of inspiration.*

If the Scriptures are indeed divinely inspired and therefore authoritative and trustworthy in conveying the revelation they were written to present, as we believe them to be, then any Spiritguided method of approaching Scripture, whether inductive or deductive, can lead us only to God's truth in terms of the proper hermeneutical principles to apply.

2. *Science, in its widest sense, is regarded as complementary to the Bible in contributing to a view of reality. The two are independent but interdependent. Thus science is autonomous and speaks for the realm of nature as the Bible does for that of faith.*

This statement expounds a commonly held confusion of categories. Science is not regarded as complementary to the Bible. The created natural world is regarded as complementary to the revealed word of the Bible. Science is a human interpretation of data derived from sense contacts with that created natural world. Its complementary category in Christian faith is not the Bible (which corresponds to the created worldthe data) but rather theology, which is the interpretation by men of the revealed word of the Bible in the light of the Bible and their experience. God made the world, and God gave the Bible. Men make science and men make theology. Men make mistakes in interpreting the natural world where their scientific hermeneutics are faulty. Men make mistakes in interpreting the Bible when their theological hermeneutics are faulty. The theology of men can no more be accepted as superior to the created world, than the science of men can be accepted as superior to the Word of God.

Any attempt to juxtapose science vs. the Bible, to plead that men accept the Bible rather than science, is a result of a misunderstanding of the categories involved. The revelation of creation and the fossil record must confront each other; the theory of fiat creation and the theory of organic evolution must each be justified. There is no more possibility, or perhaps I should say there is just as much possibility of "accepting the Bible" as there is of "accepting the natural world." Both can be accepted in that we accept their trustworthiness. What we then apply to ourselves on the basis of this trustworthiness depends on the results of our hermeneutics and our exegesis.

(a) Modern science is regarded as the study of natural revelation and its best authenticated results (according to its own standards of course!) are considered to represent a valid picture of that revelation.

This statement in itself is not a condemnation. Its counterpart would be that modern theology (orthodox fundamental Protestant, if you will) is regarded as the study of the special revelation and its best authenticated results (according to its own standards of course!) are considered to represent a valid picture of that revelation. I do not consider this latter statement offensive, and so I do not consider the statement of Gordon Keddie offensive either.

On the other hand, modern science is not regarded as presenting a "valid picture" of the natural revelation, if by "valid picture" is meant any more than a relative pragmatic approach to the nature of physical reality. The Christian man of science regards his scientific understanding as a helpful guide, not as a source of ultimate truth. As his science is more and more faithful to the created structure of the physical universe, he believes that his scientific picture is more and more like that of the reality of that created universe. Also, although his scientific picture may not be able to describe completely what physical reality is like, it may well be able to give a definite indication as to what it is not like.

(b) Any conflict between "traditional Biblical interpretation" and modern scientific discoveries must result in a re-examination of the former with a view to reinterpretation.

I would agree that when traditional Biblical interpretations come into conflict with any type of experience, whether scientific or personal, it is good to re-examine them. I demand the same of traditional scientific interpretations. Not to be willing to re-think interpretations is to forsake the pursuit of truth. Because interpretations may be faulty never implies that the ground of revelation is faulty. Because a scientific interpretation of the physical world is in error casts no doubt on the trustworthiness of the revelation contained in that created physical world. Because a theological interpretation of the Bible is in error casts no doubt on the trustworthiness of the revelation contained in the Bible.

(c) No distinction is recognized in principle between the science practiced by Christians (true science) and that of unbelievers (apostate science). There is no consistently Christian philosophy of science.

There is a consistent Christian philosophy of science. The Christian believes in the structure of reality, a created structure given by God Himself. Truth is that which is in conformity with this structure of reality. Thus it is possible for an unbeliever to know partial truth, but never total truth.

The Christian philosophy of science is this: Christian science is good science. And good science is science that is faithful to the structure of reality. Science that is honest, open, seeking to capture and to reflect the structure of the world that is really there-that is good science, and that is Christian science. The unbeliever is successful in science, successful in apprehending partial truths of the universe, when and only when he appropriates for himself the Christian approach to the world without recognizing that he is doing it, and without acknowledging the ground of reality that makes it possible for him to do it successfully. Thus the success of the nonChristian in science can be attributed to his use of Christian principles of scientific investigation, principles which seek above all else to be faithful to the created structure of the world.

It is becoming for the followers of Christ to be informed and perceptive as well as militant. We spend so much time and effort in battles that need not be fought that we do not have the strength left to engage where we are desperately needed. If we recognize that we have trustworthy revelation from Cod both in the natural world and in the Bible, can we not then cease from pursuing these false dichotomies: science or Scripture, evolution or creation, natural or God caused, chance or providence? R.H.B.

—6—
Evangelical Attitudes to the Environment

Here are two contrasting declarations on by *Evangelical Environmental Network* published in 1994 and the second *The Cornwall Declaration* of 2000 by ISA (Interfaith Stewardship Alliance). Aspects are discussed in Chapter 9. Note the difference in understanding of the Fall and the Curse. These two documents illustrate the on-going disagreement between evangelicals on the environment.

CREATION: AN EVANGELICAL DECLARATION ON THE CARE OF CREATION

The Earth is the Lord's, and the fulness thereof—Psalm 24:1

As followers of Jesus Christ, committed to the full authority of the **Scriptures**, and aware of the ways we have degraded creation, we believe that biblical faith is essential to the solution of our ecological problems.

Because we worship and honor the Creator, we seek to cherish and care for the creation.

Because we have sinned, we have failed in our stewardship of creation. Therefore we repent of the way we have polluted, distorted, or destroyed so much of the Creator's work.

Because in Christ God has healed our alienation from God and extended to us the first fruits of the reconciliation of all things, we commit ourselves to working in the power of the Holy Spirit to share the Good News of Christ in word and deed, to work for the reconciliation of all people in Christ, and to extend Christ's healing to suffering creation.

Because we await the time when even the groaning creation will be restored to wholeness, we commit ourselves to work vigorously to protect and heal that creation for the honor and glory of the Creator—whom we know dimly through creation, but meet fully through Scripture and in Christ. We and our children face a growing crisis in the health of the creation in which we are embedded, and through which, by God's grace, we are sustained. Yet we continue to degrade that creation.

These degradations of creation can be summed up as 1) land degradation; 2) deforestation; 3) species extinction; 4) water degradation; 5) global toxification; 6) the alteration of atmosphere; 7) human and cultural degradation.

Many of these degradations are signs that we are pressing against the finite limits God has set for creation. With continued population growth, these degradations will become more severe. Our responsibility is not only to bear and nurture children, but to nurture their home on earth. We respect the institution of marriage as the way God has given to insure thoughtful procreation of children and their nurture to the glory of God.

We recognize that human poverty is both a cause and a consequence of environmental degradation.

Many concerned people, convinced that environmental problems are more spiritual than technological, are exploring the world's ideologies and religions in search of non-Christian spiritual resources for the healing of the earth. As followers of Jesus Christ, we believe that the Bible calls us to respond in four ways:

First, God calls us to confess and repent of attitudes which devalue creation, and which twist or ignore biblical revelation to support our misuse of it. Forgetting that "the earth is the Lord's," we have often simply used creation and forgotten our responsibility to care for it.

Second, our actions and attitudes toward the earth need to proceed from the center of our faith, and be rooted in the fullness of God's revelation in Christ and the Scriptures. We resist both ideologies which would presume the Gospel has nothing to do with the care of non-human creation and also ideologies which would reduce the Gospel to nothing more than the care of that creation.

Third, we seek carefully to learn all that the Bible tells us about the Creator, creation, and the human task. In our life and words we declare that full good news for all creation which is still waiting "with eager longing for the revealing of the children of God," (Rom. 8:19).

Fourth, we seek to understand what creation reveals about God's divinity, sustaining presence, and everlasting power, and what creation teaches us of its God-given order and the principles by which it works.

Thus we call on all those who are committed to the truth of the Gospel of Jesus Christ to affirm the following principles of biblical faith, and to seek ways of living out these principles in our personal lives, our churches, and society.

The cosmos, in all its beauty, wildness, and life-giving bounty, is the work of our personal and loving Creator.

Our creating God is prior to and other than creation, yet intimately involved with it, upholding each thing in its freedom, and all things in relationships of intricate complexity. God is transcendent, while lovingly sustaining each creature; and immanent, while wholly other than creation and not to be confused with it.

God the Creator is relational in very nature, revealed as three persons in One. Likewise, the creation which God intended is a symphony of individual creatures in harmonious relationship.

The Creator's concern is for all creatures. God declares all creation "good" (Gen. 1:31); promises care in a covenant with all creatures (Gen. 9:9-17); delights in creatures which have no human apparent usefulness (Job 39-41); and wills, in Christ, "to reconcile all things to himself" (Col.1:20).

Men, women, and children, have a unique responsibility to the Creator; at the same time we are creatures, shaped by the same processes and embedded in the same systems of physical, chemical, and biological interconnections which sustain other creatures.

Men, women, and children, created in God's image, also have a unique responsibility for creation. Our actions should both sustain creation's fruitfulness and preserve creation's powerful testimony to its Creator.

Our God-given, stewardly talents have often been warped from their intended purpose: that we know, name, keep and delight in God's creatures; that we nourish civilization in love, creativity and obedience to God; and that we offer creation and civilization back in praise to the Creator. We have ignored our creaturely limits and have used the earth with greed, rather than care.

The earthly result of human sin has been a perverted stewardship, a patchwork of garden and wasteland in which the waste is increasing. "There is no faithfulness, no love, no acknowledgment of God in the land . . . Because of this the land mourns, and all who live in it waste away" (Hosea 4:1,3). Thus, one consequence of our misuse of the earth is an unjust denial of God's created bounty to other human beings, both now and in the future.

God's purpose in Christ is to heal and bring to wholeness not only persons but the entire created order. "For God was pleased to have all his fullness dwell in him, and through him to reconcile to himself all things, whether things on earth or things in heaven, by making peace through his blood shed on the cross" (Col. 1:19–20).

In Jesus Christ, believers are forgiven, transformed and brought into God's kingdom. "If anyone is in Christ, there is a new creation" (II Cor. 5:17). The presence of the kingdom of God is marked not only by renewed fellowship with God, but also by renewed harmony and justice between people, and by renewed harmony and justice between people and the rest of the created world. "You will go out in joy and be led forth in peace; the mountains and the hills will burst into song before you, and all the trees of the field will clap their hands" (Isa. 55:12).

We believe that in Christ there is hope, not only for men, women and children, but also for the rest of creation which is suffering from the consequences of human sin.

Therefore we call upon all Christians to reaffirm that all creation is God's; that God created it good; and that God is renewing it in Christ.

We encourage deeper reflection on the substantial biblical and theological teaching which speaks of God's work of redemption in terms of the renewal and completion of God's purpose in creation.

We seek a deeper reflection on the wonders of God's creation and the principles by which creation works. We also urge a careful consideration of how our corporate and individual actions respect and comply with God's ordinances for creation.

We encourage Christians to incorporate the extravagant creativity of God into their lives by increasing the nurturing role of beauty and the arts in their personal, ecclesiastical, and social patterns.

We urge individual Christians and churches to be centers of creation's care and renewal, both delighting in creation as God's gift, and enjoying it as God's provision, in ways which sustain and heal the damaged fabric of the creation which God has entrusted to us.

We recall Jesus' words that our lives do not consist in the abundance of our possessions, and therefore we urge followers of Jesus to resist the allure of wastefulness and overconsumption by making personal lifestyle choices that express humility, forbearance, self restraint and frugality.

We call on all Christians to work for godly, just, and sustainable economies which reflect God's sovereign economy and enable men, women and children to flourish along with all the diversity of creation. We recognize that poverty forces people to degrade creation in order to survive; therefore we support the development of just, free economies which empower the poor and create abundance without diminishing creation's bounty.

We commit ourselves to work for responsible public policies which embody the principles of biblical stewardship of creation.

We invite Christians—individuals, congregations and organizations—to join with us in this evangelical declaration on the environment, becoming a covenant people in an ever-widening circle of biblical care for creation.

We call upon Christians to listen to and work with all those who are concerned about the healing of creation, with an eagerness both to learn from them and also to share with them our conviction that the God whom all people sense in creation (Acts 17:27) is known fully only in the Word made flesh in Christ the living God who made and sustains all things.

We make this declaration knowing that until Christ returns to reconcile all things, we are called to be faithful.

THE CORNWALL DECLARATION (2000)

The past millennium brought unprecedented improvements in human health, nutrition, and life expectancy, especially among those most blessed by political and economic liberty and advances in science and technology. At the dawn of a new millennium, the opportunity exists to build on these advances and to extend them to more of the earth's people.

At the same time, many are concerned that liberty, science, and technology are more a threat to the environment than a blessing to humanity and nature. Out of shared reverence for God and His creation and love for our neighbors, we Jews, Catholics, and Protestants, speaking for ourselves and not officially on behalf of our respective communities, joined by others of good will, and committed to justice and compassion, unite in this declaration of our common concerns, beliefs, and aspirations.

Our Concerns

Human understanding and control of natural processes empower people not only to improve the human condition but also to do great harm to each other, to the earth, and to other creatures. As concerns about the environment have grown in recent decades, the moral necessity of ecological stewardship has become increasingly clear.

At the same time, however, certain misconceptions about nature and science, coupled with erroneous theological and anthropological positions, impede the advancement of a sound environmental ethic. In the midst of controversy over such matters, it is critically important to remember that while passion may energize environmental activism, it is reason—including sound theology and sound science—that must guide the decision-making process. We identify three areas of common misunderstanding:

1. Many people mistakenly view humans as principally consumers and polluters rather than producers and stewards. Consequently, they ignore our potential, as bearers of God's image, to add to the earth's abundance. The increasing realization of this potential has enabled people in societies blessed with an advanced economy not only to reduce pollution, while producing more of the goods and services responsible for the great improvements in the human condition, but also to alleviate the negative effects of much past pollution. A clean environment is a costly good; consequently, growing affluence, technological innovation, and

the application of human and material capital are integral to environmental improvement. The tendency among some to oppose economic progress in the name of environmental stewardship is often sadly self-defeating.

2. Many people believe that "nature knows best," or that the earth—untouched by human hands—is the ideal. Such romanticism leads some to deify nature or oppose human dominion over creation. Our position, informed by revelation and confirmed by reason and experience, views human stewardship that unlocks the potential in creation for all the earth's inhabitants as good. Humanity alone of all the created order is capable of developing other resources and can thus enrich creation, so it can properly be said that the human person is the most valuable resource on earth. Human life, therefore, must be cherished and allowed to flourish. The alternative—denying the possibility of beneficial human management of the earth—removes all rationale for environmental stewardship.
3. While some environmental concerns are well founded and serious, others are without foundation or greatly exaggerated. Some well-founded concerns focus on human health problems in the developing world arising from inadequate sanitation, widespread use of primitive biomass fuels like wood and dung, and primitive agricultural, industrial, and commercial practices; distorted resource consumption patterns driven by perverse economic incentives; and improper disposal of nuclear and other hazardous wastes in nations lacking adequate regulatory and legal safeguards. Some unfounded or undue concerns include fears of destructive manmade global warming, overpopulation, and rampant species loss.

The real and merely alleged problems differ in the following ways:

1. The former are proven and well understood, while the latter tend to be speculative.
2. The former are often localized, while the latter are said to be global and cataclysmic in scope.
3. The former are of concern to people in developing nations especially, while the latter are of concern mainly to environmentalists in wealthy nations.
4. The former are of high and firmly established risk to human life and health, while the latter are of very low and largely hypothetical risk.
5. Solutions proposed to the former are cost effective and maintain proven benefit, while solutions to the latter are unjustifiably costly and of dubious benefit.

Public policies to combat exaggerated risks can dangerously delay or reverse the economic development necessary to improve not only human life but also human stewardship of the environment. The poor, who are most often citizens of developing nations, are often forced to suffer longer in poverty with its attendant high rates of malnutrition, disease, and mortality; as a consequence, they are often the most injured by such misguided, though well-intended, policies.

Our Beliefs

Our common Judeo-Christian heritage teaches that the following theological and anthropological principles are the foundation of environmental stewardship:

1. God, the Creator of all things, rules over all and deserves our worship and adoration.
2. The earth, and with it all the cosmos, reveals its Creator's wisdom and is sustained and governed by His power and lovingkindness.
3. Men and women were created in the image of God, given a privileged place among creatures, and commanded to exercise stewardship over the earth. Human persons are moral agents for whom freedom is an essential condition of responsible action. Sound environmental stewardship must attend both to the demands of human well being and to a divine call for human beings to exercise caring dominion over the earth. It affirms that human well being and the integrity of creation are not only compatible but also dynamically interdependent realities.
4. God's Law—summarized in the Decalogue and the two Great Commandments (to love God and neighbor), which are written on the human heart, thus revealing His own righteous character to the human person—represents God's design for shalom, or peace, and is the supreme rule of all conduct, for which personal or social prejudices must not be substituted.
5. By disobeying God's Law, humankind brought on itself moral and physical corruption as well as divine condemnation in the form of a curse on the earth. Since the fall into sin people have often ignored their Creator, harmed their neighbors, and defiled the good creation.
6. God in His mercy has not abandoned sinful people or the created order but has acted throughout history to restore men and women to fellowship with Him and through their stewardship to enhance the beauty and fertility of the earth.
7. Human beings are called to be fruitful, to bring forth good things from the earth, to join with God in making provision for our temporal well being, and to enhance the beauty and fruitfulness of the rest of the earth. Our call to fruitfulness, therefore, is not contrary to but mutually complementary with our call to steward God's gifts. This call implies a serious commitment to fostering the intellectual, moral, and religious habits and practices needed for free economies and genuine care for the environment.

Our Aspirations

In light of these beliefs and concerns, we declare the following principled aspirations:

1. We aspire to a world in which human beings care wisely and humbly for all creatures, first and foremost for their fellow human beings, recognizing their proper place in the created order.

2. We aspire to a world in which objective moral principles—not personal prejudices—guide moral action.
3. We aspire to a world in which right reason (including sound theology and the careful use of scientific methods) guides the stewardship of human and ecological relationships.
4. We aspire to a world in which liberty as a condition of moral action is preferred over government-initiated management of the environment as a means to common goals.
5. We aspire to a world in which the relationships between stewardship and private property are fully appreciated, allowing people's natural incentive to care for their own property to reduce the need for collective ownership and control of resources and enterprises, and in which collective action, when deemed necessary, takes place at the most local level possible.
6. We aspire to a world in which widespread economic freedom—which is integral to private, market economies—makes sound ecological stewardship available to ever greater numbers.
7. We aspire to a world in which advancements in agriculture, industry, and commerce not only minimize pollution and transform most waste products into efficiently used resources but also improve the material conditions of life for people everywhere.

References

Ackermann, Paul. (1991). *It's a Young World After All.* Grand Rapids, MI: Baker.

Albritton, Claude C. (1980). *The Abyss of Time.* San Francisco, CA: Freeman, Cooper.

Alexander, T. Desmond, and Baker, David W. (eds.). (2003). *Dictionary of the Old Testament: Pentateuch.* Leicester, UK: Inter-Varsity Press.

Altholz Josef L. (1988). The Warfare of Conscience with Theology. In G. Parsons (ed.), *Religion in Victorian Britain*, Vol 4. Manchester: Manchester University Press.

Anderson, Katherine (2005). *Predicting the Weather: Victorians and the Science of Meteorology.* Chicago: University of Chicago Press.

Armstrong, Patrick (2000). *The English Parson-Naturalist.* Leominster, UK: Gracewing.

Ashton, John (ed.). (1999). *In Six Days.* Australia: New Holland. A useful introduction on why scientists adopt YEC.

Astore, William J. (2001). *Observing God.* Aldershot, UK: Ashgate.

Babinksi, Edward. (2005). *From YEC (Young Earth Creationist) to Evolutionist* (http://www.edwardtbabinski.us/articles/yec.html

Balmer, Randall (ed.). (2002). *Encyclopedia of Evangelicals.* Louiville, UK: Westminster.

Barrett, David (2001). *World Christian Encyclopedia.* Oxford: Oxford University Press.

Bartholomew, Craig (ed.). (2003). *The Futures of Evangelicalism.* Leicester, UK: Inter-Varsity Press.

Baylee, Joseph (1857). *Genesis and Geology.* Liverpool: Publisher unknown.

Bebbington, David W. (1987). *Evangelicalism in Modern Britain.* London: Unwin.

Bebbington, David, and Noll, Mark (eds.). (2004). *A History of Evangelicalism*, 5 vols. Leicester, UK: Inter-Varsity Press: Noll, Mark (2004). *The Rise of Evangelicalism, Vol. 1: 1730s–1790s.* Leicester, UK: Inter-Varsity Press; *Projected volumes*: Wolffe, John. *The Expansion of Evangelicalism, Vol. 2: 1790s–1840s*; Bebbington, David. *The Dominance of Evangelicalism, Vol. 3: 1850s–1900s;*

Treloar, Geoff. *The Disruption of Evangelicalism, Vol. 4: 1900s–1940s;* Stanley, Brian. *The Global Diffusion of Evangelicalism, Vol. 5: 1940s–1990s.* Essential for understanding the history of evangelicals.

Beisner, E. C. (1990). *Prospects for Growth: A Biblical View of Population, Resources, and the Future.* Westchester, IL: Crossway.

———. (1994). *Man, Economy, and Environment in Biblical Perspective.* Moscow, ID: Canon Press.

———. (1997). *Where Garden Meets Wilderness: Evangelical Entry Into the Environmental Debate.* Grand Rapids, MI: Eerdmans. An anti-global warming book on the environment.

Berry, R. J. (1972). *Ecology and Ethics.* London: Inter-Varsity Press.

———. (ed.). (2000). *The Care of Creation.* Downers Grove, IL: Inter-Varsity Press. Very useful on a green evangelical approach.

———. (2003). *God's Book of Works: The Nature and Theology of Nature.* Edinburgh: Continuum. A useful serious theological reflection.

Birks, Thomas R. (1862). *The Bible and Modern Thought.* London: Religious Tract Society.

———. (1872). *The Scripture Doctrine of Creation.* London: SPCK.

———. (1876). *Modern Physical Fatalism and the Doctrine of Evolution.* London: Macmillan.

Blocher, Henri (1984). *In the Beginning.* Leicester, UK: Inter-Varsity Press.

Bowler, Peter J. (1988). *The Non-Darwinian Revolution.* Baltimore, MD: Johns Hopkins University Press.

Bray, Gerald (1996). *Biblical Interpretation: Past and Present.* Leicester, UK: Inter-Varsity Press.

Brewster, D. (1862). The Facts and Fancies of Mr Darwin. *Home Words* (Edinburgh).

Brooke, John Hedley (1991). *Science and Religion: Some Historical Perspectives.* Cambridge, UK: Cambridge University Press. A classic general work.

———. (2001). The Wilberforce-Huxley Debate: Why Did It Happen? *Science and Christian Belief* 13: 127–141.

Brooke, J. H., and Cantor, Geoffrey (1998). *Reconstructing Nature.* Edinburgh: T & T Clark.

Brown, W. B, Murphey, N. C., and Malony, H. N. (1998). *Whatever Happened to the Soul? Scientific and Theological Portraits of Human Nature.* Augsburg: Fortress Publishers.

Browne, Stanley G. (1990) *Heralds of Health.* London: Christian Medical Fellowship.

Bryant, J., and Searle, J. (2003). *Life in our Hands.* Leicester, UK: Inter-Varsity Press.

Bube, R. (ed.). (1968). *The Encounter between Christianity and Science.* Grand Rapids, MI: William B. Eerdmans Publishing Company.

———. (1971). *The Human Quest: A New Look at Science and the Christian Faith.* Waco, TX: Word.

———. (1995). *Putting It All Together: Seven Patterns Relating Science and Christian Faith.* Lanham, MD: University Press of America.

Buckland, William (1820). *Vindiciae geologicae.* Oxford: Oxford University Press.

———. (1836). *Geology and Mineralogy, Considered with Reference to Natural Theology.* London: John Murray.

Buffon, G. L. (1778). Des Époques de la nature. In Buffon *Histoire Naturelle, Supplement* 5, 1–254.
Bugg, G. (1826–1827). *Scriptural Geology*, 2 vols. London: Hatchard.
Bultmann, R. (1984). *New Testament and Mythology and Other Writings.* Philadelphia, PA: Forstress.
Burchfield, Joseph D. (1976). *Lord Kelvin and the Age of the Earth.* London: Macmillan.
Burkhardt, F. (1993). *The Correspondence of Charles Darwin, Vol. 8: 1860.* Cambridge, UK: Cambridge University Press.
———. (1994). *The Correspondence of Charles Darwin, Vol. 9: 1861.* Cambridge, UK: Cambridge University Press.
Burkhardt, F., and Smith, S. (1985). *The Correspondence of Charles Darwin, Vol. 1: 1821–36.* Cambridge, UK: Cambridge University Press.
———. (1986). *The Correspondence of Charles Darwin, Vol. 2: 1837–43.* Cambridge, UK: Cambridge University Press.
———. (1991). *The Correspondence of Charles Darwin, Vol. 7: 1858–9.* Cambridge, UK: Cambridge University Press.
Burnet, Thomas (1684–1690). *The Sacred Theory of the Earth.* London: Norton.
Calvin, John (1847) *Commentary on Genesis.* London: Banner of Truth. (Original published in 1554.)
Cannon, S. F. (1978). *Science in Culture: The Early Victorian Period.* New York: Dawson.
Chadwick, Owen (1965). *The Victorian Church*, Vol. I. London: A & C Black.
———. (1970). *The Victorian Church,* Vol II. London: A & C Black.
———. (1975). *The Secularization of the European Mind in the 19th Century.* Cambridge, UK: Cambridge University Press.
Chalmers, T. (1814). *The Evidence and Authority of the Christian Revelation.* Edinburgh: William Blackwood.
———. (1833). *On the Power, Goodness and Wisdom of God as Manifested in the Adaptation of External Nature to the Moral and Intellectual Constitution of Man.* London: Pickering.
Chambers, Robert (1844). *The Vestiges of the Natural History of Creation.* London: John Churchill.
Chateaubriand, F.-R. (1966). *Génie du Chrsitianisme.* Paris: Garnier-Flammarion.
Clark, J. W., and Hughes, T. Mc. (1890). *The Life and Letters of the Reverend Adam Sedgwick.* Cambridge, UK: Cambridge University Press.
Clark, Robert E. D. (1945). *Scientific Rationalism and Christian Faith.* London: Inter-Varsity Press.
———. (1945). *Darwin, Before and After.* Exeter, UK: Paternoster.
Close, Francis (1826). *The Book of Genesis Considered and Illustrated.* London 1826, 6th edn. 1841).
Close, Francis J. (1859). *Hugh Miller's "Testimony of the Rocks"—God in His Word and in His Works.* Lectures delivered before the YMCA, Nov 1857 to Feb 1858, London, 1859, 239–272.
Cohen, I. B. (1990). *Benjamin Franklin's Science.* Cambridge, Mass.: Harvard University Press.
Cole, Henry (1834). *Popular Geology Subversive of Revelation! Being a Letter to Adam Sedgwick.* London: Hatchard.

Colenso, Joseph (1861). *St Paul's Epistle to the Romans.* London: Longman.
———. (1862–1879). *The Pentateuch and the Book of Joshua.* London: Longman.
Collins, Francis S. (2003). Faith and the Human Genome. *Prospectives on Science and Christian Faith* 55.3: 142–153.
Collins, Francis (2006). *The Language of God.* London: Free Press. Excellent evangelical apologetic.
Conybeare, William D., and Phillips, William (1822). *Outline of the Geology of England and Wales.* London: Phillips.
Cumming, J. (1851). *"God in Science" in Letters to Young Men.* London: James Nisbet.
———. (1860). *Church before the Flood.* London: Arthur Hall.
Dana, James D. (1863). *The Manual of Geology.* Philadelphia: Theodore Bliss.
Darwin, Charles (1839). *Voyage of the Beagle.* London: Henry Colburn.
———. (1868). *The Variation of Plants and Animals under Domstication,* 2 vols. London: John Murray.
———. (1983). *Autobiography of Charles Darwin.* Oxford: Oxford University Press.
Davies, L. Merson (1934). *The Bible and Modern Science.* London: Constable.
Davis, Edward B (1991). A Whale of a Tale: Fundamentalist Fish Stories. *Perspectives on Science and Christian Faith* 43: 224–237. Both amusing and erudite!
———. (1995). *The Anti-Evolution Pamphlets of Harry Rimmer.* New York: Garland.
Davis, F. B. (2003). The Word and the Works: Concordism and American Evangelicals. In Miller, Keith (ed.), *Perspectives on an Evolving Creation.* Michigan: Eerdmans.
Dawson, John W. (1860). *Archaia.* London: Hodder and Stoughton.
———. (1872). *The Story of the Earth.* London: Hodder and Stoughton.
———. (1877). *The Origin of the World according to Revelation and Science.* London: Hodder and Stoughton.
Derham, W. (1713). *Physico-Theology.* London: Innys.
Dewar, Douglas (1931). *Difficulties of the Evolution.* London: Arnold.
———. (1938). *More Difficulties of the Evolution Theory.* London: Thynne.
DeWitt, C. (1991). *The Environment and the Christian.* Grand Rapids, MI: Baker.
Dixon, A. C. (1910). *Fundamentals, The; A Testimony to the Truth, 1910–5,* 12 vols. Chicago, IL: Testimony Publishing Company. An Excellent Source for Evangelicalism in 1910.
Draper, J. W. (1874). *History of the Conflict between Science and Religion.* New York: D. Appleton.
Drummond, Henry (1883). *Natural Law in the Spiritual World.* London: Hodder and Stoughton.
Dudley-Smith, Timothy (1999). *John Stott, the Making of a Leader.* Leicester, UK: Inter-Varsity Press.
———. (2001). *John Stott, a Global Ministry.* Leicester, UK: Inter-Varsity Press.
Duncan, Ligon III, Hall, David W., Ross, Hugh, and Archer, Gleason L. (2000). *The Genesis Debate: Three Views on the Days of Creation.* Orlando, FL: CruPress. Very useful for this question.
Dupree, A. H. (1988). *Asa Gray.* Cambridge, MA: Belknap.
Edwards, Jonathan (1980). *Works, Vol. VI: Scientific and Philosophical Writings.* In Anderson, Wallace E. (ed.). New Haven, CT: Yale University Press.

Eldredge, Niles (1982). *The Monkey Business.* New York: Pocket Books.
Eliot, George (1973). *Scenes of Clerical Life.* Harmondsworth: Penguin.
Everest, F. A. (ed.) (1950). *Modern Science and Christian Faith.* Wheaton, IL: Van Kampen.
Faber, George S. (1802). *Horae Mosaica.* Oxford: Publisher unkown.
———. (1816). *The Origin of Pagan Idolatry.* London. Publisher unknown.
———. (1823). *A Treatise on the Genius and Object of the Patriarchal, the Levitical and the Christian Dispensations.* London: Rivington.
Fairholme, George (1833). *General View of the Geology of Scripture.* London: James Ridgeway.
Farrington, Benjamin (1953). *Greek Science.* Baltimore, MD: Penguin Books.
Ferngren, Gary B. (ed.). (2000). *The History of Science and Religion in the Western Tradition: An Encyclopedia*: New York: Garland.
Fields, W.W. (1976). *Unformed and Unfilled.* Nutley, NJ: Presbyterian and Reformed.
Fitzroy, Robert (1839). *Narrative of the Surveying Voyages of His Majesty's Ship Adventurer and Beagle,* Vol. 2. London: Henry Colburn.
Fleming, Fergus (1998). *Barrow's Boys.* London: Granta.
Fleming, John (1826). The Geological Deluge, as Interpreted by Baron Cuvier and Professor Buckland, Inconsistent with the Testimony of Moses and the Phenomena of Nature. *Edinburgh Philosophical Journal* 14: 205–239.
Fouts, David M. (1997). A Defense of the Hyperbolic Interpretation of Large Numbers in The Old Testament. *Journal of the Evangelical Theological Society* 40(3): 377–387.
Frost, David (1997). *Billy Graham: Personal Thoughts of a Public Man.* London: Hodder.
Fuller, J. G. C. M. (2001). Before the hills in order stood: the beginning of the geology of time in England. In Lewis, C. L. E., and Knell, S. J. (eds.), *The Age of the Earth from 4004 BC to AD 2002.* Geological Society, London: Special Publications, Number 190, 15–23.
———. (2005). A Date to remember: 4004BC. *Earth Sciences History* 24(1): 5–14.
Funkhouser, J. G. and Naughton, J. J. (1967). Radiogenic Helium and Argon in Ultramafic Inclusions from Hawaii. *Journal for Geophysical Research* 73(14): 4601–4607.
Fyfe, Aileen (2004). *Science and Salvation.* Chicago: Chicago University Press.
Galileo, Galilei (1953) *Dialogue Concerning the Two Chief World Systems: Ptolemaic and Copernican.* Berkeley: University of California Press.
Galloway, William Brown (1872). *Science and Geology in Relation to the Universal Deluge.* London: Thynne.
Gardner, Martin (1957). *Fads and Fallacies in the Name of Science.* New York: Dover.
Geikie, Archibald (1923). *A Life's Long Work.* London: Macmillan.
Geisler, N. and Anderson, K. (1987). *Origin Science, a Proposal for the Creation-Evolution Controversy.* Grand Rapids, MI: Baker.
Giberson, Karl W., and Yerxa, Donald. A. (2002). *Species of Origins.* New York: Rowman and Littlefield. A superb exposition of "origins" in a wider sense.
Gillispie, Charles C. (1951). *Genesis and Geology a Study in the Relations of Scientific Thought, Natural Theology and Social Opinion in Great Britain, 1790–1850.* Cambridge, MA: Harvard University Press.

Gingerich, Owen (2006). *God's Universe.* Cambridge, MA: Belknap. Yet another fine apologetic.

Gisbourne, Thomas (1818). *The Testimony of Natural Theology to Christianity.* London: T. Cadell.

González, Roberto (2001). *Zapotec Science: Farming and Food in the Northern Sierra of Oaxaca.* Austin: University of Texas Press.

Gosse, Edmund (1907/1949). *Father and Son.* Harmondsworth: Penguin.

Gosse, Philip H. (1857). *Omphalos.* London: Van Voorst.

Gould, Stephen J. (1987). *Time's Arrow. Time's Cycle.* Harmondsworth: Penguin.

———. (1991). *Bully for Brontosaurus.* Harmondsworth: Penguin.

———. (1999). *Rocks of Ages: Science and Religion in the Fullness of Life.* New York: The Ballantine Publishing Group.

Graebner, Theodore (1932). *God and the Cosmos.* Grand Rapids, MI: Eerdmans.

Gray, Asa (1963). *Darwiniana.* Cambridge, MA: Belknap. (Originally published in 1876). The classic response to Darwin in the 1860s.

Grotius, H. (1719). *The Truth of the Christian Faith in Six Books* (Clark, John, trans.). London: Publisher unknown.

Grudem, Wayne (1994). *Systematic Theology.* Downers Grove, IL: Inter-Varsity Press. A standard evangelical theology, but favoring YEC.

Grundmann, Christoffer H. (2005). *Sent to Heal: Emergence and Development of Medical Missions.* Lanham, MD: University Press of America.

Haas, Jack W. (1995). John Wesley's Vision of Science in the Service of Christ. *Prospectives on Science and Christian Faith* 47: 234ff.

Ham, Ken, and Taylor, Paul (1988). *The Genesis Solution.* Grand Rapids, MI: Baker. Like anything by Ham (and AIG) clearly gives the rationale of YEC.

Hanna, William (1852). *Memoirs of the Life and Writings of Thomas Chalmers, D.D., LL.D.* Edinburgh: Constable.

Hart, David G. (1991). The Fundamentalist Origins of the ASA. *Perspectives on Science and Christian Faith* 43.4: 238–248.

Hepp, Valentine (1930). *Calvinism and the Philosophy of Nature.* Grand Rapids, MI: Eerdmans.

Herbert, Sandra (2005). *Charles Darwin, Geologist.* Ithaca, NY: Cornell University Press.

Hitchcock, Edward (1847). *Elementary Geology,* 8th ed. New York: Newman. (Originally published in 1847.)

———. (1851). *The Religion of Geology.* Glasgow: William Collins.

Hobbs, W. (1981). *The Earth Generated and Anatomized.* Ithaca, NY: Cornell University.

Hodge, Charles (1870). *Systematic Theology.* New York: Scribners.

———. (1874/1994). *What Is Darwinism?* Grand Rapids, MI: Baker.

Holmes, Arthur (1913). *The Age of the Earth.* New York: Harper.

Hooykaas, Reijer (1959). *Natural Law and Divine Miracle.* Leiden, The Netherlands: E. J. Brill.

———. (1972). *Religion and the Rise of Science.* Edinburgh: Scottish Academic Press.

Houghton, John. T. (1994). *Global Warming.* Oxford: Lion.

Israel, Charles. (2004). *Before Scopes: Evangelicalism, Education, and Evolution in Tennessee, 1870–1925.* Athens: University of Georgia Press.

James, Frank A. J. L. (2005). An "Open Clash between Science and the Church"?: Wilberforce, Huxley and Hooker on Darwin at the British Association, Oxford, 1860. In Knight D. M., and Eddy M. D. (eds.), *Science and Beliefs.* Aldershot: Ashgate, pp. 171–193.

Jeeves, Malcolm, and Berry, R. J. (1998). *Science, Life and Christian Belief.* Downers Grove, IL: Inter-Varsity Press. A good exposition of proevolutionary evangelicalism.

———. (ed.). (2004). *From Cells to Souls.* Downers Grove, IL: Inter-Varsity Press.

Johnstone, Patrick, and Mandryk, Jason (2001). *Operation World.* Carlisle: Paternoster.

Johnson, P. (1991). *Darwin on Trial.* Downers Grove, IL: InterVarsity Press.

———. (1995). *Reason in the Balance.* Downers Grove, IL: InterVarsity Press.

Jones, D. G. (2005). Responses to the Human Embryo and Embryonic Stem Cells: Scientific and Theological Assessments. *Science and Christian Belief* 17(2): 199–222.

Jones, James. (2003). *Jesus and the Earth.* London: SPCK.

Jowett, B. (1860). On the Interpretation of Scripture. In *Essays and Reviews.* London: Parker.

Kant, I. (1755/1981). *Universal Natural History and Theory of the Heavens* (Jaki, S., trans.). Edinburgh: Scottish Academic Press.

Karl, F. (1995). *George Eliot.* London: HarperCollins.

Keith, Arthur (1935). *Darwinism and its Critics.* London: Watts.

Kellogg, John (1879). *Harmony of Science and the Bible.* Battle Creek, MI: Review and Herald.

Kelly, Douglas F. (1997). *Genesis 1.1–2.4 in the Light of Changing Scientific Paradigms.* Fearn, Ross-shire: Mentor. The theological argument for YEC. Very Accessible.

King-Hele, D. (1999). *Erasmus Darwin: A Life of Unequalled Achievement.* London: De la Mere.

Kinns, Samuel (1883). *Moses and Geology.* London. Cassell.

Kirby, W. (1835). *On the Power, Wisdom and Goodness of God as Manifested in the Creation of Animals.* London: John Murray.

Kitcher, Philip (1982). *Abusing Science.* Cambridge MA: MIT Press.

Kline, Meredith G. (1958). "Because it had not rained." *Westminster Theological Journal* 20: 146–157.

———. (1996). Space and Time in the Genesis Cosmogony. *Perspectives on Science and Christian Faith* 48: 2–15.

Kuhn, Thomas S. (1962). *The Structure of Scientific Revolution.* Chicago: University of Chicago Press.

Larsen, Timothy (ed.). (2003). *Biographical Dictionary of Evangelicals.* Leicester, UK: Inter-Varsity Press.

Larson, E. (1997). *Summer of the Gods.* Cambridge, MA: Harvard University Press.

Lindberg, David C., and Numbers, Ronald L. (eds.). (1986). *God and Nature.* Berkeley: University of California Press.

———. (eds.). (2003). *When Science and Christianity Meet.* Chicago: University of Chicago Press.

Lindsell, Harold (1976). *The Battle for the Bible.* Grand Rapids, MI: Zondervan.

Lippard, J. (2007). *Trouble in Paradise: Answers in Genesis Splinters*, http://www.ncseweb.org/resources/rncse_content/vol26/6646_trouble_in_paradise_answers_i_12_30_1899.asp

Litton, E. A. (1882/1960). *An Introduction to Dogmatic Theology.* London: James Clarke.

Livingstone, David N. (1987). *Darwin's Forgotten Defenders.* Michigan: Eerdmans. A good account of early evangelical defenders of Darwin.

Livingstone, David N., Hart, D. G., and Noll, M. A. (1999). *Evangelicals and Science in Historical Context.* New York: Oxford University Press. A very useful selection of historical essays.

Lord, Eleazer (1851). *The Epoch of Creation: The Scripture Doctrine contrasted with the Geology Theory.* New York: Charles Scribner.

Lord, David (1855). *Geognosy, or the Facts and Principles of Geology against Theories.* New York: Franklin Knight.

Lucas, Ernest (2001). *Can we Believe Genesis Today?* London: Inter-Varsity Press. An accessible non-YEC theologian on Genesis.

Lyell, Charles (1863). *The Antiquity of Man.* London: John Murray.

Lynch, John M. (2002). Introduction. In Lynch, J. M. (ed.), *Creationism and Scriptural Geology, 1817–1857*, Vol. 1., Bristol: Thoemmes, pp. ix–xxiv.

MacArthur, John (2003). *Think Biblically!: Recovering a Christian Worldview.* Wheaton, IL: Crossway Books.

McCalla, Arthur (2006). *The Creationist Debate: The Encounter between the Bible and the Modern Mind.* London: Continuum.

McGrath, Alister E. (1998). *The Foundations of Dialogue in Science and Religion.* Oxford: Blackwell.

———. (1998b). *Science and Religion, An Introduction.* Oxford: Blackwell.

McGrath, A. (2001). *Christian Theology, An Introduction*, 3rd ed. Oxford: Blackwell. The standard "moderate" evangelical theology.

———. (2002–2003). *Scientific Theology*, 3 vols. Grand Rapids, MI: Eerdmans. The most heavyweight science and theology work by an evangelical!

———. (2004). *Dawkins' God.* Oxford; Blackwell.

McIver, Tom (1992). *Anti-Evolution. A Reader's Guide to Writing before and after Darwin.* Baltimore, MD: Johns Hopkins University Press. An excellent bibliographical source.

MacKay, Donald (1988). *The Open Mind.* Leicester, UK: Inter-Varsity Press.

McKeown, John. The Use of Genesis in Debates About Environmental Stewardship among Conservative Christians in the USA. Unpublished paper. Paper given at *Critical Perspectives on Religion and the Environment*, Birmingham, 18–19 September 2006.

McMullen, E. (1998). Galileo on Science and Scripture. In *The Cambridge Companion to Galileo.* Cambridge: Cambridge University Press.

Marsden, George (1980). *Fundamentalism and American Culture.* New York: Oxford University Press.

———. (1987). *Reforming Fundamentalism.* Grand Rapids, MI: Eerdmans.

———. (1991). *Understanding Fundamentalism and Evangelicalism.* Grand Rapids, MI: Eerdmans.

———. (2003). *Jonathan Edwards: A Life.* New Haven, CT: Yale University Press.

Marshall, H. (1982). *Biblical Inspiration.* London: Hodder and Stoughton.
Martin, William (1992). *The Billy Graham Story.* London: Hutchinson.
Mattson, James, and Merrill, Simon (1996). *The Pioneers of NMR and Magnetic Resonance in Medicine: The Story of MRI.* New York: Bar-Ilan University Press.
Miller, Hugh (1857). *The Testimony of the Rocks.* Edinburgh: Thomas Constable.
———. (1881). *Footprints of the Creator.* New York: Carter.
Miller, Keith (ed.). (2003). *Perspectives on an Evolving Creation.* Michigan: Eerdmans.
Millhauser, Milton (1954). The Scriptural Geologists. *OSIRIS* 11: 65–86.
Mixter, Russell (1959). *Evolution and Christian Thought Today.* Michigan: Eerdmans.
Moberg, David (1973). *The Great Reversal.* London: Scripture Union.
Monty White, A. J. (1985). *How Old is the Earth?* Welwyn: Evangelical Press.
Mooney, Chris (2005), *The Republican War on Science.* New York: Basic Books.
Moore, James R. (1979). *The Post-Darwinian Controversies: A Study of the Protestant Struggle to Come to Terms with Darwin in Great Britain and America, 1870–1900.* Cambridge, UK: Cambridge University Press.
Moore, John A. (2002). *From Genesis to Genetics.* Berkeley: University of California Press.
Moreland, John P., and Reynolds, John M. (eds.). (1999). *Three Views on Creation and Evolution.* Grand Rapids, MI: Zondervan.
Moreland, J. P., and Rae, S. B. (2000). *Body & Soul: Human Nature & the Crisis in Ethics.* Downers Grove, IL: Inter-Varsity Press.
Morgan, Llwyd D. (1988). *The Great Awakening in Wales.* London: Epworth.
Morrell, J., and Thackray, A. (1981). *Gentlemen of Science.* Oxford: Clarendon.
———. (1984). *Gentlemen of Science, Early Correspondence.* London: Royal Historical Society.
Morris, Henry M. (1982). *Men of Science—Men of God.* San Diego: Master Books.
———. (1984a). *The Biblical Basis of Modern Science.* Grand Rapids, MI: Baker.
———. (1984b). *History of Modern Creationism.* San Diego: Master Books.
Morris, Henry M., and Parker, Gary E. (1974). *Scientific Creationism.* San Diego: Master Books.
Morris, Henry M., and Whitcomb, John C. (1961). *The Genesis Flood.* Nutley, NJ: Presbyterian and Reformed. The book that started the recent YEC movement. Essential to read.
Mortenson, Terry (2001). Boundaries on Creation and Noah's Flood. Available at http://www.zondervanchurchsource.com/convention/parallel.htm#DB
———. (2004). *The Great Turning Point: The Church's Catastrophic Mistake on Geology—Before Darwin.* Green Forest, AR: Master Books. An idiosyncratic history.
Moule, H. C. G. (1889). *Outlines of Christian Doctrine.* London: Hodder and Stoughton.
Murphy, H. R. (July 1955). The Ethical Revolt against Christian Orthodoxy in Early Victorian England. *American Historical Review.*
Nares, Edward (1805). *A View of the Evidences of Christianity.* Oxford: Publisher unknown.
———. (1834). *Man, as Known to us Theologically and Geologically.* London: Rivington.
Nelkin, Dorothy (1977). *Science Textbook Controversies and the Politics of Equal Time.* Cambridge, MA: MIT Press.

Nelson, Byron C. (1927). *After its Kind.* Minneapolis: Bethany Fellowship.
Newman, Robert C., and Eckelmann, Herman Jr. J. (1977). *Genesis One & the Origin of the Earth.* Downers Grove, IL: Inter-Varsity Press.
Newton, B. W. (1882). *Remarks on "Mosaic Cosmogony" and Genesis II.5.* London: Houlston.
Newton-Malony, H. (1995). John Wesley and the Eighteenth Century Therapeutic Uses of Electricity. *Perspectives on Science and Christian Faith* 45 (December 1995): Pp. 244ff.]
Nolan, Frederick (1833). *The Analogy of Revelation and Science Established.* Oxford: Publisher unknown.
Noll, Mark (1986). *Between Faith and Criticism.* Leicester, UK: Inter-Varsity Press.
———. (1992). *A History of Christianity in the United States and Canada.* Grand Rapids, MI: Eerdmans.
———. (1994). *The Scandal of the Evangelical Mind.* Leicester, UK: Inter-Varsity Press. A good self-critical work.
———. (2004). *The Rise of Evangelicalism.* Leicester, UK: Inter-Varsity Press. The first of an excellent five volume series on the history of evangelicalism. Indispensable.
Numbers, Ronald L. (1992/2006, 2nd ed.). *The Creationists.* Berkeley: University of California Press [2nd ed., Cambridge, MA: Harvard University Press]. The best historical survey.
———. (1998). *Darwinism Comes to America.* Cambridge, MA: Harvard University Press
Oelschlaeger, M. (1994). *Caring for Creation.* New Haven, CT: Yale University Press.
Olson, Richard G. (2004). *Science and Religion, 1450–1900: From Copernicus to Darwin.* Westport, CT: Greenwood Press.
Orr, James (1897). *The Christian View of God and the World.* Edinburgh: Andrew Elliot.
Paley, William.1804 & many editions. *Natural Theology.*
Patrick, S., Louth, Whitby (1834/1764). *Critical Commentary and Paraphrase on the Old and New Testament.* 4 vols. London: Thomas Tegg.
Pennock, Robert T. (1999). *The Tower of Babel.* Cambridge, MA: MIT Press.
Pennock, R. and Ruse, M. (eds.). (2007). *But Is It Science?: The Philosophical Question in the Creation/Evolution Controversy.* Amherst, NY: Prometheus Books.
Phillips, John. (ed. Torrens, Hugh). (2003). *Memoirs of William Smith, LL. D.* by Phillips, John, with introduction by Torrens, Hugh. Bath, UK: Bath Royal Literary and Scientific Institution.
Pratt, John H. (1871/1856). *Scripture and Science Not at Variance,* London: Hatchard.
Preston, James (2001). *Genetic Turning Points.* Cambridge: Eerdmans.
Price, George McCready (1902). *Outlines of Modern Christianity.* Los Angeles: Modern Heretic.
———. (1906). *Illogical Geology.* Los Angeles: Modern Heretic.
———. (1923). *The New Geology.* Mountain View, CA: Pacific Press.
———. (1926). *Evolutionary Geology.* Mountain View, CA: Pacific Press.
Prince, Thomas (1717). *An Account of the First Aurora Borealis.* Boston: Publisher unknown.

———. (1755). *Earthquakes the Works of God and Tokens of His Just Displeasure.* Boston: Publisher unknown.

Pye Smith, John (1839). *The Relation between the Holy Scriptures and Some Parts of Geological Science.* London: Jackson and Walford.

Ramm, Bernard (1954). *The Christian View of Science and Scripture.* Grand Rapids, MI: Eerdmans. The book which guided most non-YEC evangelical scientists in the 60s and 70s. Highly significant.

Ray, John (1693). *Three Physico-Theological Discourses.* London: Smith.

Rea, Michael (2007). *World without Design: The Ontological Consequences of Naturalism.* New York: Oxford University Press.

Redford, George (1837). *The Holy Scriptures Verified.* London: Publisher unknown.

Roberts, M. B. (2001). Revising the Age of the World. In McCready, S. (ed.), *The Discovery of Time.* London: MBQ Publications.

———. (2002). The Genesis of Ray and His Successors. *Evangelical Quarterly* 76(2): 143–165.

———. (1997). Darwin's Doubts about Design. *Science and Christian Belief* 9: 113–127.

———. (1998). Geology and Genesis Unearthed. *The Churchman* 112(3): 225–256.

———. (1999). Design Up to Scratch? A Comparison of Design in Buckland (1832) and Behe. *Perspectives on Science and Faith* 51.4: 244–257.

———. (2004). Intelligent Design: Some Geological and Historical Perspectives. In Ruse, Michael, and Dembski, William (eds.), *Debating Design: From Darwin to DNA.* Cambridge, UK: Cambridge University Press.

———. (2007). Genesis Chapter One and Geological Time from Hugo Grotius and Marin Mersenne to William Conybeare and Thomas Chalmers (1620 to 1825). In Piccardi, L., and Masse, W. B. (eds.), *Myth and Geology* (Special Publication 273). London: Geological Society of London.

Rogers, Jack B., and McKim, Donald K. (1979). *The Authority and Interpretation of the Bible.* San Francisco: Harper and Row.

Rogerson, John (1984). *Old Testament Criticism in the Nineteenth Century England and Germany.* London: SPCK.

Ross, H. (2004). *A Matter of Days: Resolving a Creation Controversy.* Colorado Springs, CO: Navpress.

Rudwick, Martin (1985). *The Great Devonian Controversy.* Chicago: University of Chicago Press.

———. (2001). Jean-André de Luc and Nature's Chronology. In Lewis, C. L. E., and Knell, S. J. (eds.), *The Age of the Earth from 4004 BC to AD 2002.* London: Geological Society, Special Publication Number 190.

———. (2004) *Bursting the Limits of Time.* Chicago: Chicago University Press.

Rupke, Nicolaas A. (1983). *The Great Chain of History.* Oxford: Clarendon.

Russell, J. B. (1991). *Inventing the Flat Earth.* New York: Praeger.

Sarfati, Jonathan (2004). *Refuting Compromise.* Green Forest, AR: Master Books.

Schaeffer, Francis (1970). *Pollution and the Death of Man.* London: Hodder.

Scott, Eugenie C. (2004). *Evolution vs. Creationism.* Westport, CT: Greenwood. A useful survey, mostly on the science of YEC.

Scott, Eugenie C. and Branch, Glenn (2003). Evolution: What's Wrong with "Teaching the Controversy." *Trends in Ecology and Evolution,* 18(10): 499–502.

Scott, Thomas (1788–1792). *Commentary of the Whole Bible.* Various editions.

Sedgwick, Adam (1831). Anniversary Address to the Geological Society, 18 Feb 1831. *Proceedings of the Geological Society* I(1834): 281–316, 302.

———. (1834/1969). *A Discourse on the Studies of the University.* Leicester: Leicester University Press.

Seely, Paul H. (1991). The Firmament and the Water above, Part I: The Meaning of raqia' in Gen 1:6–8. *Westminster Theological Journal* 53: 227–240; available at www.thedivinecouncil.com/seelypt1.pdf

———. (1992). The Firmament and the Water above, Part II: The Meaning of 'The Water above the firmament' in Gen 1:6–8. *Westminster Theological Journal* 54 31–46; available at www.thedivinecouncil.com/seelypt2.pdf

———. (2005). Concordism Lacks Concord with Both Scripture and Jesus (56.3 [2004]: 235–236). *PSCF* 57.1: 74–75.

———. (1969). The Three-Storied Universe. *JASA* 21.1: 18–21.

———. (1997). The First Four Days of Genesis in Concordist Theory and in Biblical Context *PSCF* 49.2: 85–95.

Seventy-Two Nobel Laureates, Seventeen State Academies of Science and Seven other Scientific Organizations. (1986) *Amicus Curiae* Brief in support of *Appelles Don Aguilard et al. v. Edwin Edwards* in his official capacity as Governor of Louisiana et al.

Sivasundaram, S. (1995). *Nature and the Godly Empire.* Cambridge, UK: Cambridge University Press.

Spufford, Francis (1996). *I May Be Some Time.* London, Faber & Faber.

Stilling, R. L. (1999). Scriptural Geology in America. In Livingstone, David N., Hart, D. G., and Noll, M. A. (1999), *Evangelicals and Science in Historical Context.* New York: Oxford University.

Stott, John (1999). *New Issues Facing Christians Today.* Grand Rapids, MI: Zondervan.

Sumner, John B. (1833). *A Treatise on the Records of Creation,* 2 vols. London: J. Hatchard.

Sutcliffe, J. (1817). *A Short Introduction to the Study of Geology.* London: Publisher unknown.

Temple, F. (1884). *The Relations between Religion and Science.* London: MacMillan.

Tertullian (1896–1903). Ad nations. (Holmes, Peter, trans.). In Roberts, Alexander, and Donaldson, James (eds.), *The Anti-Nicene Fathers,* Vol. 3. New York: Charles Scribner's Sons.

Thaxton, C. B., Bradley, W. I., and Olsen, R. L. (1984/1992). *The Mystery of Life's Origin.* Dallas: Lewis and Stanley.

Theophilus of Antioch (1970). *Ad Autolycum.* Oxford: Clarendon.

Thomas, K. (1983). *Man and the Natural World.* Harmondsworth: Penguin.

Thomas, W. H. Griffith (1922). Evolution and the Supernatural. *Bibliotheca Sacra.*

———. (1943). *The Catholic Faith.* London. Longmans.

Torrens, H. S. (2001). Timeless Order: William Smith (1769–1839) and the Searching for Raw Materials 1800-1820. In Lewis, C. L. E., and Knell, S. J. (eds.), *The Age of the Earth from 4004 BC to AD 2002.* London: Geological Society, Special Publications, Number 190, 61–83.

Toumey, Christopher (1994). *God's Own Scientists: Creationists in a Secular World.* New Brunswick, NJ: Rutgers University Press.

Townsend, Joseph (1813). *The Character of Moses Established for Veracity as an Historian.* Bath, London: Longman, Hurst, Rees, Orme & Brown.

Ure, A. (1829). *A New System of Geology.* London: Publisher unknown.

Ussher, James (1650–1654). *Annales Veteris et Novi Testamenti.* London: Publisher unknown.

Van de Fliert, J. R. (1969). Fundamentalism and the Fundamentals of Geology. *JASA* 21.3: 69–81.

Van Riper, A. Bowdoin (1993). *Men among the Mammoths.* Chicago: Chicago University Press.

Van Till, H. J. (1986). *The Fourth Day.* Grand Rapids, MI: Eerdmans.

———. (1996). Basil, Augustine and the Doctrine of Creation's Functional integrity. *Science and Christian Belief* 8: 21–38.

Van Till, H. J., Snow, R. E., Stek, J. H., Young, D. A. (1990). *Portraits of Creation.* Grand Rapids, MI: Eerdmans.

Warfield, B. B. (1951). *The Inspiration and Authority of Scripture.* Philadephia: Presbyterian and Reformed.

Wells, Jonathan (2002). *Icons of Evolution.* Washington, DC: Regnery.

Wesley, John. (1755). *Serious Thoughts on the Earthquake at Lisbon* in Vol. 11 of Jackson *The Works of John Wesley.* Thomas Jackson, ed. (14 vols.: 3rd ed.). London: Wesleyan Methodist Book Room, 1872; reprint ed., Grand Rapids: Zondervan 1958–1959.

———. (1760). *The Desideratum: Or, Electricity Made Plain and Useful by a Lover of Mankind and of Common Sense.* London: Bailliere, Tindall and Cox.

———. (1763–1777). *Survey of the Wisdom of God.* 3rd ed. London: J. Fry and Co.

———. (1781). *The Great Deliverance.* In Vol. 2, pp. 448–449, *The Bicentennial Edition of the Works of John Wesley.* Editor in Chief, Frank Baker. Nashville, TN: Abindgon, 1984.

———. (1987). *Notes on the Bible.* Grand Rapids, MI: Francis Asbury Press.

Wheatley, Phillis (1994). *The Norton Anthology of American Literature.* New York: Norton.

Whiston, William (1696). *A New Theory of the Earth.* London: Roberts.

White, Andrew D. (1896/1955). *The Warfare of Science with Theology.* London: Arco. The epitomy of the conflict thesis of science and religion. Of immense influence.

White, Ellen G. (1864). *Spiritual Gifts: Important Facts of Faith.* Battle Creek, MI: Seventh-day Adventist Publishing Association.

———. (1890). *Patriarchs and Prophets.* Battle Creek, MI: Seventh-day Adventist Publishing Association.

White, Lynn. (1967). The Historical Roots of Our Ecologic Crisis. *Science* 155: 1203–1207.

Wiester, John (1983). *The Genesis Connection.* New York: Thomas Nelson.

Wilberforce, S. (ed.). (1861). *Replies to Essays and Reviews.* London

Wilkinson, Loren (1980/1990). *Earth-keeping.* Grand Rapids, MI: Eerdmans.

Williams, A. (1948). *The Common Expositor.* Chapel Hill: University of North Carolina Press.

Williams, William (Pantecelyn) (1764). *Golwg ar Deymas Crist (A View of Christ's Kingdom).* (Jones, R., trans.) London (1878).

Winchester, S. (2001). *The Map that Changed the World.* London: Viking.
Wonderly, Daniel E. (1977). *God's Time-Records in Ancient Sediments.* Flint, MI: Crystal Press.
Wright, George Frederick (1882) *Studies in Science and Religion.*
———. (1899). *Man and the Glacial Period.* New York: Appleton.
———. (1913). *The Origin and Antiquity of Man.* London: John Murray.
Wright, Richard. (1971). Responsibility for the Ecological Crisis. *Bioscience* 20: 851–853.
———. (1995). Tearing Down the Green: Environmental Backlash in the Evangelical Sub-Culture. *PSCF* 47: 80–91.
———. (1989). *Biology through the Eyes of Faith.* Leicester, UK: Inter-Varsity Press.
Young, Davis (1977). *Creation and the Flood.* Grand Rapids, MI: Baker.
———. (1995). *The Biblical Flood.* Grand Rapids, MI: Eerdmans.
Zimmer, Carl. (2001). *Evolution, the Triumph of an Idea.* London: Heinemann.

WEB SITES

The number of Web sites on evangelicals and science is absolutely vast, and those by evangelicals range across the whole spectrum. These are just a few important ones, and are the most useful to start on. Between these one will get a clear picture of where evangelicals are with science today.

www.cis.org.uk: The Web site of the British *Christians in Science.* A wide range of material and links to the journal *Science and Christian Belief.*

www.asa3.org: The Web site of the *American Scientific Affiliation.* A useful and comprehensive evangelical science and religion site, varying in perspective from evolutionary to creationist. Most of their journal is on-line going back to the forties.

http://www.cornwallalliance.org/docs/Climate_Change_and_the_Responsibility_of_Civil_Society.pdf

http://graphite.st-edmunds.cam.ac.uk/faraday/Seminars.php: Web site of *The Faraday Institute,* Cambridge, UK. An intellectual evangelical science site.

Young Earth Creationist Web sites

There are innumerable sites and these are those of the two most influential groups.

www.answersingenesis.org: Web site of Answers in Genesis. A tremendous amount of material, much very popular.

www.icr.org: Institute of Creationist Research. A tremendous amount of material, but less popular than AIG.

Two Secular Sites on YEC

http://www.talkorigins.org: A useful technical anti-Young Earth Creationist site. Comprehensive. Anti-christian at times!

http://www.ncseweb.org/: National Center for Science Education, Oakland. Very useful.

Index

About the Author

MICHAEL ROBERTS has degrees in both geology and theology, worked as an exploration geologist, was ordained in the Church of England, and is currently a vicar of parishes near Lancaster. He has written numerous articles on the history of geology and on evangelical connections to science.